WATER IN CENTRAL ASIA

WATER
IN
CENTRAL ASIA

PAST, PRESENT, FUTURE

Viktor A. Dukhovny
Interstate Commission for Water Coordination
Tashkent, Uzbekistan

Joop L.G. de Schutter
UNESCO – IHE Institute for Water Education
Delft, The Netherlands

CRC Press
Taylor & Francis Group
Boca Raton London New York

CRC Press is an imprint of the
Taylor & Francis Group, an **informa** business
A BALKEMA BOOK

UNESCO-IHE
Institute for Water Education

Cover Illustrations:
Where the Aral Sea water is retiring salty soils are left, which will dry out and become the cause of regular salt and dust storms. Former islands now show as hills in the landscape of the flat Aral Sea bottom landscape. Water management infrastructure for the irrigated agriculture of Central Asia, from a simple water wheel to complex pumping stations, has been developed ever since civilization started in this region.

CRC Press
Taylor & Francis Group
6000 Broken Sound Parkway NW, Suite 300
Boca Raton, FL 33487-2742

First issued in paperback 2019

Typeset by Vikatan Publishing Solutions (P) Ltd, Chennai, India

ISBN-13: 978-0-415-45962-4 (hbk)
ISBN-13: 978-0-367-38308-4 (pbk)

Library of Congress Cataloging-in-Publication Data

Dukhovnyi, Viktor Abramovich.
Water in Central Asia : past, present, future / Viktor A. Dukhovny, Joop L.G. de Schutter.
p. cm.
Includes bibliographical references.
ISBN 978-0-415-45962-4 (hard cover : alk. paper)—ISBN 978-0-415-47555-6 (paper back (limited ed) : alk. paper)—ISBN 978-0-203-84145-7 (e-book)
1. Water-supply—Asia, Central—Management. 2. Water resources development—Asia, Central. I. Schutter, Joop de. II. Title.
HD1698.A783D85 2010
333.9100958–dc22

2010043727

Visit the Taylor & Francis Web site at
http://www.taylorandfrancis.com

and the CRC Press Web site at
http://www.crcpress.com

Highlands of Tadjikistan

Former Island in the Northern Part of the Aral Sea

Contents

Sudoche Wetlands before Restoration

About the Authors

Viktor A. Dukhovny

Professor Viktor A. Dukhovny has many years of experience with water in Central Asia and is one of the leading specialists in the development of complex water management and irrigation systems. He was involved with the rehabilitation program of the Hunger and the Karshi Steppe and of the Karakum Canal. At present, he leads the future water development policies plan for the region. Viktor Dukhovny is the director of the Scientific Information Centre of the Interstate Commission for Water Coordination of Central Asia in Tashkent, Uzbekistan.

Joop L.G. de Schutter

Ir. Joop de Schutter has dealt with many water development projects in the Central Asian region. He was involved with the integrated management schemes for the Amudarya and Syrdarya deltas as part of the regional Aral Sea Basin management model. He also led the implementation of the Sudoche Wetlands Restoration project. At present, he is the deputy director of the UNESCO-IHE International Institute for Water Education in Delft, the Netherlands.

Water diversion structure
in the highlands

Acknowledgements

The authors want to thank UNESCO-IHE, Prof. Andras Szolossi-Nagy and Prof. Bart Schultz for their guidance and generous support to the production of this book. Our thanks are also due to Prof. Goga Khidoyatov for providing outstanding input into the chapters one and two; for Anatoli Sorokin and his team for the data and analysis materials used in the chapters four and five and to the colleagues at SIC-ICWC, namely G. Poltarev, D. Abdurakhmanov and L. Rojenko who have done a great job on the pictures and maps. Special thanks go to Nikolai Goroshkov for his excellent and critical translation work and to Raya Kadyrova, who took care of the many requirements for preparation and production of this book.

Typical regulator structure in the upstream part of the rivers

Preface

"Each streamlet and each drop of water are treasures
in this country; and all efforts and all possible measures
should be aimed at their preservation and proper use..."
V. Masalsky

Water is the basis for life and essential for all life forms. The blue planet, as our globe is called, is not very rich in renewable fresh water resources. Fresh water reserves amount to just 43,219 km^3, of which 10% or 4,200 km^3 are already in use (UNESCO 2006). The world's fresh water reserves make up only 2.5% of the total available water resources and they are steadily reducing. If one takes into account that between 9.07 to 10.7 bilion people will live on Earth by 2050 according to various forecasts, the need for a careful and responsible attitude to water becomes obvious. This is especially important because water is extremely unequally distributed: some countries have about 30,000 m^3/year per capita while others must do with less than 300 m^3/year per capita. Water problems are on the rise, and not only as a result of population growth. Industrial water consumption is also growing. Water is the major ingredient of all foodstuffs which support life and congest supermarkets. Five giant food and drink manufacturers, Nestlé, Unilever, Coca-Cola, Groupe Danone and Anheuser-Busch InBev, together consume up to 575 million m^3 of fresh water annually, which compares to the daily water consumption of the total world population.

Many reservoirs and other sources of potable water are subject to rapidly increasing pollution from all sorts of waste. Inefficient and wasteful use of water can be seen in many regions all over the world. Increasingly intense drought events as a result of climate change, which are expected to lead to a drop in agricultural production, pose another threat to mankind. Water professionals are rightfully raising the alarm, calling upon societies to take measures to prevent water-related disasters. Under increasing anthropogenic pressure and the effects of climate change, water, just like organic fuel, will become a scarce resource, and it may also become a subject of major financial speculation in the twenty-first century. Moreover, putting water on a par with resources such as oil and gas, could increase water deficits, because, for example, hydropower generation will compete ever more intensively for water with agriculture production and that needed for the preservation of nature.

This book focuses specifically on the problems related to water use in Central Asia. Based on the experiences from this region, we show ways to consume and preserve water wisely and demonstrate how to use water as an ally to turn arid lands into "Gardens of Eden" in a sustainable way.

Preface

Humans living in Central Asia have understood and demonstrated the art of efficient water use since the earliest periods of recorded history. They used water not only as a means of life support but also as a natural means of defence for towns and settlements. They used it to turn arid steppe and desert lands into flourishing oases. Water, in return, has also been a great teacher to farmers, whose patience and diligence allowed them to change the potential of the region. An amazing symbiosis of humans and nature, based on the use of water, existed here. Nowhere in the ancient world had similar irrigation systems and the water management infrastructure created in Central Asia. The people of ancient Egypt used flood water from the Nile for agriculture and during the Achaemenid Dynasty of ancient Persia people used *karez* (underground conduits). People in ancient India waited for the summer monsoons to flood their lands. But there was nothing to compare with the practices in Central Asia, where people understood to the need to protect and take care of water and make water shape their culture. This needed no specific teaching: by its very behaviour, water taught people the required laws that shape sustainable use and protect nature.

The monumental historic Silk Road cities of Samarkand and Bukhara are masterpieces of medieval architecture and they are well-known throughout the world. However, the world is less acquainted with other monuments in this region, which can serve as a model of how to use natural resources and preserve them for future generations. The noted Russian scientist, Central Asia explorer, geographer and traveller, V.I. Masalsky, wrote: "*Among all monuments of hoary antiquity in Central Asia, grandiose water infrastructure in the form of irrigation canals that often look like rather large rivers in accordance with their length and water abundance attracts the greatest attention. The enormous significance of irrigation water, which provides life and creates civilizations in lifeless deserts, is well realized by the local population that from times immemorial has considered land reclamation by means of irrigation as a God-pleasing activity; the memory [of] khans*[1] *and all other persons whose efforts were aimed at irrigating waterless lands was surrounded with a aura of holiness; and the true believers are visiting their graves for worship.*" (Masalsky 1913).

During every period in history the fight for water was a struggle for survival, and water has therefore become the basis for the development of the civilizations of this region. The need for advanced water resource management and agriculture practice stimulated the development of studies of earth and sky as well as fundamental mathematics. The names of those who have enriched science include Al-Fargoni, Al-Khorezmi, Imam Al-Bukhari, Al-Beruni, Ulugbek, and Ibn Sino (Avicenna), and many other scholars from previous generations are now well-known and are remembered all over the world.

Russian colonization and Soviet realism brought a completely new perspective and lead to the development of irrigated agriculture in the region on an unprecedented scale. This resulted from the efficient combination of the ancient water management practices of Turkestan with the fundamentals of new European engineering practice. The disintegration of the Soviet Union resulted in important geopolitical changes. Six independent nations now share the resources of the Aral Sea Basin, the five former Soviet Union republics of Central Asia – Kazakhstan, Kyrgyzstan, Tajikistan, Turkmenistan and Uzbekistan – and Afghanistan in South Asia. Today, the international community has still not completely grasped

---

[1]Khan (Turkish khān, "lord") is a title used by rulers of Turkic peoples in Central Asia and was adopted also by the Mongols.

the global political and economical importance of this region, and Central Asia is still terra incognita for many. However, the role and influence of these nations in the world is rising.

The region is at the crossroads between Europe and Asia. This region can be rightfully considered to be the eastern edge of Europe or, equally, the western border of Asia. With the ongoing process of globalization of the world economy, the region is a main link between Europe and Asia. There is no doubt that the region is an important part of the Eurasian continent. It not only is extremely rich in natural energy resources (oil, gas, hydropower) but it also has a unique place in history. Central Asia made a considerable contribution to world civilization through its scientific geniuses, great military commanders and statesmen, and talented poets and philosophers. There were never famines in these arid steppes and deserts because prosperous and productive oases were created using sound and well-established scientific practices. The people and nations of this region never needed assistance to raise them out of poverty, nor to overcome food shortages or political turmoil. After the collapse of the Soviet Union, the countries in Central Asia are in transition. Their main requirements are to modernize their social and political life and to introduce state-of-the-art technologies and new socioeconomic ideas and values that will allow them to speed up the transition progress.

Countries in this region differ considerably from other Asian countries. They should not be called "developing countries". Almost 100% of the population is literate, and there exists a well-developed infrastructure, a modern public health system and a high level of spiritual culture. They aspire to co-operate on an equal basis with developed, civilized and democratic nations rather than enter into a relationship of development assistance. The authors dedicate this book to the indefatigable toilers, patient irrigators and modest farmers of Central Asia, hoping to bring a better understanding of their present problems and their significance for developing world civilization to Western readers.

Fisherman at work in the restored Sudoche Wetlands in the Amudarya Delta

Foreword

In December 1993 Prof. Victor Dukhovny presented, during the Water Resources Seminar of the World Bank in Richmond USA, an excellent overview of the history of the Aral Sea Basin and the expected developments after the independence of the five States - Kazachstan, Kirgistan, Tajikistan, Turkmenistan and Uzbekistan - in the basin. He presented a broad overview of the hydrology of the area, the economic and social developments and the on-going degradation of the environment. In addition he presented the developments during the Soviet period and his ideas how the countries would have to cooperate in sharing the limited water resources of the basin, while by that time the Aral Sea had already significantly reduced in size due to long-term hydrological fluctuations and over exploitation, primarily by irrigation and hydropower. He also showed the importance of optimal sharing of the water resources among the various uses: hydropower, irrigation, drinking water supply, etc. Since then many developments have taken place in the region in which he generally was involved as an expert with an in depth insight in the various relevant processes.

Over the years Joop de Schutter has been involved as an expert in a wide range of environmental projects in various countries, non the least in the countries in the Aral Sea Basin. Based on this he is also in the position to place the problems and achievements in the basin in a global context and has used his expertise and this insight in the various projects in the region where he was involved in.

About five years ago the authors decided to combine their decades of experience. Since then they have actively worked on the preparation of this book *Water in Central Asia - past, present and future*. The book is a must for those who are professionally involved in the development and management of the land and water resources in the region, as well as for those who have to deal, or students that may have to deal, in future with similar issues in other river basins.

I therefore like to congratulate the authors wholeheartedly with the result of their hard work over the past years. I strongly like to recommend this very interesting book to the potential readers. You will enjoy it and it will undoubtedly significantly improve your insight in the region where those in charge have to be so careful in the development and management of the scarce water resources to the benefit of their societies.

Prof. Bart Schultz
Prof. of Land and Water Development UNESCO-IHE
Pres. hon. International Commission on Irrigation and Drainage (ICID)

Moscow
RUSSIA
UKRAINE
Black Sea
Caspian Sea
Aral Sea
Amuc
Georgia
Tbilisi
ARMENIA
Yerevan
AZERBAYDJAN
Baku
TURKEY
TURKMENI
SYRIA
IRAQ
Baghdad
Teheran
IRAN

Map of Central Asia

Accumulation of freshwater in the
highlands of Central Asia

1

Introduction

I

Introduction

The joint declaration signed by the presidents of five Central Asian states (the Republic of Kazakhstan, Kyrgyz Republic, Republic of Tajikistan, Turkmenistan and Republic of Uzbekistan) at their summit held in Dashhowuz on March 3, 1995 was devoted to the problems of the Aral Sea. It was based on the very old understanding that the Nations of Central Asia are consolidated by their common historical and cultural development and that ancient traditions and common moral values bring them together. The nations of Central Asia have always lived as good neighbors peacefully and in a spirit of mutual respect. Along with these moral values, they are united by traditional trade and economic relations that are based on joint use of regional water and mineral resources.

Chapter one is dealing with the very old history of Central Asia with a special focus on how the nomads of the region became sedentary and developed their special and unique relationship with the waters that feed their land. Scant precipitation (less than 350–400 mm a year), extremely low humidity, high evaporation rates and abundant solar radiation are the major features of the climate of this very arid region that covers an area of more than 300 million hectares. Geology and climate here have always been very dynamic and the main rivers Amudarya and Syrdarya have regularly changed location and direction. Sometimes the rivers discharged through the channels of the Jana Darya and the Uzboy into the Caspian Sea and sometimes they came back to their original riverbeds discharging into the Aral Sea. Ever since the Mongol invasion during the 1320s, anthropogenic influences have played an important role in water management in Central Asia. The understanding of these actions and processes remain their value up to today.

1.1 Challenges and Problems

The joint declaration signed by the presidents of five Central Asian states (the Republic of Kazakhstan, Kyrgyz Republic, Republic of Tajikistan, Turkmenistan and Republic of Uzbekistan) at their summit held in Dashhowuz on March 3, 1995 was devoted to the problems of the Aral Sea. It contains these statements:

"Nations of this region are consolidated by their common historical and cultural development. Ancient traditions and common moral values bring them together. These nations live as good neighbours peacefully and in a spirit of mutual respect. Along with moral values, we are also united by traditional trade and economic relations that are based on joint use of regional water and mineral resources.

We, the heads of new independent states, have to remember this reality and, in practice, facilitate further development of the rich heritage of our noble ancestors. We need to start from this heritage and realize the great natural and economic potential that exists in our countries. This potential should be also used for solving the most urgent environmental issues including the rehabilitation of the Pre-Aral[1] region."

Central Asia is a vast region, with densely populated oases located mainly along the upper and middle reaches of two large rivers, and new and old irrigated areas in the lower reaches and deltas of these rivers. These areas are surrounded by deserts, which often retreat or encroach further due to natural processes that may change the direction of streams or as a result of destructive human intervention. Layers of sediment have buried traces of the ancient origin of life and the first stages of human existence in the depths of centuries-old deposits. It is no coincidence that so many archaeological finds have been discovered during excavations in Tashkent (ancient Shash), Samarkand, Bukhara, Osh, ancient Davani in the Fergana Valley and ancient Nisa in Turkmenistan. These ancient settlements are all located close to fresh water sources that provided opportunities for life and well-being: the Chirchik River for Tashkent, Zeravshan River for Samarkand and Bukhara, Khojibakirgan River for ancient Khodjent and Vakhsh River for Tutkabul in Tajikistan. Evidence of human activity, which were initially buried as "vertical traces of history", were later dispersed along the edges of irrigation areas and over deserts as a result of the disintegration of sediment layers. The description of these "horizontal traces of history" by S.P. Tolstov in his famous work *The Ancient Khorezm* is typical:

"The desert surrounding the Khorezm oasis in the west and in the east is a strange desert. Between barkhans[2], on the tops of the motley rocks of the Sultan-uiz-Dag ridges, on the cliffs of the Usturt Plateau, and on the flat, pink, surfaces of takyrs – everywhere over the area covering hundreds of thousands of hectares – we find traces of human activity. These are for example the double lines of weather-beaten hills that stretch out over dozens of kilometres, which are the remains of ancient irrigation canals, and silt deposits on takyrs shaped in a checkerboard pattern typical for irrigation networks. These traces also are countless ceramic fragments covering takyrs over dozens of square kilometres. Sometimes they are red, smooth and shining but sometimes they

---

[1]Pre-Aral or Pre-Araliё is used to indicate the dried out coastal zones of the Aral Sea.

[2]Barkhan is a crescent-shaped shifting sand dune, convex on the windward side and steeper and concave on the leeward side.

Figure 1.1 Ruin of Ancient Khoresm fortress.

are also reddish-brown with a rough texture. There are also fragments of many-coloured slip glaze, bronze and iron, three-edged bronze arrowheads, ancient earrings, pendants, bangles, and rings among which cameos with portraits of riders, gryphons and Ammon's horns are found. Furthermore there are small terracotta statues of men and women in typical clothes, figurines of horses, camels, sheep and bulls and coins with portraits of ḳings in magnificent dresses on one side and of riders surrounded with characters of the ancient alphabet on the other side. All these are reminders of ancient cities, settlements and dwellings that existed long ago. Sometimes they show as almost invisible traces on the bright surface of taḳyr – as the layout of ancient houses or the reddish rings of pithoi[3] *that were dug into ground cut level with the taḳyr surface. Sometimes these traces are dead towns, settlements, fortresses, castles or ruins of whole areas that were once populated. Remains may show as walls with narrow, arrow-shaped loopholes, large towers and round or conical arches that rise ten to twenty metres above the dry channels of irrigation canals now dispersed by the wind or filled with sand.*

I never forget the magnificent view, when once, after our difficult march across the sandy dunes with my fellow travellers – the Kazaḳh-worḳer and the photographer E. Polyaḳov – I came to the site of the Angḳa-Kalin taḳyrs. A smooth clay plain covered with the crimson scatterings of antique ceramics was spreading out at the bottom of the sandy hills and under the legs of our camels. The square of greyish-pinḳ adobe walls with tall lancets of loopholes and rectangular towers in the corners was rising above the plain all contributing to this magnificent view.

[3]Pithos (pithoi–pl.) is a large crock (clay barrel).

It almost seemed that the fortress that has existed there for more than one and a half thousand years was left only yesterday. Our small caravan passed through between the pylons of the gates into the passage and into the yard of the fortress accompanied by a hollow sound. The takyr soil of the yard looked like cobblestone with desert vegetation coming out of the cracks." (Tolstov 1948)

The same impressions occured to one of the co-authors when he visited the site of ancient Merv for the first time in 1964. Located in the middle reach of the Murgab River, a new land reclamation project was being started at the time of the author's first visit. According to Al-Beruni (Figure 1.1 taken from Tolstov S.P. 2004), the lands of the Kyzyl Kum Desert near Kyrkyz-Kala and Ayaz-Kala looked similar almost a thousand years prior to the expedition of Tolstov. A question remains about the origins of the water that supported the civilization in the Aral Sea Basin. The Silk Road, and all other routes of which crossed ancient Turkestan, could not have emerged without reliable fresh water sources. The corridors that these routes followed had to be sufficiently supplied with water.

Most of Central Asia exists under very arid conditions. Scant precipitation (less than 350–400 mm a year), extremely low humidity, high evaporation rates and abundant solar radiation are major features of the climate. This arid region covers an area of more than 300 million hectares. However, in places where small amounts of water emerge, nature blossoms. Life always arises near water sources such as lakes, springs and temporary bodies of water in depressions that accumulate rainwater or water from melting snow. Here, humans not only get drinking water but also obtain food from the abundance of various plants and animals found near these water sources in the middle of desert areas. Because it is relatively scarce, the people in Central Asia idolize water; they glorify it in numerous legends and stories as the fountain of life and as the means for the continuation of life. In this region, water is identified with life and life is identified with water.

Today, in spite of the enormously increased technical and socioeconomic potential of a modernizing society, the importance of water has not decreased in Central Asia. The region is one of the areas in the world where the conflict between socioeconomic development and environmental demands has become critical due to the growing deficit of water resources. The environmental disaster that has transformed the Aral Sea and its surrounding basin is well known throughout the world. This situation is aggravated by demographic pressures and by the urgent need to improve the social conditions of people while at the same time preserving and rehabilitating ecosystems.

Significant experience of survival and adaptation under new political and social conditions has been obtained during the twenty years of independence of the Central Asian nations. All five states have declared the political will and intention to collaborate, which is reflected in many joint agreements and the establishment of mutli-country organizations to govern the management and use of water resources in the Aral Sea Basin. This is a rather unique example of transboundary river basin management. The lesson is that by agreeing common principles for sharing and using water resources and by establishing an appropriate management organisation serious conflicts can be avoided, even under very difficult conditions with both floods and droughts occurring in the last fifteen years. However, expanding commercial interests, attempts to use water resources as a market commodity in a similar fashion to oil and gas supplies, and the speculator-driven rise in hydroelectric power prices have created new destabilizing

factors in the water arena in Central Asia. Inter-country tensions over water seen in the growing season of 2008 should be marked as "an outburst of hydro-egoism", and they show that the joint agreements signed by the Central Asian states for water resource allocation cannot serve anymore as the single management tool for collaboration on water in the region.

Globalization and the transformation of Central Asia into a sphere for the geopolitical games of mega-actors such as the USA, Russia, China, India and the European Union (Dukhovny 2007) make the need for a reassessment of the role of transboundary water resource management in the region ever more important. All riparian states in Central Asia should understand their mutual dependence and the need for joint and equitable water resource governance and use. In order to do this, they should draw on their thousands of years of common history and their common experience of water resource management.

The history of water resource management in Central Asia – one of the ancient regions from where humanity originates – is of interest not only because it teaches us about the importance of water but also because it allows us to understand the development of concepts for combining a water-based economy with sociocultural development. From time immemorial, water and irrigation have been important drivers for progress, the development of science and culture, and social cohesion in Central Asia. Water resource management always required strict implemention of both written and unwritten principles of mutual respect. These principles were often anchored in traditions, rules and customs, as well as in the minds of the people who had to appreciate, protect and care for water and everything related to water. According to all local traditions, and especially the spiritual and moral principles of ancient relations between people, water was never a source of profit. From these original religions and beliefs, and later through the adoption of Muslim traditions and values, water has been (and should continue to be) recognized as the main basis for the survival and well-being of humanity.

In Turkestan, since the dawn of time people have considered farming as one of the most honourable occupations. The Muslim religious doctrine states that farming has divine origins. The first plough was made of a tree from paradise by the Archangel Gabriel, who made a few first furrows and then handed over this plough to Adam. The Sharia calls farmers who cultivate their land as "*ashraf-ul-ashraf*" ("the noblest among the nobles").

When we attempt to understand the key role of water resources management based on the experience of our ancestors, we should equally understand that the present transition towards a market economy in this arid region is impossible without government attention, oversight and support. Historical analysis shows that because of the specific natural conditions, economic progress in Central Asia depended on whether societies and governments could manage these conditions. They needed the ability to administrate their vital water-based systems to protect the prosperity of the people and avoid degradation. Wars and human aggression cause destruction, but not as much that which results from loss of water management capacity due to carelessness, lack of will and weakness of the central administration. This is one of the fundamental lessons of history for us and for this region.

The potential of society has grown incredibly since the days of the wooden plough. Today, as thousands of years ago, the selfish aspirations of human beings are still often aimed at achieving short-term results. However, in these days of ballistic missiles and rapid economic and technological development, the consequences of large-scale mistakes by mankind become potentially ever more dreadful and can threaten the very existence of civilization.

In speaking about the relationships of civilizations with their water resources, we apply the term *civilization* as meaning "*a level of social development and material culture achieved by a given social and economic entity*" (Bongard-Levin 1989). Our target is to show how these relationships have developed in the region by describing attitudes to water, methods of water use and the struggle for water resources. In doing so, we reiterate that water always was a driving force of sociocultural development.

There is no common view on the initial realm and origin of the hydrographic networks in Central Asia. Geologically the Aral-Caspian Sea Basin was a part of Turkistan in the middle of the Tertiary period, but already during the Pliocene only the western part of Turkistan (the Sarmat Sea) was covered with water. The Aral-Caspian Basin continued to gradually dry up with the separation of the Caspian Sea from the Aral Sea by the Usturt divide. After this, the Sarakamish depression became landlocked within the Aral Sea side of the divide. One feature is certain, at that time the hydrographic networks of Central Asia carried much more water.

Today's researchers base their understanding of the hyprographic features of Central Asia in earlier times on an interpretation of descriptions by the Chinese traveler Zhang Qian, who refers to the tenth century BC and his own epoch (the second century BC). They also use works by Strabo[4], which contain descriptions of the trade routes from India to Girkania[5] along the great rivers in Central Asia. Reference is also found in Vedic literature and in Chapter 21 of the Book of the Prophet Ezekiel that coincides with decscriptions by Herodotus (see Bekchurin 1950, Konshin 1883 and Solovyev 1989). These descriptions of the river networks in Central Asia mention the Vedi-Darya – "the Sarasvati and the Sure" – the river now known as the Amudarya. This river then flowed from the mountain ridges of "the Charaiti" (Tian Shan), together with "water off-spring" such as the Chu River (running through the Buam Canyon), "the Soghd" and "the Mug" – "Sogdiana" – the Zeravshan and Murgab rivers and the Tejen River, into the "Zahra Kosha" (the Caspian Sea). The country between the Syrdarya and Amudarya rivers was the location of ancient settlements occupied by Aryans who engaged into agriculture and practised Zoroastrianism. These people left after the water flow through the upper reaches of the Vedi-Darya – the Sarasvati and Suna (Sarisu and Chu) – stopped following a major earthquake. Records of this disaster in the Buam Canyon can be found in the *Avesta* and *Rigveda*, as well as in the works of Herodotus and the Book of the Prophet Ezekiel. These descriptions enabled A. Chaykovsky to produce the sketch map of the ancient river network in Central Asia shown in Figure 1.2. This network in its geological setting has been reproduced with the help of presently available GIS data in the map shown in Figure 1.3.

This explanation fits with the hypothesis that during the Aryan period the Aral Sea did not exist. The *Avesta* contains a detailed description of the entire region and if the Aral Sea existed in that time,

[4]Strabo (63BC–24AD) was a Greek geographer and historian, born in Amasya, Pontus (now in Turkey).
[5]Girkania corresponds to the modern city of Gorgan (Iran) located near the Caspian Sea.

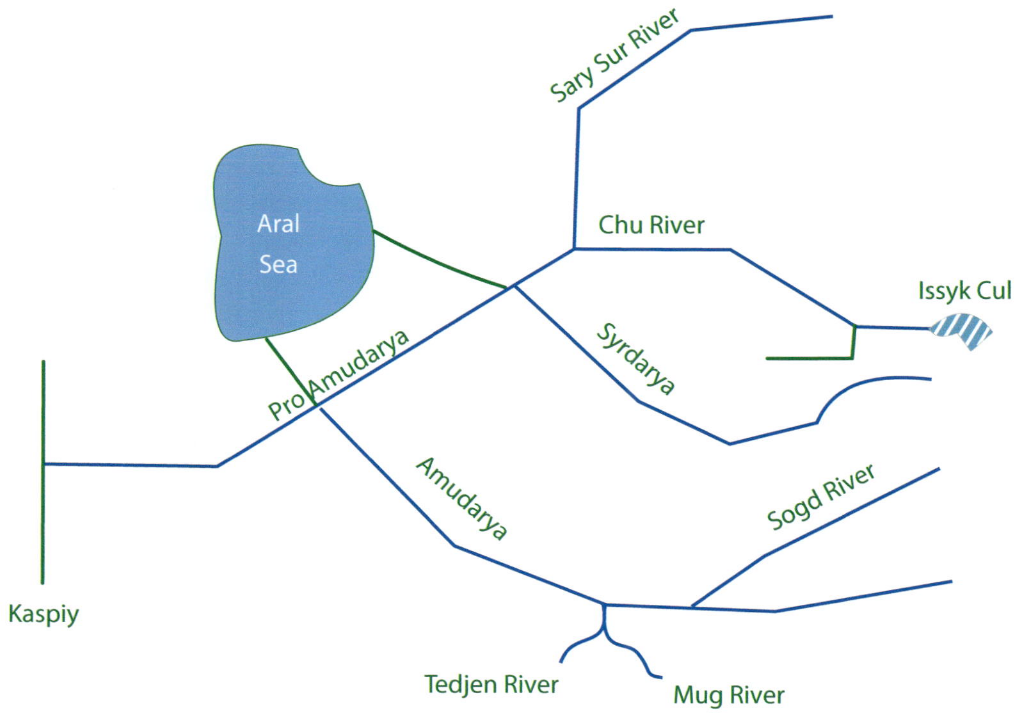

Figure 1.2 Sketch map of the Aral-Caspian Basin prior to the creation of Issyk-Kul (after Chaikovsky (in UNESCO 2006)).

this would have been definitely mentioned. After the loss of the links with the Chu and Saris rivers, the system consisting of seven rivers transformed into a system of five rivers (the Amudarya, Syrdarya, Zeravshan, Murgab and Herirud (Tejen)). This system gradually disintegrated – most likely because of the extension of irrigated areas – through formation of large water bodies in the river deltas. Ptolemy's map shows that the Murgab and Tejen rivers were tributaries to the Amudarya, but in photographs made by Kornilov in 1834 both rivers run in parallel and drain into one big lake and do not reach the Amudarya River.

Over their long period of existence, both the Amudarya and the Syrdarya have regularly changed their location and direction. Sometimes the rivers discharged through the channels of the Jana Darya and the Uzboy into the Caspian Sea and sometimes they came back to their original riverbeds discharging into the Aral Sea. Al-Khorezmi, the outstanding mathematician and astrologer, of whom it was said he could not make a mistake, mentions the surface area of the Aral Sea in his work – giving a figure that is three times smaller in size than the area of the sea in 1960. B. Bardold, the great historian and explorer of ancient Turkestan, has also pointed to the changes in direction of flow of the Amudarya and Syrdarya in the past (Bardold 1966).

After the Mongol invasion of the 1320s, the Amudarya River was again located in the Sarakamish depression and flowing onwards through the Uzboy riverbed drained into the Caspian Sea. Kyat village

Figure 1.3 Map of the Aral Sea and Caspian Basin prior to the creation of Issyk-Kul (reconstructed with recent GIS data).

was located at that time near the river mouth at the Caspian coast and boats went upstream from there. However, there was no continuous waterway from Kunya-Urgench down to the Caspian Sea due to the existence of waterfalls and rapids along the Uzboy River. There is also information that at the beginning of the eleventh century the Syrdarya was a tributary to the Amudarya. As a result, the towns that existed in the fourteenth century along the lower reaches of the Syrdarya near its confluence with the Jana Darya faced desolation, although they were neither affected by Tamerlane's military campaign nor other military events. In the seventeenth century, after the final return of the Amudarya towards the Aral Sea, the capital of the Khanate of Khorezm moved from Kunya-Urgench to Khiva.

It is clear that as a result of the interaction of nature (including climate change) and human intervention the landscape of the region changed many times from a land with fertile soils and rich in water to semideserts and deserts. We will return to this subject later in this book. Human societies changed too: here we can cite the main conclusion of the work of S. Tolstov that "in Central Asia, society went

through many political reforms from the ancient closed communities to a slave-owning system and a centralized state" (Tolstov 1948). With reference to Karl Marx, Tolstov added: "*I doubt whether there is a need to remind ourselves that these are the features which Marx and Engels considered as most essential prerequisites for the flowering of irrigation.*"

In an article entitled "The British Rule in India" published on June 10, 1853, Karl Marx wrote: "*The elementary need for joint and economical use of water, which in the West stimulated private entrepreneurs to team up into associations on a voluntary basis as in Flanders and Italy; in the East, where civilization was at a lower level and where territories are too large to allow to call associations into being on a voluntary basis, this has required interference by the consolidating forces of the government. Therefore organizing community-based works became the leading economic function that had to be implemented by all Asian governments*" (Marx 1969). In his letter to Marx dated June 6, 1853, Engels wrote: "*Here, farming is based on man-made irrigation, and irrigation is already a business of communities, provinces and central government*" (Engels 1956). In these articles and letters, Marx and Engels clearly show us the causes of the collapse of ancient Oriental cultures that were based on irrigation.

A system of artificial land reclamation that depends on a strong government will immediately fall into decay if neglected by this central authority. This explains why today we can find large areas of barren and deserted land that were previously excellently cultivated irrigated areas, such as Palmyra[6], the ruins in Yemen and large areas of Egypt, Persia and Hindustan. Lack of water management can render a land barren in the same way that just one disastrous war can depopulate a country for many centuries or even destroy its civilization completely.

Central Asia has experienced numerous natural and, particularly, manmade disasters in its history. Each military invasion can be considered anthropogenic "transgressions", where the raids by conquerors with many thousands of troops have created waves of disaster. Alexander the Great, the Huns[7], Genghis Khan and many others inflicted great destruction on the civilizations of Central Asia. Each "wave" lead to disaster, as they resulted in the collapse of the irrigation and water supply systems through weak governance and disorder in the societies that followed.

A century and a half of Russian colonization in Central Asia followed by the Soviet period was a time of strong governance in the region, and it brought step-by-step rehabilitation, enhancement and development of water systems. This created one of most technically advanced and developed water infrastructure systems of its time, certainly by comparison with those in other arid regions of the world. Unique waterworks, such as the longest canal in the world (Kara-Kum Canal), the highest dam (Nurek), the Karshi Pumping Cascade with the highest elevation and largest capacity, and the largest single area irrigation project (Golodnaya Steppe), were developed. These projects applied new principles (economically, institutionally and technically) and left a heritage of irrigation infrastructure that the newly independent nations of the region are using and developing today.

[6]An ancient city in central Syria said to have been built by Solomon.

[7]Huns are nomadic Asian people, probably of Turkish, Tataric or Ugrian origin, who spread from the Caspian steppe to make repeated incursions into the Roman Empire during the fourth and fifth centuries AD.

However, some important adverse features inherent to the Soviet water sector have also been inherited by Central Asia. Important are the lack of attention to preserving the environment, which caused the Aral Sea crisis, and centralized systems of water resources management. There is a legacy of "top down" governance, based on central government investment policies for water infrastructure and central government managed budgets for operation and maintenance. There is insufficient public participation in the decision-making processes related to both investment strategies and the operation and maintenance of the water infrastructure. This is causing many inefficiencies and is slowing down developent and modernization processes in the sector.

At present, the newly independent nations face a choice: to either learn lessons from history and find ways towards a system of equitable water resource management based on sound government support, or to ignore these lessons and invite disaster, which may lead to large-scale violence and the suffering of millions of people. Many destructive factors are expected to lead to greater competition for water resources in the future, such as the impact of climate change, reduction of river flows (30% for the Amudarya), population growth, increased water demand and higher energy prices. In addition, cultural changes as a result of modernization processes may lead to a reduction in solidarity or, as we say, "growing water market egoism".

Forward planning based on very well-understood integrated challenges used in development scenarios for the region should guide decision-makers and society as a whole. The choice is between the possible loss of the essential values of Central Asia civilization or the rise of new levels of understanding to meet the needs of the "*Central Asian Homo Sapiens*" of the future.

1.2 A Geographical Review

Central Asia covers the territory between the Ural Mountains in the north and the Hindu Kush in the south, and between the Caspian Sea in the west and the Tien Shan mountains in the east near the border with China. The region covers an area of 4,000,000 sq km (10% of the Asian continent and is two times larger than the combined areas of Germany, France, Italy, Spain and Great Britain). It stretches over a distance of 2,400 km from west to east, and over 1,280 km from north to south. There is a total population of around 54 million people. In literature, the region has several different names. It is known as Central Asia, Middle Asia and Turkestan. The idea of Central Asia as a distinct region of the world was first introduced in 1843 by the well-known German scientist, traveller and geographer Alexander von Humboldt (1769–1859). He coined the name Central Asia for the first time. In 1869 the Russian colonial administration administered the region as "Turkestan", because the territory is populated by Turkic people. A more limiting definition was the official one used by the Soviet Union: the authorities defined "Middle Asia" as consisting solely of Uzbekistan, Turkmenistan, Tajikistan and Kyrgyzstan. The concept of Central Asia and Kazakhstan was also used. Soon after independence, the leaders of the five former Soviet Central Asian republics met in Tashkent and declared that Central Asia should include Kazakhstan as well as the other four countries included in the Soviet definition. Since then, this

has become the most commonly accepted definition of what constitutes Central Asia. The region represents a unique part of the globe. There are some very peculiar natural features inherent only to this place and environment. However, more important even, is the imprint made by the people of this region at the crossroads of Europe and Asia, with their unique history, lifestyle, mentality and culture.

The different territories of Central Asia can be distinguished according to their size and specific natural conditions. Waterless deserts and steppes cover hundreds of thousands of square kilometres, lakes are so large that they are called seas, and nature still breathes eternity and greatness despite the impact of human activities on the region over the ages. An ocean, part of the Genesis flood, covered the region millions of years ago, with the surrounding ridge formations serving as a coastline. Some of the highest mountains in the world are found here. The region is isolated geographically – all the rivers that originate in Central Asia drain into landlocked lakes or sink into desert sands. Wind is a major force driving the development of geological formations. With few obstacles to stop it, wind can carry tons of sand and dust over the steppes, valleys, deserts and plains, and then deposit this material elsewhere as dunes or barkhans. Heat and wind are the engines of the weathering process that breaks up rocks and causes dust storms. Larger particles are moved more slowly and collect as agglomerations of loose sands. These created deserts such as the Kyzyl Kum and Kara Kum.

Over thousands of years, the winds have also been instrumental in the creation of other soils. By lifting light dust high up in the air, and by carrying clouds of dust over the whole region and depositing the dust in thin layers, wind action has created a thick soil cover that has buried ancient towns and settlements. Wind action also inevitably spreads sand widely throughout the region. Sand intrudes into fertile oases and waves of yellow sand have ruined cultivated fields, irrigation canals and wells, and buried gardens, houses and whole towns. Numerous towns, castles, temples, caravan roads and even kingdoms have disappeared because human beings were powerless against these forces of nature. The ruins of Carthage and Palmyra are evidence of this phenomenon. During the long cloudless scorching summers, the soil disintegrates thus producing the material for dust storms. Steppe and desert now cover over 75% of the territory in Central Asia.

Along its southern, eastern and north-eastern borders Central Asia is surrounded by high mountain ranges. These play an essential role in making the region suitable for farming. The mountain chains of Tarbagatay, Tien Shan, Pamir-Alay, Kopet Dagh and the Hindu Kush[8] are exposed to arctic cyclones that bring the frosts, abundant rainfall and snow that contribute to the formation of the large rivers that are the source of life for the region. The Pamirs, "the roof of the world", are among the highest mountains of the world and the peaks of the Tien Shan mountain range reach 6,000 metres above sea level. In ancient times, these mountains carried the Turkic name *Oķ doglar*, which means "white mountains". Chinese warriors who occupied these territories in the tenth century renamed them the Tien Shan ("heavenly mountains"). The ice caps on the mountains and the constant low temperatures produce abundant snowfall in the valleys, resulting in snow cover and wet soils in these valleys. Glaciers slowly move from the ice caps into the valleys where they thaw. The melt water feeds the rivers. Glaciers are

[8]Some western geographers (such as Glantz) consider that Tzinzjan (part of China) and Western Mongolia also belong to Central Asia.

natural reserves of fresh water, and they provide 10–30% of the river run-off. The Fedchenko Glacier, which is 25 km long and 2 km wide, is the largest glacier in the region, with the thickness of the ice layer reaching up to 5 metres. More than 6,000 rivers longer than 10 km in length originate in the mountains, including the two great Asian rivers, the Amudarya and Syrdarya.

The Amudarya is the largest river of Central Asia considering discharge. From the headwaters of Pandj to the Southern Aral Sea, the river is 2,540 km in length. The Syrdarya River is longer. It runs for about 3,000 km, from the headwaters of Naryn to the Northern Aral Sea. The vast Turan lowlands stretch between these rivers. A major portion of these lowlands, within which the Kyzyl Kum Desert is also located, is now part of the Republic of Uzbekistan. The ridges of the Tien Shan and Hissar-Alay mountain systems surround these lowlands to the north-east and to the south. Ancient authors used the older name "Oxus" for the Amudarya River and called the whole lowland area "Transoxiana", which means "beyond the Oxus River". Following the Arab conquest of this area, it became known as *Mawara'un-Nahr* (Arabic for "what is beyond the river"). It is interesting to note that in ancient Chinese literature this region was also called "what is beyond the river" (*Maure-en-nagar*), but the Chinese referred to the Syrdarya River. Note also that the phonetic transcriptions *Ma wara'un-Nahr* and *Maure-en-nagar* are surprisingly similar. The well-known explorer of Mongolia and Central Asia, N. Bichurin (Father Yakim) noticed that Chinese authors were the first to call this territory "Turan" ("the country of Turkic peoples") (Bekchurin 1950). The first agricultural communities of Central Asia settled in these lowlands and the first manmade infrastructure for irrigating agricultural fields and orchards was developed here.

The climate of Central Asia is unique. Each spring, streams rush through the mountains into the valleys and, under a moderately warm sun, grasses and wild flowers blossom, showing all colours of the rainbow. Spring starts in February, when the fields turn emerald green and the flowers in the orchards fill the valleys with a sweet fragrance like the Gardens of Eden. The hot summer lasts four months, and ripens all heat-loving plants. Some vegetation has already dried out by the beginning of July under the hot sun, but some remains verdant until September, when autumn arrives and sometimes brings rain. The "velvet" season – the warm autumn months from September to December – provide the opportunity to harvest crops and to carry out any necessary repair and cleaning work on the irrigation canals Snowfall starts at the beginning of December and covers the winter crops with a layer of snow. Winter is cold, and typically this cold weather lasts for about 40 days. After that, the cycle of the seasons begins anew.

It is possible that these climatic conditions allowed Central Asia to become one of the world's first regions to cultivate cereals. According to the outstanding academician N. Vavilov (1887–1943), some 15% of cultivated plants known to the world are indigenous to Central Asia (Vavilov 1967). Vavilov claims that cultivation of cereals started in this region far back in history, first in the foothills of the mountains and later in the valleys of the great rivers where the first farming communities formed. As crop production became established, pastoral agriculture developed in parallel through tribes of herders who owned sheep and goats. Wild horses (*Equus przewalskii*) inhabited these lands, and they were domesticated by humans for use as work horses in agriculture and transport. Since about 1.8 million years ago, the mountain slopes have been covered by nut-bearing groves and forests, while apple, cherry-plum and almond trees have grown at lower altitudes.

Nature created all conditions necessary for life in Central Asia, and a key task for humans is to use these properly. Nature has also fostered in humans a care for the environment and awakened their love for the land and for water – its most significant feature. People became farmers, livestock breeders and irrigators. In Central Asia a *dekhkan* (peasant) is a creature of nature. Dressed in white shirt and with a skullcap on his head, he can work all day long in the fields under the merciless sun. He seems to be permanently tied in with nature, communicating with the land, thanking God for this invaluable gift that feeds him and his family, whispering gently to the water and sometimes reproaching it flowing in the wrong direction, while the water babbles gently in response. Humans, land and water are intimately connected. This union has formed important features inherent to the national character of the people living in Central Asia. It has inspired poets and writers, composers and painters, who have glorified their love for their native soil. The outstanding Uzbek poet Hamid Alimjan (1908–1944) in his poem "Zeynab and Omon" wrote:

> "*When I roamed over valleys and groves*
> *I have seen amazing and marvellous orchards,*
> *They awakened my affectionate feelings,*
> *And I kissed my darling land...*"

The work of peasants in Central Asia is very hard but, in contrast to the labour of Sisyphus, it provides results. A major challenge is salinity. While 10% of the total area of Central Asia are natural oases, saline soils cover about 35% of this territory. The soils emerged after the retreat of the oceans, and salts accumulate towards the bottom of the the top soil layers. In the southern part of the region, along the borders with the Kyzyl Kum and Kara Kum deserts, there are vast areas of land where the soil contains considerable amounts of soluble salts. These soils are called *takyrs* and they are unfit for farming. There are also areas named *shors* (an Uzbek word meaning salty soil) where the salt content in the soils is so high that the ground surface is covered with a white crust resembling snow cover. *Shors* look like huge white snow lakes and are absolutely devoid of vegetation.

Solonchaks usually occupy depressions, especially in steppe areas, that drain rainwater from the surroundings. True solonchaks usually have soils that are unfit for vegetation. However, in Central Asia, *dekhkans* have proved that not only solonchaks but also shifting sands can be transformed into quite productive soils. Through subsurface drainage and leaching operations, hundreds of thousands of hectares have been won back from solonchaks and previously unproductive soils have been turned into fertile arable lands and pastures.

The natural conditions in Central Asia "gave birth" to one other group of toilers who made a contribution to make these boundless steppes habitable. These are the nomads. A gigantic strip of steppe stretches from the banks of the Amur in the Russia Far East to Hungary in the west. The Great Steppe crosses over Altai, southern Siberia, northern Kazakhstan, the northern coast of the Black Sea and ends in Pannonia[9] (Hungary). In the north of Kazakhstan, the Great Steppe connected to steppe lands running south forming the Kazakh steppes. Nomadic tribes adapted themselves to the mainly similar

[9]Pannonia is the ancient country of Illyria, bounded to the north and east by the Danube River, to the south by Dalmatia, and to the west by Noricum and parts of upper Italy. It covered parts of contemporary Austria, Hungary, Croatia and Slovenia.

living conditions that existed along the whole length of the Great Steppe. They were mainly Turkic peoples, and were endurable, strong and viable. Their major occupation was cattle breeding. Horses, sheep, cows, bullocks and camels were the basis of their economies. Nomads created their own empires, teamed up into confederations and unions, and played an enormously important role in the world history. Attila and Genghis Khan created great steppe empires. They are known as conquerors that ruined ancient civilizations, and although this is largely true, they also headed tribes that developed the steppes and created their own style of life. Without this development, the Great Steppe would have remained lifeless terrain. In his fundamental work *The Empire of the Steppes: A History of Central Asia* that was first published in 1939 and has since been republished many times, the well-known French historian René Grousset has clearly shown that these "children of steppes" played a creative role to develop the steppes and to assist nations to emerge and build civilizations (Grousset 1996). It is not by mere chance that Grousset chose Central Asia as the historical scene where nomads were allies and assistants to farmers in the development and use of local natural resources.

Three confederations of nomads (*juses*) emerged on the land of present-day Kazakhstan in the fifteenth century. They roamed over a territory of 2,700,000 sq km, equal in size to the area of Western Europe. Each *juse* had use of strictly-specified districts, including agreed grazing areas, camping grounds and winter quarters. In spring, summer and autumn, they grazed their livestock near rivers, and in the winter they moved to warmer areas in the south. *Juses* were subdivided into tribes, and tribes into clans. While roaming from place to place, they made long marches, often travelling 60–70 km a day during the 12 hours of daylight. Camels could manage without water for 10–12 days, but horses and livestock had to be watered each six or seven hours to keep them in good shape. Each head of a tribe or clan had to know the locations of wells along their migration routes. The wells served as "lighthouses" in the steppe. They were jointly dug by tribes and clans for common use, and nobody could have exclusive right of use. Water was the basis for nomadic life and also the regulator of the nomads' social, economic and political relations. Water bodies were "temples", which were piously maintained, and the major responsibility of each tribe or clan was to leave wells or other water bodies clean and in good working order. They always had to have special buckets for lifting water, and trays for watering livestock.

Nomads were an integral part of the regional economy. Their presence and activities have created sustainable mechanisms for relations between the nomadic and settled populations. The nomadic lifestyle endured until the Soviet period, when nomadic tribes were forced to settle, but even today numerous nomadic traditions remain in use.

There were established days when nomads visited the markets in towns to sell their products. They sold cattle and sheep, wool, skins and pelts, poultry and game, adobe for construction material, and hooves and bones to make glue. In return, nomads purchased horse tackle and saddles, sabres, garments and textiles, jewellery, tableware, timber doors for yurts (circular shaped nomad tents), silk and soap. The collaboration between the settled and nomadic population was shaped on this basis.

Trade relations, both within Central Asia and with neighboring nations, played an essential role in developing the region and its water infrastructure. The great overland trade route linking the West and the East dates from around the second century. Credit for establishing this trade route is usually attributed to Chinese diplomats of the Han dynasty. In 135 BC, the envoy Zhang Qian returned to China after ten years spent travelling through Central Asia. He brought news to the Chinese Emperor Wudi (who ruled 141–87 BC) amazing information on the great potential of this region. Zhang Qian reported that farming areas in Central Asia were abundant in rice and wheat and rich in fruits and vegetables, and wine production also prospered. He especially admired the horses, in which were not present in China at that time. He wrote: "These horses are blood red and originate from the horses of heaven."

Intensive trade relations were established between China and Central Asia. Over time, trade extended further. The trade routes reached the Mediterranean and goods were delivered to the Roman Republic (and later the Roman Empire). In 60 BC, Gaius Julius Caesar (100–44 BC) had a shirt made of red silk that came for the first time from China via the Silk Road. Later, soap made in Samarkand was traded in Europe. The French word for soap, *savon*, also means soap in the Uzbek language. The first paper made from pure cotton using clean spring water was manufactured in Samarkand. This paper soon supplanted the use of parchment in Europe. In time, a trade route also reached India, creating a unique triangle (China – Central Asia – India) and the basis for the interchange of cultural and material values between these three regions. For example, the Buddhist religion was brought into Central Asia, China and Japan along these trading routes.

Apart from commerce and cultural exchange, the Silk Road also played an essential role in the development of water resource management. It was necessary not only to guard this road but also to organize a water supply, since most part of the route run through waterless steppes. There could be no water for hundreds of kilometres along the route. Local rulers started to build *sardobas* (see Figure 1.4) – covered storages for collecting rainwater in the rainy periods and meltwater in the winter and keeping this water under guard over the hot summer period for the use of travellers. A long chain of these covered reservoirs stretched from the Tien Shan mountains to the Amudarya (Suleymanov 2000).

Figure 1.4 *Sardoba* means "roof above water". These structures were placed around hand dug wells such as at Mirza-kuduk in the Hunger Steppe (left) and Malik Rabat (right).

One of these structures, the Malik *sardoba* near Kershme, is described by Nemtzova. This was built in the form of a water reservoir with a dome-shaped roof. The structure dated from the eleventh century and unfortunately is no longer standing. The water reservoir was covered with a dome of 13 m in diameter and sunk below the surface to a depth of 12 m, keeping water cool during all summer. It was located near the caravansary. A sixteenth century author compared this structure with that housing the Zam-Zam spring in Mecca. Three windows for light were made in parts of the dome, and a small portal with steps leading towards the water was located near the entrance.

The Sardoba is a typical well structure from ancient Central Asia

2

The Cost of Water is Life

2

The Cost of Water is Life

The individual character of a nation is mainly formed by its environment, and by the ways people adapt to this environment in an attempt to survive and this principle has shaped the communities of Central Asia from the very beginning of their sedentary life. In Central Asia for example, the late Stone Age was marked by the Keltiminar and first traces of this culture were discovered not far from Urgench in the Amudarya delta. The Keltiminar culture represents one clear stage in the development of the region's civilization. Similar sites were discovered in South Xinjiang, Western Kazakhstan, the Lower Volga Region, the Southern Pre-Ural, North-Eastern Europe and South Siberia. This confirms that Central Asia was the starting point for the development of a human society that later on spread northwards.

Turkmenistan is the cradle of irrigated farming in Central Asia. Twenty seven mountain streams flow down from the northern slopes of the Kopet Dagh Ridge, which became the basis for settled life in the two oases at Akhal-Tekin and Atrek. Using small barrages, farmers created a manmade microrelief for irrigating their fields. There was no need to build any special structures for flooding fields, since it was sufficient to build small embankments along the edges of the fields to retain water for a period of time. These rather primitive methods initiated the development of irrigation. The iron age introduced a new era in the development of irrigated farming and irrigation systems. Iron tools, such as the spade, hoe and ploughshare, played a revolutionary role in driving technological development in the agricultural sector. Iron tools allowed agricultural production to extend over larger areas, by helping humans construct deep and wide irrigation canals and clear woodlands for cultivation.

Eventually irrigated agriculture in Central Asia developed around numerous oases where ever water resources were adequate and it was possible to establish a governing structure strong enough to develop, operate and maintain the systems. The oases became the places where civilization would prosper and science would develop. The chapter describes the history and present condition of many of these oases and their irrigation systems that have played such an important role in the development of the culture and structure of Central Asia. Emphasis is put on the role of water as an educator for how a civilization is shaped and how traditions and religion derive many of their characteristics from the special role that water plays in society.

2.1 A Revival of the Will

The individual character of a nation is mainly formed by its environment, and by the ways people adapt to this environment in an attempt to survive. This character shows itself at an early stage of a nation's development. A nation can simply remain subject to environmental forces or it can attempt to bring them under control. People can accept the role of primitive hunter gatherer, but they can also become master, demiurge[1] and reformer by forcing the environment to serve the needs of human beings. A nation may remain at the mental level of primitive societies and lead a life in the desert, or it can transform arid areas into prosperous oases as happened in Central Asia.

One cannot agree with Arthur Schopenhauer (1788–1860), who believed that a person inherits all virtues and vices from his parents and cannot go beyond the path inherited at birth. Nature awakens the will to live and the will to improve life, and this leads people not to wait for favours from nature but to take what they need. However, ancient history clearly testifies that humans need to be on friendly terms with nature, to take care of the environment in order to benefit from "reciprocal feelings". We all know that this is needed, but few people know how this can be achieved.

The famous Russian anthropologist V. Bartold made valuable contributions to the study of the social and cultural history of the Turkic peoples of Central Asia. Bartold has shown the significance of past irrigation experience in this region for its present and future development. In his article "The History of Irrigation in Turkistan", he wrote: "In this region, Russians discovered the centuries-old culture of farming based on irrigation that aroused the interest of both orientalists and archaeologists and business people." Furthermore he emphasized: "The close link between studying the past of this region and [planning] activities to achieve a prosperous future seems to be more obvious here than in other places" (Bartold 1966).

The Russian administration created a school of orientalists specializing in the history of Central Asia. The founders of this school included V. Vyatkin, V. Bartold. N. Veselovsky and A. Semenov, and their works and those of others established a basis for developing the policy of the tsarist administration in this region. The past became the adviser of the present, and helped to prepare for the future. Let us also look back at the history of irrigation in Central Asia in order to realize the significance of contemporary water problems and to learn lessons for the future.

Chronologically, three principal epochs can be distinguished in human history – the Stone Age, the Bronze Age and the Iron Age. The Stone Age, in turn, is usually divided into three successive stages[2] – the Old Stone Age (Paleolithic), the Middle Stone Age (Mesolithic) and the New Stone Age (Neolithic). The Paleolithic stage is the longest period in the history of *Homo sapiens* and covers the period from 2.5 million years ago to 10,000 BC. It can be subdivided into the pre-Chellean period, which

[1]In Gnostic and some other philosophies the creator of the universe.

[2]The stages also imply broad time frames and are perceived as stages of human cultural development; the Paleolithic is considered to be the period in which stone tools were chipped or flaked.

dates from approximately 2.5 million years ago until 700,000 years ago, the Chellean (Abbevillian) period between 700,000 and 300,000 years ago, the Acheulian period between 300,000 and 100,000 years ago, and the Mousterian period between 100,000 and 35,000 years ago. We will now briefly describe some major features of these periods in Central Asia.

2.1.1 The Old Stone Age

In 1985, the archaeologist U. Islamov discovered Sel-Ungur, a primitive cave site near the village of Khaydarkan in the Fergana Valley which dates back to the Chellean period (700,000 years ago). Based on this archaeological find, it was hypothesized that Central Asia belongs to the zone where hominoids (the early forerunners of man) evolved during the Stone Age into forms such as *Homo habilis*, *Homo rudolfensis*, *Homo erectus*, *Homo neanderthalensis* (also called *Neanderthals*) and, finally, *Homo sapiens*, modern humans. Over time, the human physical constitution developed with improved diet and through undertaking more diverse activities, and the parts of the brain responsible for labour and speech grew. The Teshik-Tash Cave in Surkhandarya Province discovered by the academician A. Okladnikov in 1938 is evidence of the evolutionary progress of the human race in Uzbekistan (Okladnikov 1966). A site where humans lived between about 100,000 to 400,000 years ago has also been found in this region as well as an ancient burial place of a Neanderthal boy who was 8–9 years old when he died. The boy had a large skull, with a high forehead, a large nasal area and big teeth, as well as other skeletal features typical of the Caucasoid race.

Investigations carried out by local historians and archaeologists during the last century have provided evidence that life in Central Asia began at approximately the same time as at other well-known Stone Age sites in Africa, the Middle East and Eastern Asia. Over 200,000 years ago, primitive people inhabited the river banks where artifacts belonging to the so-called "stone culture" (when stone tools[3] were in use) have been found. Archaeological sites dated to the Old Stone Age (Paleolithic) period include a site near the town of Naryn in the valley of the Ok-Archa River in Kyrgyzstan investigated by Okladnikov (1966), sites in Tajikistan on the plain of the Vakhsh River and sites in the Khoji-Bakirgan region in the Syrdarya Valley (Gafurov 1989).

The archaeological finds from the Mousterian culture (from around 40,000 BC) include a collection of stone tools found in the Teshik-Tash cave in the Kuhitang Mountains in Uzbekistan close to Baysun. This discovery comprises an enormous number of stone tools (more than 3,000 pieces) that were used for processing wood and skins as well as for hunting. More recently, Lvov discovered an Old Stone Age site in present-day Samarkand that shows evidence of a more advanced culture. The dwellings are made of clay and reed, and the tools relate not only to hunting and gathering but also to the domestication of animals (Masson 1950).

[3]The term *tool* is used to refer to something that has been used by a human or a human ancestor for some purpose regardless of whether the *tool* has been modified or not. Anthropologists commonly use the term *culture* to refer to a society or group in which many or all people live and think in the same ways. The earliest physical evidence of culture are rough stone tools produced in East Africa over two million years ago.

Professor E. Barker (1932) has asserted that humans of the Caucasoid race originate from Central Asia and then migrated to Russia and Europe. In a museum in Urumchi, the capital of the Xinjiang Uygur Autonomous Region, are mummies of primitive humans, which were found in the Takla Makan Desert. These were very well-preserved when discovered because of salt deposits in the soil. Following studies of these mummies, Barker concluded that the ancient Celts actually moved to Europe from Central Asia in the late Bronze Age.

Similar Neanderthal sites have been discovered in Uzbekistan, including at the Amir-Temir cave close to Teshik-Tash, at the Aman-Kutan site 25 km from Samarkand and the Khodjikent site 70 km from Tashkent. More than 100 monuments from this period have been discovered in Central Asia and archaeologists are sure that this region was one of the important centres in the development of human civilization (Shishkin, 1969).

Evidence of further development in human societies can be found at Mesolithic archaeological sites (dating from between 10,000 and 15,000 years ago) located in the foothills of the Kopet Dagh and near the Caspian Sea (Jebel, Dam-Dam Chashma). These provide evidence to show that fishing and (in the Neolithic epoch) hunting and gathering were replaced with cattle breeding and farming (of the Jeytun culture) (Lisitsina 1965).

It is interesting that monuments of the New Stone Age (Neolithic) are located in piedmont areas. Okladnikov considered development in this period in terms of the Hissar culture, the most significant monuments of which are located at Tatkaul near Naryn on the Vakhsh River. Excavations of archaeological sites dated between the tenth and eighth millennia BC show an ancient culture with high level of development. Gafurov characterized this important stage of development as "incipient farming."

Why are all these Stone Age monuments located in mountains or piedmont areas? Masson (1950) and Mushketov (1886) assume that at the beginning of the Quaternary Period most of Turkestan was covered by the Aral-Caspian Sea. Human settlements in Central Asia therefore arose beyond this huge sea, concentrating in piedmont areas with suitable conditions for life support.

2.1.2 The Keltiminar (Khorezm) Culture

The late Stone Age was the period of human history between 10,000 and 3,000 years ago. Large well-polished stone tools, sometimes with drilled orifices, were in use. Tribes had started to domesticate animals and form families. In Central Asia, the late Stone Age was marked by the Keltiminar culture (dating from the period between 5,500 and 3,500 years ago). In 1939, the first traces of this culture were discovered by academician S. Tolstov not far from Urgench in the Amudarya delta. This site was inhabited between the turn of the fourth and third centuries BC. In total, eighteen sites of Keltiminar culture have been discovered in Uzbekistan. These reveal similar features and Keltiminar culture has been recognized as a specific stage in the development of mankind. Keltiminar humans made their tools using stones and bones. They mainly ate fish, but their diet also included meat, which they got from hunting, and vegetables, which they gathered in forests. They cooked their meals in pottery that was made without a potter's wheel. The surface of this pottery was painted with a red paint. Keltiminar shelters were made of reed and wood, and were rather big structures (24 m by 17 m) with bearing members (pillars), rafters and lathing. A fireplace with room for numerous simple fires for preparing meals was located in the centre of each shelter. Another living space (for sleeping) was constructed separately. The whole clan consisting of some 100 to 120 people lived in these structures. If a family pair lived (or slept) separately, they did not yet have a separate family fireplace.

The Keltiminar culture represents one clear stage in the development of the region's civilization. Similar sites have also discovered outside of Uzbekistan, in South Xinjiang, Western Kazakhstan, the Lower Volga Region, the Southern Pre-Ural, North-Eastern Europe and South Siberia. However, these sites only show evidence of Keltiminar culture one or two thousand years later than the sites in Uzbekistan. This confirms that Central Asia was the starting point for the development of a human society that then spread northwards (Tolstov 1948).

2.1.3 The Bronze Age

Central Asia had one important advantage in comparison to other regions. These are its loess soils, which contain up to 20 different minerals that facilitate the growth of plants and make them resistant against various diseases. Loess (aeolian soil) is one of the most fertile organic (original) surface formations, though its fertility varies depending on its specific texture and chemical composition. Loess soils cover mountain slopes and inter-mountain valleys and troughs, and they are one of main natural riches of the region. The creation of "agricultural oases" is closely related to the formation and presence of loess soils. Undoubtedly, a major precondition for the emergence of tribes that engaged mainly in farming was the combination of abundant water resources and loess soils. These early farming-based tribes settled in the valleys of the Amudarya and Syrdarya, Zeravshan, Kashkadarya and Surkhandarya, as well as the Fergana and Hissar valleys.

At the end of the third millennium and at the beginning of the second millennium BC, wheat, barley and oats were already being cultivated in these locations. Archaeological excavations in Sapaly-Tepe have revealed remains of granaries, rounded stones for grinding grain and hoes in ancient houses. Archaeologists concluded that from that time – and possibly as early as the end of the fourth millennium BC – irrigated farming was practised in Central Asia. Irrigated farming started in the two most favourable geographical zones, piedmont valleys and the floodplains and deltas of the great rivers.

The environment has played an essential role in the development of human culture, especially during the period that saw the formation of the first farming communities. Similar cultures were usually formed under the same geographical and natural conditions, which explains the similarity (and parallel development) of the known ancient agricultural centres.

Information collected by the outstanding Soviet scientist Nikolay Vavilov and his followers allowed them to conclude that major ancient farming centres mainly arose in arid and tropical zones as shown in Figure 2.1 (Vavilov 1967). They indicate that Central Asia was one of the six most important areas.

Development in these early centre of agriculture is not uniform, but occurs in different time periods and with differences in the origin of domesticated plants. B.V. Adrianov (1969) wrote that the south-western agriculture areas, which date back to 7000–6000 BC, are the most ancient of the early agricultural lands in the Middle East and Central Asia. Numerous crops (such as wheat, rye, cucumber, carrot and cotton)

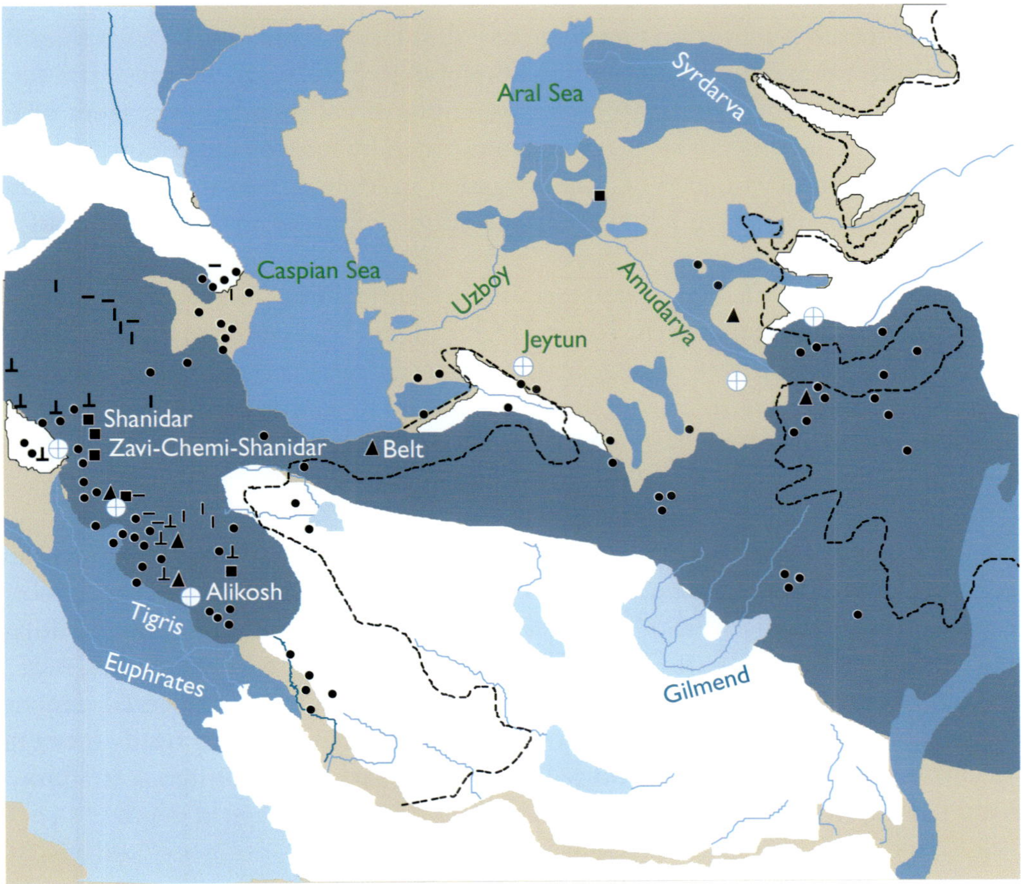

Legend

- ■ Pre-agriculture sites (11th–9th millenium BC)
- ▲ First agricultural settlements (8th–6th millenium BC)
- ⊕ Sites of flood irrigation in piedmont valleys (4th–3rd millenium BC)
- ● Sites of irrigated farming in the river deltas (2nd millenium BC)
- ⊥ Dry farming lands
- ı Sites with the remainders of ancient grain crops
- ⁄ Sites with the first irrigated crops other than cereals
- - - Desert zone boundary
- Alluvial river plains

Figure 2.1 Map of ancient agricultural centers in Central Asia and Asia Minor according to Vavilov[4] (1967).

[4]Vavilov, N.I. 1967. The agriculture. Selected works in two volumes, Leningrad.

were grown. Many other scientists have proved that centres of ancient civilization developed in Asia Minor, Mesopotamia, Central Asia and the Mediterranean Region in the same period (Bartold 1966, Herodotus 1968, Zhukovsky 1894, Masson 1950, Okladnikov 1962 and Yakubovsky 1947).

Areas cultivated with barley and wheat, adobe houses with wattles, as well as sickles with flint blade segments have been found during archaeological excavations in these zones. It is wrong to think that each region – and each zone within these regions – developed in isolation (Tzinzerling 1927). There was a permanent cultural exchange between the developing agricultural centres of the ancient world, which enabled knowledge and experience to be accumulated and handed over from one generation to another. In Central Asia, this can be clearly demonstrated by looking at the cultural heritage of the Geoksur agriculture dating from 4000 BC, the irrigation in Mohandir at the beginning of the second millennium BC and the irrigation in Khorezm in the middle of the second millennium BC.

Turkmenistan is the cradle of irrigated farming in Central Asia. Twenty seven mountain streams flow down from the northern slopes of the Kopet Dagh Ridge. These were the basis for settled life in the two oases at Akhal-Tekin and Atrek. Using small barrages, farmers created a manmade microrelief for irrigating their fields. There was no need to build any special structures for flooding fields, since it was sufficient to build small embankments along the edges of the fields to retain water for a period of time. These rather primitive methods initiated the development of irrigation.

The development of these ancient irrigation techniques in Southern Turkmenistan and their spread throughout Central Asia is a consequence of rather favourable conditions for developing irrigation (rich coniferous forests and brooks on the slopes of Kopet Dagh, fertile soils, a climate that was milder than today) and proximity to the centres of ancient civilization in Mesopotamia, Asia Minor[5] and India. Thanks to the resilient environment, a significant number of well-preserved remnants of these ancient cultures based on irrigation have been found and many remained practically intact until the last two decades. Prominent archaeologists and paleontologists such as Bukinitch, Okladnikov, Masson, Lisitsina, Kuften, Hlupin and Sarianidi have studied this cultural heritage in depth.

The well-studied caves of Jebel, Kailso and Dam-Dam-Cheshme in Turkmenistan, as well as the Kzil-lay and Cuba-Sangjr caves located on the Krasnovodsk Peninsula are cultural heritages from the Middle Stone Age and Old Stone Age. Bukinitch (1924) was the first to discover these sites, and he concluded that in the eighth and seventh millennia BC the first humans settled in these piedmont valleys, which had enough rain to sustain agriculture, warm winters and an abundance of wild animals and fish.

The practice of a mountain brook type of irrigation started in the Mesolithic period in areas where the harvesting of wild cereals was well-organized (Andrianov 1969). First, humans began to reap and thrash grain before later introducing farming activities such as tillage and sowing. According to Andrianov: "... *when in piedmont valleys, gatherers of wild cereals understood that plants are growing much*

[5]Some of the earliest Neolithic settlements in the Middle East have been found in Asia Minor.

better on naturally flooded plots, they arrived at the idea of diverting water to the plots under cereals using small guide dams made of tree trunks, stones, etc. It was rather easy to divert water from small mountain creeks into ditches with sufficient gradient and then towards irrigated plots flanked by levees." Okladnikov (1962) and Bukinitch (1924) came to the same conclusion based on their studies of sites in the Hissar Valley (such as Tutkaul and Cub-Burien) as did G.N. Lisitsina (1965) who researched Kopet Dagh's historical monuments (Jelum and Bami) dated to the Late Stone Age.

In Southern Turkmenistan, there is well-preserved evidence of one of the most ancient settled farming communities called the Annu culture. Excavations made by Komarov, the American archaeologist Pampelli, Litvinsky, Kuftinsky and Masson discovered rather large ancient settlements. The most significant of these settlements is Namazga Tepe. This settlement covered about 70 hectares and was surrounded by irrigated plots that were planted with a diverse set of crops. At that time, bronze agricultural tools were already employed for cultivation of fields. Masson (1935) considered that Southern Turkmenistan was the most advanced area in Central Asia, belonging to a belt of first time and early settling farming communities that adjoined the ancient civilizations of the East.

Another community of ancient farmers existed in the Fergana Valley. This was the Chust culture, and it was similar to the Late Annu culture. Excavations in Kuchuk Tepe and Sapali Tepe provide evidence that primeval-agricultural settlements with quite highly-developed socioeconomic and farming systems existed at these locations from the end of the third millennium BC (Gafurov 1989).

Karez[6], the subsurface water supply systems used by local inhabitants for delivering water to distant steppe settlements, were used for the first time in Central Asia in the Akhal-Tekin oasis. Traces of more advanced irrigation systems remain in the Tejen delta. This system consisted of two irrigation canals each about 2.5 km long that were constructed almost at right angles to the river channel. Irrigated agriculture was introduced considerably later in the central and eastern parts of Central Asia (around the middle of the second millennium BC). It was more developed in the floodplains and deltas of large rivers such as the Amudarya and Zeravshan. Local farmers could not divert water from the river channels or train the river flows because the riverbanks were high and they only had a low level of engineering knowledge. Therefore, they used the floods of lateral deltaic streams. Flood water was stored in natural depressions or by means of manmade diking along the borders of the field. As a result, the water content of the soil was sufficient for growing wheat, barley and millet.

One more centre of manmade irrigation was established in the lower reaches of the Syrdarya River. It utilized a form of flood irrigation. During high water, river water flooded riverside depressions, causing a natural circulation of the water. As the lakes formed during high water gradually dried up in the summer, farmers sowed crops at the edges of the receding water. This activity was repeated for a few years at one location before the farmers moved to another place. This practice was unreliable, and it completely depended on the flood flow rates, the heights of embankments and the unpredictable behaviour of the river.

[6] Karez (also *kanat* in the Middle East) are gently sloping underground channels or tunnels constructed to lead water from the interior of a hill to a village below. Some remain in existence today.

The Keltiminar culture was a step forward in developing irrigated farming but it did not go beyond primitive irrigation methods. After the Keltiminar culture, new cultures can be considered to have made progress in developing manmade irrigation systems. The first of these cultures were the Tazabag-Yab and Amir-Abad cultures in the Khorezm region of the Amudarya delta. They were first discovered and studied by academician S. Tolstov. Farmers began to use horses as working animals in combination with wooden ploughs and hoes. These farming tools were made of hardwood (cherry or elm wood), and they were the basic implements for tillage and the construction of manmade water flow passages.

In the lower reaches of the rivers, 6–20 m wide barrages were built within a river channel to divert water into small irrigation ditches a few tens of metres in length. Some larger irrigation systems were also coming into use. One of the largest irrigation systems discovered that dates from the second half of the second millennium BC had a length of 2.5 km and served 70–90 hectares of arable lands. The lateral canals, each 0.5–0.7 m wide, branched off the main 2–3 m wide irrigation canal at right angles at intervals almost along its whole length (Dingelshtadt 1895, 1893).

2.1.4 The Iron Age

The Iron Age started at the beginning of the first millennium BC, and iron began to be used by humans in different parts of the world at around the same time. The origin of the first iron tools and weapons can traced back to the military campaigns of Cyrus the Great and Darius I, kings of Persia in the sixth and fifth centuries BC. Later the Greeks headed by Alexander the Great brought many diverse iron weapons and housewares on their campaigns. The first emergence of iron and iron tools in Central Asia occurred at that time as well. In 1959, an ancient settlement dating to the sixth and fifth centuries BC was found in the Dingilje oasis located in the eastern part of the Amudarya delta, close to the Kyzyl Kum Desert. It is still known as the Dingilje settlement. Agriculture was the basis of the Dingilje economy, but some craft workshops were also found. The earliest tools and housewares made of iron, such as cauldrons, pans, knives, sickles, needles, nails and pins, discovered in Central Asia were found here (Tolstov 1948). Tolstov discovered some similar settlements elsewhere in the region, including Koy Krylgan-Kala, Ayaz-Kala, Bazar-Kala and Kyuzeli-gir. These findings produced convincing evidence that iron tools were used in the economy of Central Asia around this time.

A new era in the development of irrigated farming and irrigation systems ensued. Iron tools, such as the spade, hoe and ploughshare, played a revolutionary role in driving technological development in the agricultural sector. Even the hardest of soils could not resist agricultural tools made from iron. Iron tools allowed agricultural production to extend over larger areas, by helping humans construct deep and wide irrigation canals and clear woodlands for cultivation. Primitive mining (to extract metallic ores), iron processing and manufacturing of iron tools emerged in the Fergana Valley.

As the population grew rates and the handicraft industry developed, demand for agricultural output increased. At the same time, it became more feasible to expand and technically upgrade irrigation systems with the establishment of city-states and larger entities such as ancient Khorezm. There was astounding growth in the use of irrigation in the period from the fourth century to the first century BC. During that period, there were drastic changes in irrigation technologies relating to improved water delivery systems as a whole and to the design of main irrigation canals. The extent of irrigation canals drastically increased, reaching hundreds of kilometres. Many small local irrigation systems were united into larger irrigation systems, and there was a shift of headwater intakes further upstream.

The progress in developing irrigation methods and irrigated farming practices resulted in vast areas being irrigated and developed on the plains and in the basins of the great Central Asian rivers. Archaeological studies show that in the fourth century BC the irrigated lands in the lower reaches of the Amudarya, Syrdarya and Zeravshan rivers exceed the present-day irrigated area several times over. In the first century BC the total irrigation area in the lower reaches of the Zeravshan and Kashkadarya rivers amounted to about 600,000 hectares, exceeding the present-day irrigated area in this region by a factor of two. In the lower reaches of the Amudarya and Syrdarya rivers, the area with a developed irrigation network was four times larger than it is at the present time. This most likely also means that the water withdrawals from these rivers were also larger than at the present time (Dingelshtadt 1993).

In the Iron Age, irrigation development shifted from the piedmont areas towards the alluvial and deltaic plains of the rivers Amudarya, Syrdarya and Zeravshan. The first large canals that diverted water directly from the main streams of the rivers rather than from their branches were built here. Head regulators were built, and there was large-scale agricultural development on the vast deltaic plains. The irrigation canal in Bazar-Kala (right-bank Khorezm) was 40 m wide and more than a kilometre long. In the process of construction of this canal about 65,000 m^3 of earth had to be excavated, which took the intense work of about 500 diggers (unskilled workers) for 35 to 40 days.

Another big irrigation canal built at the section from Adamly-Kala to Janbas-Kala (right-bank Khorezm) was 15 km long. The irrigation network covered an area of 2,000 hectares. According to our estimates, the total volume of earthworks to construct this canal would amount to about 400,000 m^3. It would have required 5,000 diggers working for some 30 days to get this done. It took about 2,000 people to clean this canal from sediments every year. This type of project was possible only under the governance of a strong centralized government. The need to protect the interests of irrigation scheme owners and to build the required irrigation systems prompted the development of oriental authoritarian forms of governance in Central Asian states. Typically a strong personality made decisions and was responsible for the welfare of his region or country. In this way, the state became the entity in Central Asia with complete authority for decisions on tackling water problems and responsibility for the sustainability of irrigation water supply and defending the water rights of the nation.

As climatic conditions were rather severe and the irrigation systems were quite unreliable, a wrong decision or a failure to safeguard irrigation systems through neglect could have fatal consequences. The state represented a regime of absolute power, and the first and principal responsibilities were to solve water issues, mobilize the population for implementing public works, collect taxes and defend its own territory. Private ownership of land was not permitted. The ruler was supreme and owned all the land.

The growth and development of ancient urban areas was closely related to the construction of the main irrigation canals. Some cities grew up around the head diversion works due to the need to guard the water, but more often irrigation systems were built near existing major settlements. The established population usually initiated the construction of water infrastructure. In this way, irrigated oases that supplied agricultural products to urban inhabitants were created around ancient cities, and water was supplied through irrigation canals to these cities to satisfy the needs of the urban population. Over time, these territories became developed states. As well as the attempts to manage water in ancient Khorezm, there were also other developments in irrigation that created the basis for future water systems in Central Asia. These events will be described in the following sections of this chapter.

2.2 The Development of Flood Irrigation

Most non-Soviet researchers consider that flood plain (liman[7]) irrigation is the oldest method of irrigation. It may be true that some primitive forms of flood plain irrigation were in use by the early Neolithic period in the sixth millennium BC. However, scientists of the former Soviet school (Aladin and Plotnikov 1995, Andrianov 1969, Dingelshtadt 1895, and others) believe that a mountain creek type of irrigation was the earliest of the ancient methods of irrigation. In piedmont valleys, irrigation using mountain creeks could be easily organized by building small guide dams.

What is clear is that step by step humans began to regulate seasonal river floods, clean stream channels from sediments and direct floodwater to downstream plots that were prepared in advance. With the strengthening of social structures in communities and the growth of technical capacities, came opportunities for constructing offtake channels and, as a result, flood plain irrigation reached the river deltas by the beginning of the fifth millennium BC. Traces of floodplain irrigation using reservoirs have been found in areas of Southern Turkmenistan (Geoksur, Annau 1, and Namazga Tepe 1). G.N. Lisitsina, who researched ancient sites in the Tejen Valley, has mentioned that some plots were separated and bordered in locations where water was delivered through irrigation ditches.

G.N. Lisitsina (1965) writes: "*During the period between the fifth and third millennia BC, Neolithic settlements have come into existence in the piedmont areas of the Kopet Dagh (Namazga Tepe 1, near the village of Kaahka). These small settlements were created on the alluvial fans of mountain streams and small rivers that flowed down from the Kopet Dagh Ridge and flooded piedmont plains. Ancient farmers were separating some flooded areas by small earth bunds[8], creating small limans, where water was being kept a little longer. Cereal crops were sown on plots with wet soils without preliminary tillage. Since soils in the piedmont zone almost didn't have a turfing, stone hoes were not really necessary because under those conditions even a digger-stick was completely sufficient.*"

[7]The term "liman" means a flooded pasture area in lower reaches of semi-desert rivers or a small coastal lake; liman irrigation is spring-flood irrigation.

[8]Bunds are low walls of earth that surround any land plot or rice field (rice paddies) to retain water.

Not only were the piedmont zones developed at the time of the Namazga settlement, but also the ancient delta of the Tejen River. In the early Neolithic period, rather large cities were built in the deltas of the Loin-Su, Dorungar, Akmazar and Cha-cha rivers and some previous settlements built on the banks of small rivers were abandoned. Agriculture output that relied on the water resources of small creeks became insufficient for satisfying the needs of growing populations (Itina 1959).

The cultural heritage of the early Neolithic period has been thoroughly studied. A large-scale study of the Geoksur oasis (now in Southern Turkmenistan) was made using aerial survey methods that provided unique evidence of ancient irrigation in the Eneolith period between the second half of the fourth millennium BC and the beginning of the third millennium BC (see Figure 2.2). The aerial

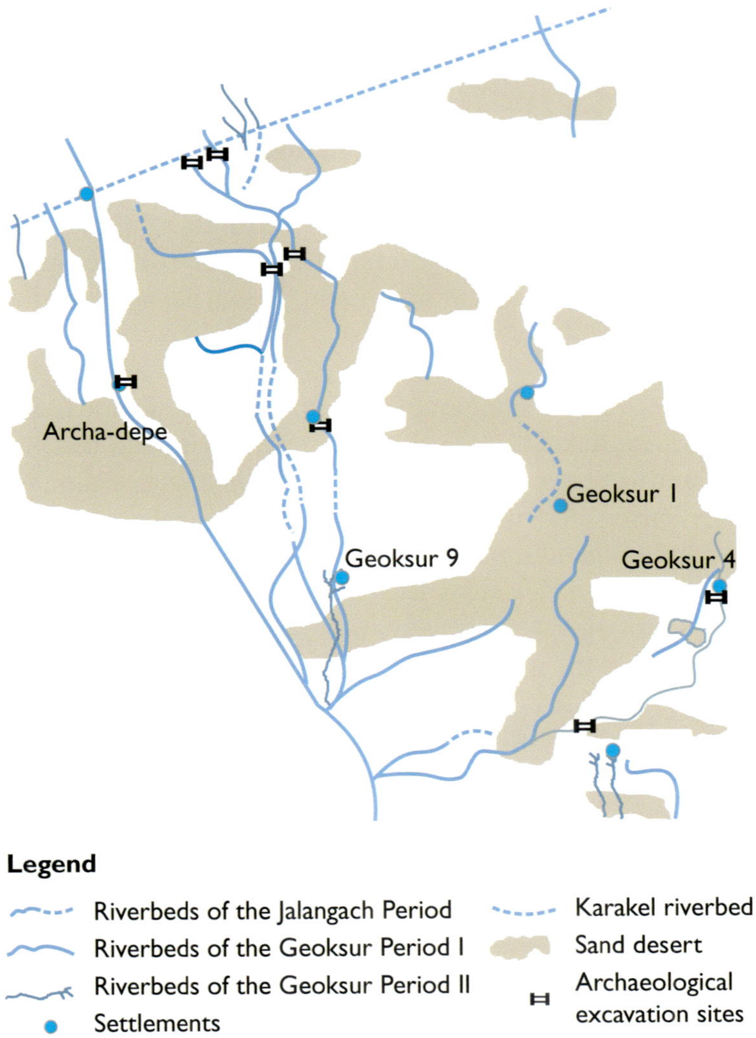

Figure 2.2 Map of the ancient hydrographic network of the Geoksur oasis (Livshits 1963).

photography revealed many distributary channels over the ancient Geoksur delta plain[9]. Still visible in this delta plain are the northern distributary channels (which were subject to rather rapid silting) and the southern distributary channels (which existed for almost 1,000 years). Ancient tribes settled along the deep distributary channels in the Tejen[10] delta (the Geoksur oasis). These river channels flooded vast areas and so created the right conditions for primitive agriculture.

An analysis of the aerial photographs together with data from field surveys in the Tejen delta plain allows the conclusion that irrigated farming was being practised in the Geoksur oasis from the beginning of the second half of the fourth millennium BC. This was based on harnessing spring floods and using regular irrigation through a system of irrigation canals. At the end of the fourth millennium BC, special water reservoirs were created by heightening the riverbanks, and these served as additional sources of water in dry seasons (in a simialr fashion to modern Indian "tanks"). The canal system had laterals at an angle to the main bends. The canal sections were typically shaped like a wash tub, with vertical or semi-vertical walls and a U-shaped bottom, 2 m wide and 1.5 m deep. Fields stretched along canals in a 250 m wide belt. Archaeologists have found three groups of buried canals, the oldest of which had a width of 5 m at the bottom and were 1.24 m deep with a slope of 20 to 50 cm/km. According to N.G. Minashina (in Khidoyatov 1990), these canals were under operation for a relatively short time, which can be explained by migration from the Tejen tributary, the volume of silting and, apparently, the difficulty of coping with these complicated delta canals because of insufficient labour and the low technical level of the structural designs. Overall, Lisitsina (1965) and Masson (1950) estimate that the Geoksur oasis existed for 1,500 years until the middle of the third millennium BC.

At the beginning of the third millennium BC, the practice of irrigation had spread over the Murgab delta, and in the second millennium BC it reached ancient Dahistan and was in use in the ancient delta areas of the Atrek and Sumbara rivers, known as the Meshed-Misrian plain. When Masson, V.M. was studying these territories during 1951 to 1956, he found a large and well-developed irrigation system that was built in the second millennium BC and existed until the Middle Ages (Figure 2.3). A series of canals constructed at designated elevations were discovered. The widest canal (with embankments 2–2.5 m high and a bed width of 30–40 m), known as the Shah-duz Canal, diverted water from the Atrek River and was apparently the main waterway in ancient Dahistan. The excavations reveal four categories or types of canal:

a) main canals: 5–9 m wide, 2.3–3 m deep
b) laterals: 1.5–3.5 m wide, 1.2–1.6 m deep
c) sub-laterals: 0.8–1.3 m wide, 0.9 m deep
d) field *ariqs*[11]: 0.5–0.7 m wide, 0.4 m deep.

[9]The delta plain is the area of extensive lowlands with active and abandoned distributary channels.

[10]Tejen River – Harīrūd River (in full *Rūdkhāneh-ye Harīrūd*), ancient Arius River in Latin (sometimes Garirud River).

[11]An ariq is an irrigation ditch.

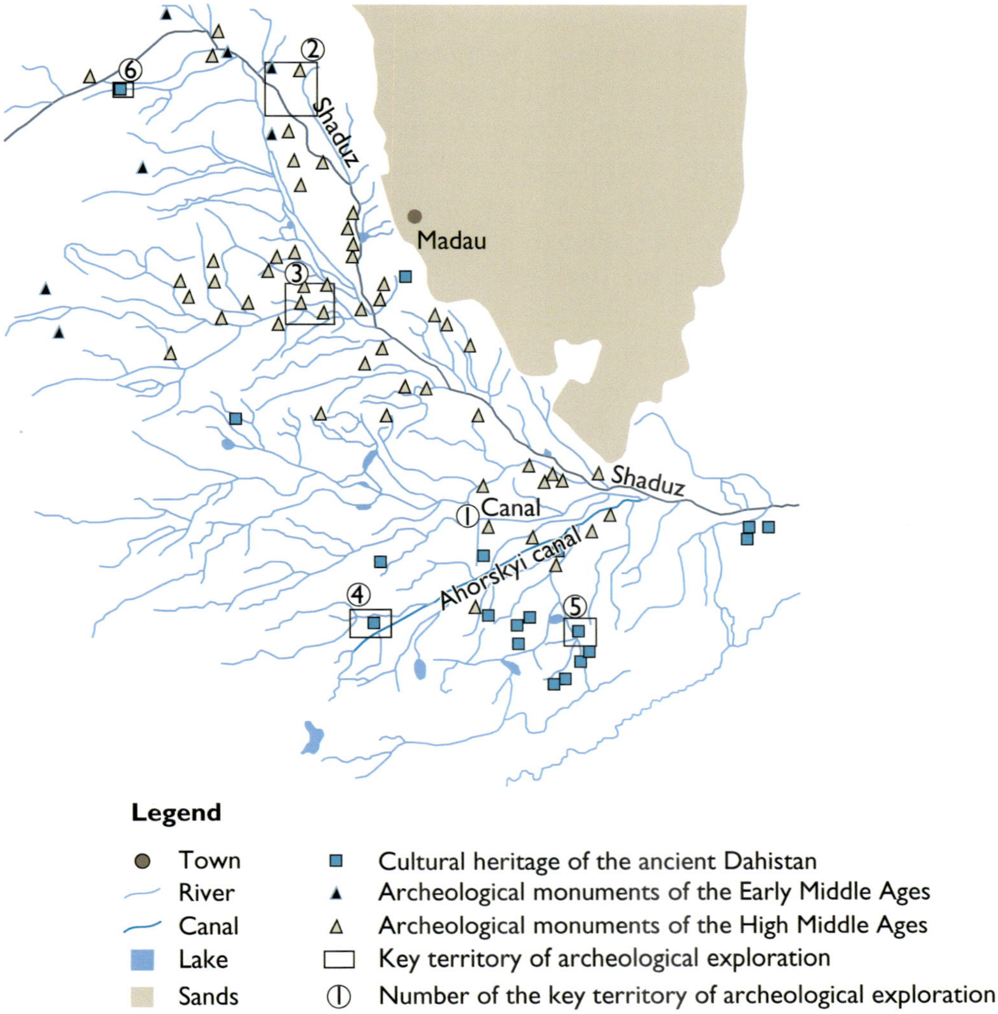

Figure 2.3 Map of the ancient irrigation networks in Dahistan (Obruchev 1951).

Based on written memoirs of ancient travellers, Bartold considered that the Shah-duz Canal existed for almost two thousand years. It diverted water from the Atrek River 60 km upstream of Chata, crossed the Sumbara riverbed and, with embankments 2.1 m high, conveyed water to the irrigated area.

Minashina, a soil scientist of worldwide reputation who participated in this archaeological research, considered that a system of farming[12] based on the principle of so-called "dry draining" was used on all these ancient irrigation lands in Turkmenistan. (this view by Minashina is supported by academician V. Kovda.) The actual area irrigated under this method of drainage amounted to no more than 20–25%

[12]Lea tillage or shifting agriculture.

of all irrigable land. Each plot was under irrigation for three to six years until salt accumulation in the topsoil reduced the crop yield too significantly. The plot was then abandoned, and farming was moved to new plots. Although first impressions might be that large areas were under irrigation, actually shift cultivation was taking place. Peter Sinnot, Director of the Caspian Project at the Columbia University, commenting on these methods of irrigation concluded: "*The tragic irony is that this region [South Turkmenistan] was home to one of the largest and most efficient irrigation systems in history until the Mogul invasion destroyed much of this network*" (quoted in Richard Stone, "A New Great Lake or Dead Sea", *Science*, AAAS, Volume 320, 23 May, pp. 1002–05).

A.V. Kaulbars (1903) described the development of liman irrigation in the Amudarya floodplain as follows: "*Downstream of Kerki, the Amudarya current steadily diverged to the northwest, towards the piedmont terrace, where it was slowing down, and the river was starting to deposit its sediments at some distance from the foot of the steep slope of the plateau. From time to time the riverbed has become silted and formed limans (depressions). People living along the riverbanks and observing how these limans were overgrown with reeds and bushes started to clear these liman areas of wild plants and cultivate different crops there. Since the river diverged further, people started to build up long levees and clean the channels, especially those which were connected with the main river stream.*" Strips of liman (flood) irrigatioan, large-scale traces of which were discovered by Tolstov in the Khorezm oasis, may have originated in this way. With a high degree of certainty, Tolstov describes the shift from liman irrigation to regular irrigation in his book *Ancient Khorezm* based on information collected during a field expedition from 1938 to 1941.

Tolstov also made a correlation between the formation of irrigation practice developed in this region and "the period of the last Aral transgression". A system of flood irrigation developed in Ancient Khorezm in much the same way as described by Kaulbars. Tolstov was firmly convinced that prior to developing this deltaic irrigation network, the reaches of the river in the ancient delta were densely populated by an agricultural population who used flood irrigation in similar way to the methods deployed in the ancient Nile floodplains. The configuration of this ancient irrigation network almost completely follows the morphology of the ancient delta. The five ancient branches of the Amudarya River (from west to east, these are the Tunkdarya-Qangadarya, Daudan-Mangitdarya, the present riverbed, Daryalyk and Su-Yargan) formed a fan-shaped configuration. (These were mentioned in ancient legends and also by Herodotus.) Ancient canals were built along the midline between these river branches. At first it is assumed that people tried to preserve (or rebuild) these existing systems of flood irrigation (as shown in Figure 2.4). Later, flood irrigation became more widespread. For example, some modified forms of flood (liman) irrigation were found in the lower reaches of the Syrdarya River.

In Khorezm, the transition from a nomadic lifesyle based on herding to a settler lifestyle based on farming started at the end of the second millennium BC. Indeed, a semi-settled lifestyle (based around cattle breeding and farming) is still followed in the lower reaches of the Syrdarya River today. The practice of irrigation became widespread throughout the whole region by the middle of the first millennium BC. Primitive systems of controlled flood irrigation were developed. Former riverbeds (and their branch channels) were used as water storage basins and irrigation was of a "tidal type", because water was supplied from the river only during the high water seasons. It was a relatively simple system of irrigation. During the high water seasons river water filled the in-channel water storage basins built

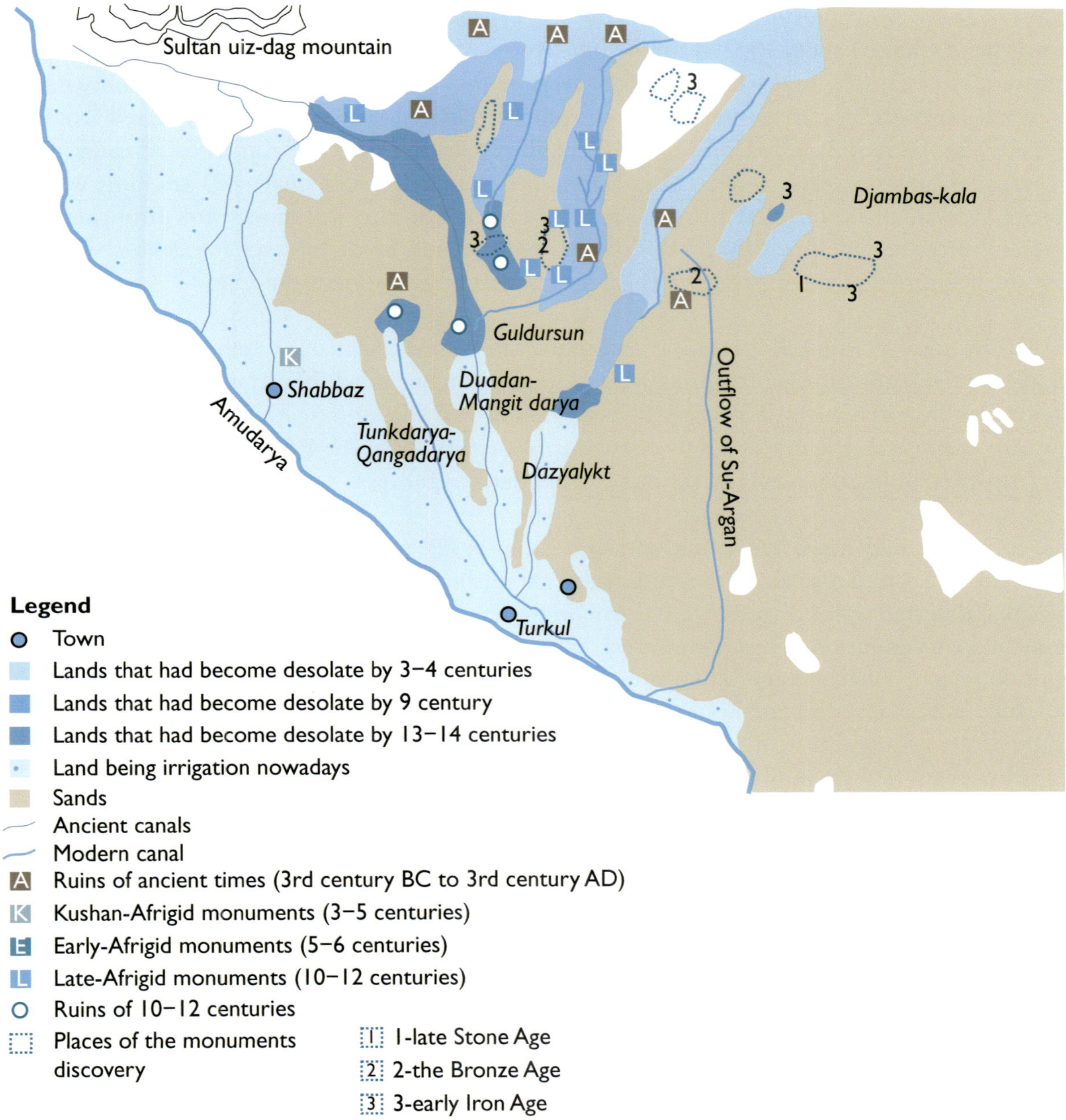

Figure 2.4 Amudarya Delta during the fourth and third Centuries BC (Tolstov 2004).

within the former riverbeds, and then water was delivered to the fields through controlled branch channels and small irrigation canals.

In the period from the fourth to the second centuries BC, along the middle reach of the Jana Darya River, near Chirchik-Rabat and Babish-Mulla, irrigation was based on the use of former riverbed branches of the so-called "inner delta". Deepened sections of former riverbeds were turned into water storage basins (a generally accepted method of irrigation in the ancient East).

The same principles of irrigation – the use of former riverbeds and in-channel water storage basins – were adopted about 1000 BC in the Jety-Asar oasis. Later, in-channel water storage basins were improved. For example, in the Upper Anka-Darya region various types of retaining dam were widely employed. Nevertheless, irrigation remained quite primitive, since water was diverted from former riverbeds that were only filled during floods, rather than fed directly from river streams. Even the biggest canal constructed along the Inkar-Darya, the Asanas-Uzyak Canal at 60 km long, was only an improved variant of a controlled former riverbed. However, around this time, water-lifting devices such as *chigirs*[13] became widespread. With the adoption of this type of water diversion device, we can speak about the beginning of a new period in the development of irrigation techniques.

Many locally adjusted irrigation techniques prevailed, such as systems with retaining dams and in-channel water storage basins as well as small-scale but much branched systems with different water controlling structures. This stage of using delta riverbeds for irrigation, as typified in the ancient Khorezm oasis (from the middle of the second millennium BC to the 800s BC), lasted almost until the end of the first[t] millennium AD in various ever modernized forms.

In the Middle Ages, beginning with the Karakhanid period (940–1212 AD), water for irrigation on the left bank areas along the middle stretch of the Syrdarya River was supplied through large canals that diverted water directly from the main Syrdarya channel. These systems had headworks and a branched configuration. They were rather extensive (30–40 km long) and resembled the mediaeval irrigation systems of Khorezm. The water supply of the big mediaeval towns located in the Syrdarya basin, such as Kyr-Uzgend and Mairam-Tobe, made use of these systems.

However, the irrigation systems in the lower reaches of the Syrdarya River were different those constructed in Khorezm and along the middle stretches of the Syrdarya River. Although there were different natural hydrographical conditions in the lower reaches of the river, there were other reasons that account for the different irrigation systems in use. In the first millennium BC, the lower reaches of the Syrdarya River were inhabited by the Saks tribe[14] who raised livestock and practised primitive farming. Therefore, there was no historical prerequisite for developing a large-scale irrigated economy, although settled communities emerged in this region from time to time. The influence of specific historical and political factors was also significant, since the Syrdarya region was bordered with areas occupied by nomadic tribes (Turs, Huns, Turks, Kimaks and Kipchaks) who often attacked farmers. The economic conditions were also a factor. Local tribes specialized in cattle breeding and used the steppes as the grazing areas for their cattle. Their economy was based on a combination of cattle breeding, farming and fishing that had changed little since the Bronze Age and provided the basis for patriarchal-tribal organization of these ancient communities. This explains the archaic irrigation techniques in use in the lower reaches of the Syrdarya River, which relied on methods such as limans, semi-stagnant water bodies and small-scale, labor extensive, irrigation systems.

---

[13]A chigir is a simple water-lifting device also known as the Persian wheel. It consists of a partly submerged vertical wheel with buckets attached to the rim. As the wheel is turned by the water stream, the buckets are filled and emptied into a reservoir and conveyor system above that carries the water to the fields.

[14]This name was given to this tribe by ancient Persians.

2.3 Irrigation in Oases

Throughout the history of Central Asia, irrigation in oases with favourable natural conditions was continuously being developed and improved. In addition, previously abandoned irrigated areas and irrigation systems were sometimes rehabilitated. Archaeological studies of ancient irrigation systems and irrigated farming areas in the deserts of ancient Khorezm and Turkmenistan (an area not coincident with contemporary Khorezm) show that some irrigation works remained unused for many centuries between periods of operation. Some fragments of the ancient irrigation infrastructure remain, although they have been abandoned and left buried for a long time. However, there are not many traces left of ancient irrigation structures, most of the ancient forms of oasis irrigation have been changed, rearranged or cruelly ruined.

This means that the centuries-old evolution of irrigation development in oases can only be evaluated from the discovered fragments. According to Dingelshtadt (1993), traces of advanced ancient agricultural activities and remnants of ancient canals, dams and waterworks were found in the valleys of Arys, Aksay and Mashatu both in the second half of the fourteenth century AD and during the Russian colonization of Turkestan. As G. Rizenkampf (1930) wrote: "*The fact that Aryan tribes have settled here can be confirmed by the development of large-scale irrigation that provided the prosperity of ancient, mediaeval, agricultural centres such as Sayram and Otar, which were ruined by the Mongol conquerors in the fourteenth century. The Zach and Khanum canals that, according to legend, were named after the lord of these lands and his wife in the earliest times, are the best preserved among other ancient canals.*"

From time immemorial, the principal oases (the centres and zones of ancient irrigation systems) emerged along the main rivers and their tributaries and were sustained by a regime of natural seasonal flooding or manmade irrigation. They appear as if threaded on the branches of these rivers. This can be seen in the Samarkand[15] and Bukhara oases located in the irrigated valley of the Zeravshan River, the Tashkent (Shah) oasis in the valley of the Chirchik River, the Khorezm oasis in the lower reaches of the Amudarya River, the Fergana and Khujand oases located along the middle reach of the Syrdarya River, the Surkhandarya oasis in the valley of the Surkhandarya River and the Katadian oasis in South Tajikistan in the valley of the Vakhsh River (Figure 2.5).

It is difficult to determine an accurate date for the beginning of these and other oases. However, information from Chinese sources (from 1149 BC), Xenophon's[16] writings (*Anabasis*, *Hellenica* and *Cyropaedia*, the latter being an idealized biography of Cyrus the Great), the writings of Pythagoras, who visited the Arius River valley during the reign of Darius the Great (son of the Persian noble Hystaspes) at the end of the sixth century BC, and Herodotus in his great work *The History* provides evidence of large-scale irrigation development in this area in the first millennium BC. Herodotus wrote that the north-eastern part of the ancient Persian Empire (Turkestan) was inhabited by Chirkans, Bactrians,

[15]Also known as Maracanda, the city was the capital of Sogdiana, an ancient Persian province.

[16]Xenophon (430–355 BC) was a Greek historian, soldier, and essayist, whose works greatly contributed to the knowledge of Greece and Persia in the fourth century BC.

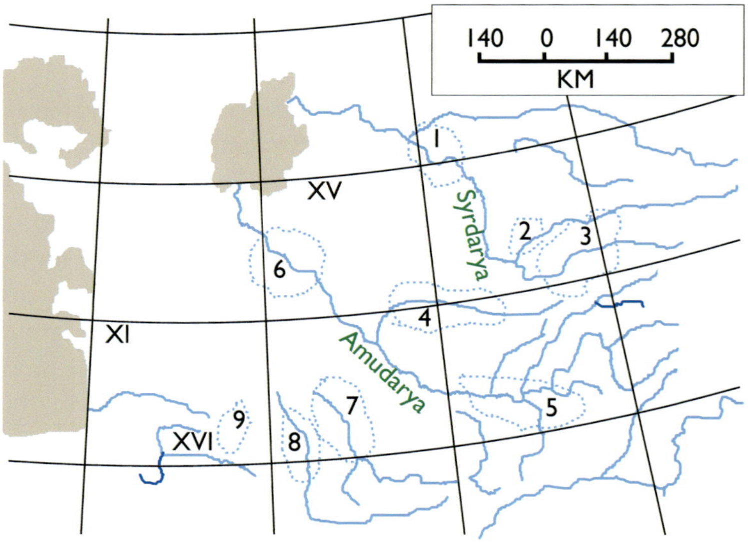

Legend

1 Otrar
2 Shash
3 Fergana and Hudjient
4 Maracanda and Bukhara
5 Katadian
6 Khorezm
7 Merv
8 Harirud
9 Geoksur

Figure 2.5 Central Asia during the fourth and third centuries BC (Massalsky 1969).

Khorezmians and Saranians (Soghds). It comprised two satrapies[17] of Persia that paid contributions to Persia of 600 and 300 talents of silver, "some 200 talents more than were paid by the prosperous Egypt" of that time (Herodotus 1968). Pythagoras witnessed that the Aryan state, which was located north of the Hindu Kush, had an abundance of water provided by the Arius River. Chinese sources such as Chjani-Kan reported that in the territory of Ansi, along the Gool-Shuri River (the Amudarya), there were many small towns and cities, and the population traded with neighbours, travelling distances of more than thousands of li[18] by land and by water. Another territory located along this river is Dakhi, which trades with Shenody (India) located at a distance of several thousand li to the south-east (Tchaikovski 1904).

Sogdiana and Shash undoubtedly were the largest oasis at that time. Sogdiana was a separate economic entity located between the Zeravshan and Kashkadarya rivers. The Zeravshan River, although it was considered as being inferior relative to the Amudarya, was of great importance for the economy of the region. Some Arabic travellers in the fourth and fifth centuries AD referred to the Zeravshan Valley as "God's own country". The Greek historians Aristobul and Strabo testified that the Zeravshan

[17]A satrapy is a province governed by a satrap.
[18]A Chinese unit of distance equal to about 0.6 km.

River (*Polytimetus*) originates from the vast Zeravshan Glacier, which is located 3,000 m above sea level and has huge reserves of fresh water. The glacier disappears among desert sands approximately 30 km before one comes to the Amudarya. The Zeravshan River has never lost its value as a source of life and well-being. The Sogdiana oasis stretches far away into the steppe in the form of emerald green orchards and fields. Two of the most famous cities of Central Asia – Bukhara and Samarkand – are located in this oasis.

The Sogdiana was bordered on the south by the Kashkadarya oasis, a manmade oasis. The Eskyangar Canal, of 100 km long, was already built here in the fifth century BC. After being restructured it has now existed for over a thousand years and the current concrete-lined canal is still kept in good condition by the local population (see Figure 2.6). Historical data provides evidence that at the beginning of the first millennium BC, the Kesh state (in Greek documents, Nautaka) became established in the Kashkadarya Valley near the foothills of the Hissar Ridge and the Nakhsab state (in Greek documents, Xenippa) occupied the middle and lower reaches of the Kashkadarya River. In the fourth century BC, these states became the XXVI satrapy of Darius I. Several ancient settlements, which later became cities, also developed along this canal, including Tallaktepa, Erkurgan, Elkendepe, Kesh, Chirakchi, and Kitab.

When the depletion of water resources of this canal system became evident, it was decided to build a link to the Dargom Canal in the Samarkand oasis. Construction of this interconnecting channel resulted in the development of a chain of new villages, fortified farmsteads and castles, as well as a line of outposts. Dry farming and cattle breeding in the vast grazing areas around these locations contributed to the economy. This interconnecting channel facilitated the establishment of a conglomerate society based around a single irrigation system with a main canal some 200 km long, which was one of the biggest ancient canals in Central Asia.

This water infrastructure is evidence of the wise and purposeful policies of well-established states. One of the main state functions consisted of developing and maintaining the irrigation

Figure 2.6 The Eskyangar Canal as it is today (2010).

systems. A quite advanced state was established in Sogdiana by the fifth to fourth centuries BC. Ancient Afrosiab, which later became known as Samarkand (although according to Chinese sources it was called Kann), was the capital of this independent state. The state encouraged construction of settlements, turning some into monitoring posts near the headworks of large irrigation systems for water sharing (Suleymanov). For example, the fortified post at Varagsar controlled water diversion into the Dargom Canal and monitored water levels and water quality became a key site. It was important to protect the dam and headworks of this canal because it supplied water for all irrigated areas around Samarkand.

The state also established a strictly controlled system of water accounting and water use monitoring. Books with records of water supply and water use in the Zeravshan oasis remain intact to this day. Water volume was measured by *ravaks*[19] (a local unit of volume where one *ravak* consists of 40 *kuraks*). One *kurak* was judged sufficient for irrigating a plot of 5 *tanabs* (about 120 hectares). In the fourth century BC, the flow rate in this canal was 21 *ravaks* per second (Mukhamejanov 1969). The potential water supply determined the development and extent of cities and urban settlements. Not many settlements existed in Sogdiana (an exeption is the large settlement at Daratepe) or expaned into cities due to the scant water supply. It required considerable labour inputs to arrange irrigation on uneven terrains and many project were abandoned over time (Suleymanov 2000).

It is worth noting here the rationale behind the irrigation system in the Karshi oasis. This irrigation system comprised a multi-branch network in the form of clusters of irrigation canals that allowed water to be redistributed between canals by means of simple operations. Usually settlements also concentrated in clusters, contained within the zones established by tail ends of small irrigation canals. A fortress encircled by moats was usually built at the river bend close to the water intake of the icanal that supplied the whole oasis with water, which allowed the status of the irrigation system to be monitored.

The Tashkent oasis was another important ancient centre of irrigated farming in Central Asia. Archaeologists have discovered 97 ancient settlements, 13 of which can be referred to as urbanized areas. Kanka, which covered 150 hectares, was the biggest city and the economic centre of the Tashkent oasis. The mainstay of the economy of this oasis was permanent farming based on irrigation, and this was supported by a complete network of canals, dams and small water reservoirs. The territory covered by the oasis included the valleys of the right-bank tributaries of the Syrdarya River – Chirchik (ancient Turk), Keles and Angren – which were covered with loess-like soils and cut deeply into the mountains. The Chirchik River was the most significant water source for irrigation in the Tashkent oasis, and some 42 main canals diverted water from this river. Masalsky (1913) wrote: "Some of these *ariqs* of ancient origin look like rather big rivers measured by water abundance and length." He described the Boz-Su, Zakh and Salar canals and others. At present, it is known for certain that these canals were constructed in the first century AD. The Zakh Canal was 20 km long with numerous barrages and laterals in the shape of *ariqs*. It delivered water for all irrigated areas east of Tashkent. The Pargostepe Fortress was built at its head section for guarding water, and there was a whole system of outposts at its laterals.

[19] 1 *ravaks* = 40 *kurags;* 1 *kurag* = 2 *toks*; 1 *tok* = 4 *choraks*; and 1 *chorak* = 2 *nimchas; etc.*

The construction of the Salar Canal help to create the Salar-Karasu-Jun irrigation system, which facilitated to shape a whole cluster of agricultural settlements. The Angren River is less important than the Chirchik River, although about 40 small canals diverted water from this river.

The first information about Tashkent is contained in ancient eastern chronicles from the second century BC, where the city is known by the name "Uni". This city was described as belonging to the Kanguy State. *Chach*, the ancient name of Tashkent oasis, was mentioned in the records of Shapur I, Sassanid king of Persia who reigned 241–272 AD. According to the transcription by Chinese sources this name is written as *Shi*, by Arabic sources as *Shash*, and by Turkic sources as *Tash*. In the second century BC, the urban settlement of Shash-Tepe had established on the bank of the Jun River and thus started the history of Tashkent.

The Tashkent oasis became the major economic centre of Central Asia owing to its advantageous geographic location. Copper and silver were mined near Almalyk, iron ore was obtained from mines in the Karamazar Ridge, and gold was found on the left bank of the Akhangaran River and in the South Chatkal Mountains. The processing of ores was done in special furnaces and the nomenclature for metalware is evidence of the development of different branches of metallurgy: blacksmith's works, metalworking and jewellery. The iron tools produced in the Tashkent oasis spread over the entire territory of Central Asia.

Archaeological excavations in ancient oases provides evidence of the dates of development of irrigation systems. For example, in the Surkhandarya oasis (Sapaliter), remnants of ancient irrigation systems date from the second millennium BC and in the Fergana oasis (Chust and Dalverzin) from the end of the second millennium BC. In 1968, archaeological excavations of large irrigation canals in the Murgab oasis were undertaken by V.M. Masson, who described the two large Guni-Yab and Gati-Akor canals (55 and 36 km long, 5 and 8 m wide, and 2–3 m deep respectively). Today, we can still find ancient irrigation canals in oases dating from the flourishing period of the Kushāna Dynasty (the 1st century BC to around 230 AD).

Detailed archaeological surveys over large areas show that during the Kushāna Period practically all the major oases in Central Asia were agriculturally developed. Irrigation was well developed and provided the basis for farming. The construction of irrigation canals in the Khorezm oasis, and in the valleys of Zeravshan and Vakhsh dates from this period. (Tajik archaeologists have also discovered the ancient main canal that diverted water from the Vakhsh River and irrigated the areas near the settlement at Uyaly, where water is still pumped for irrigation today.)

The Zang Canal in the Surkhandarya oasis, the Zakh, Bozsu and Salar canals in the Tashkent oasis, the Eskyangar and Dargom canals in the Samarkand oasis, and the Shahrood and Ramitanrood canals in the Bukhara oasis all date from around this period and some have remained in use until the present time. According to Bartold (quoted in Andrianov 1969), the total area under irrigation in these oases in ancient times was approximately 3.5–3.8 million hectares. However, no more than 10–15% of this land was in actual use at any one time. Several research works are devoted to irrigation in the Samarkand Oasis and adjacent regions (see, for example, Bekchurin 1950), and these describe the physical, political and economic conditions that existed in the oases around the time of the Kushāna Dynasty.

Samarkand really was the cultural and economic centre of Central Asia in that period. The ancient Greeks were very impressed by Sogdiana and in Greek literature we find the quote: "*Samarkand, I do not know another place all over the world where a bird's eye view would be so pleasant. The city is located on the right bank of the Soghd River and [Sogdiana] stretches out from the Bukhara border up to the Buttan border. Its size is impressive and can be measured by eight days of travel along green fields and orchards. Flourishing orchards are surrounded by canals with permanent running water; and houses are located among meadows, fields, and ponds … Sogdiana is the most fertile land among all God-given countries. The best trees and fruits grow here and ditches with flowing waters are crossing all homesteads. It is very seldom when a street or a house does not have a canal with flowing waters.*" In the first century AD, the Zargasar Dam on the left bank of the Zeravshan River was the key component of the Dargom irrigation system. First-hand acquaintance with the many research documents concerning ancient Bukhara allows us to complete the review of irrigation development in the Zeravshan valley. To do this, we need to draw on the work of medieval historians such as Abu Bakr (Abū Bakr as-Siddīq) and Mukhammad Ibn Djafar Narshakhi, the author of *The History of Bukhara* in the tenth century, and Arabic historians, Istakhri Ibn Haukal (Masdlik ul-Mamalik of al-Istakhri), Mahsidi and others, as well as the research work on ancient irrigation done by Russian scientists (Levi 1955).

Mukhamedjanov (1978, 1984) has purposefully devoted his monograph to the history of the Bukhara irrigation systems, enabling us to outline a general picture of irrigation development in the Bukhara oasis. Until 4000 BC the Zeravshan River was a tributary to the Amudarya; at the very least, the rivers were linked by two channels the Makhan-Darya and the Karakul-Darya. In the same period, the Kashkadarya River was most likely also linked with the Karakul-Darya and, through a channel, with the Amudarya. With step-by-step development of irrigation in the region and anthropogenic impacts on the upper watersheds of these rivers through deforestation and other factors, together with a certain degree of natural aridization, these two channels dried up in the third and second millennia BC.

In the Zeravshan Valley, remnants of ancient irrigation systems, which resembled the archaic irrigation systems in Khorezm, have been found in the lower reaches of the Vashkent-Darya Canal. The length of these canals did not exceed 5–7 km. The Kizil Kir Canal had a length of 6 km, a width of 20 m and a few laterals 2.5–4 km long. This system existed from the beginning of the first millennium BC until the third century BC. The maximum irrigation area reached 50,000 hectares; much later, in the ninth century AD, it had decreased to 15,000 hectares.

Material remnants of large irrigation systems were also discovered along the Vashkent-Darya Canal near ancient Varahsha. A large irrigation canal was found here, 30 m wide with multi-headed water intakes adapted for diverting water from the Vashkent-Darya River under varying water levels. The canal had three branches with numerous laterals. Archaeological monuments such as the Varahsha fortress and the ancient settlement of Khojihatia that date from the second and third centuries AD were discovered near this canal. However, the irrigation network predates these monuments. Chronicles written by the Greek historian Arrian of Nicomedia confirm this fact (Arrianus 1908). In *Anabasis*

of Alexander, his authoritative biography of Alexander the Great, Arrian wrote that Alexander had undertaken a campaign in the valley irrigated by the Polytimetus (Zeravshan) River, crossed over this river and reached the borders of the desert after a three-day march, having travelled more than 280 km. Here, his warriors killed more than 120,000 local people and ruined their fortresses. Apparently, this area was densely populated and already irrigated in the fifth century BC.

Mukhamedjanov, who states that the beginning of irrigation here can be dated back to the second millennium BC, is right. There was already developed agricultural production in the fourth century BC. The total irrigated area was 80,000–100,000 hectares. Due to developing irrigation in the upstream parts of the river basin, the water supply to the downstream irrigated areas became less sustainable and these lands gradually became desolate. According to Bartold (1966), at the end of the fourth and at the beginning of the fifth centuries AD there was an abrupt reduction in irrigation areas over all the low-lying areas in the Zeravshan Valley, including Varahsha and the territories within the contemporary borders of the Bukhara oasis.

In the eighth and ninth centuries AD, the irrigation area in the Bukhara oasis was again expanded and it reached its maximum extent in the tenth and eleventh centuries. Irrigation canals with high carrying capacity, such as those at Burantepa, Shafurkan and Naukanda (in the Kziltepa District), were built during this period. In the south-western part of the oasis the Kami Daimun and Paikend irrigation canals were constructed, the latter more than 20 km long. As a result, the irrigated area in the lower reaches of the Zaravshan River reached about 300,000 hectares. Haukal of Varksar wrote: "*From the city of Varḳsar the river waters have reached Buḳhara after a six-day trip among blooming gardens and villages*." This was one of three main canals existing in this area at that time. Another big canal that was built in the Zeravshan Valley was the Narpay Canal that ran close to Samarkand. The Shapyrkam Canal (later Shapirkan) was according to the legend created by Shapur II, Sassanid king of Persia, during his reign to irrigate the northern part of the Bukhara oasis.

From the beginning of the twelfth century, the canals gradually become desolate due to the devastating conquest by the Mongols and neglect during their 150-year reign, and the area of land under irrigation was drastically reduced.

2.4 Ancient Irrigation in Khorezm

The vast terrains of the ancient Khorezm[20] oasis that encompassed the present-day regions of Karakalpakstan, Khorezm Province and Dashhowuz Province were the most important centres of agriculture in the lower reaches of the Amudarya River.

[20]Khorezm means "land of the sun".

The Khorezm oasis (Khorazmia City and "country of Khorezmians") was first mentioned in Greek sources by Herodotus, who was the predecessor of Hecataeus of Miletus. It was described as a quite populated area, a partly plain and partly mountainous country located east of Parthia. Perhaps Herodotus was referring to the region where Khorezmians had political power. Khorezmians together with Soghds, Aryans and Parthians had stayed true to the traditions of "the Big Khorezm" prior to unification under the ruling of the Achaemenid dynasty. However, it is also possible that Khorezmians had controlled vast territories all over the Central Asian region for a long time (Lifshits 1963).

Tolstov was among the first explorers to discover and investigate this vast region. Work in the Khorezm oasis was started by archaeologists of the USSR Academy of Sciences under his leadership. They studied a territory which exceeds two million hectares in area, where the remnants of ancient cultures can be found in numerous ruined fortresses, ancient settlements, irrigation canals and irrigated fields. The description that follows is based on Tolstov's (2004) last publication, *Following the Tracks of Ancient Khorezmian Civilization*.

Descriptions of the Khorezm oasis in pre-Arabic times are rather limited. There are a few records in Persian documents, in the Avesta and in the religious literature of the Pahlavi dynasty period. More references can be found in Greco-Latin, Chinese and Armenian sources. Khorezm was mentioned by the Greek historian Arrian at the end of fourth century BC and in the *History of the Tang Dynasty*, which was written in the middle of the eighth century AD. Some fairly negligible data presented by the Arabic historians Tabari and Balazuri are the only information available about the period prior to the tenth century AD, when several minor records about the Khorezm state appeared in Arabic geographical literature.

In the tenth century AD and at the beginning of the eleventh century AD, the Khorezm khanate became the centre of the greatest oriental empire yet seen, spreading from the border of Georgia up to Ferghana and from the Indus River to the northern Aral steppes. This attracted the attention of eastern authors, who described the breathtaking rise of formerly unknown territories situated at the edge of the Muslim world.

Tolstov wrote that J. Marquart searched for the mysterious country of *Airyanem Vaējah* (supposedly the first inhabited land, which was created by the supreme deity Ahura Mazda) in the Khorezm oasis. It was traditionally described as the most northern region and a very cold country. According to the legend, the prophet Zarathushtra was born there.

The so-called lands of ancient irrigation are now a vast wilderness, and there are only traces of irrigation systems and ruins of castles in the Kyzyl Kum Desert that surrounds the present-day Khorezm oasis. However, ancient maps show considerable numbers of structures in this land. Surveying these lands and searching for the reasons for desertification have engaged not only historians but also geographers and geologists in Central Asia. There have been many theories that have tried to explain the decline of the ancient irrigation cultures, including changes in the direction of riverbeds due to erosion, the inevitable invasion of sands, the salinity of the soild and he general aridization of Central Asia.

Tolstov focused on these "lands of ancient irrigation" south-east of Karakalpakstan because he already had some historical records that suggested their early abandonment. Local historical traditions that were established by Al-Beruni[21] and first studied and published by E.D. Sachau (1873) provide only a rough outline of the Khorezmian history since the beginning of the Middle Ages.

In 1938, Tolstov discovered the archaeological site Janbas Kala 4. This belongs to an earlier unknown Neolithic culture that, according to existing archaeological tradition, was called the Kaltaminar Culture after the name of the nearest settlement. This site can be dated back to the beginning of the third millennium BC or even as early as the fourth millennium BC. After excavations on this site, Tolstov's expedition undertook a study along a new archaeological route. Over two weeks they advanced along the ancient 150 km canal (the main canal of the so-called "lands of ancient irrigation" in the south-western part of the Khorezm oasis) towards the ruins of Dev-Kala and investigated twelve newly found archaeological monuments of the ancient and mediaeval period. These were buried by loamy sediments to a depth of 1.5 m. These sediments accumulated due to the rather strong depletion of the river flows towards the delta in the third and second millennium BC.

The period from the beginning of the fourth millennium BC until the fourteenth century AD (that is, about four and half millennia) is represented by an almost continuous chain of archaeological monuments. These enabled Tolstov to trace the development of the ancient Khorezmian civilization and historical agricultural trends in this area. It also provided an insight into the political borders of southern Khorezm before and during the Middle Ages, both on the right bank and the left bank of the Amudarya. Tolstov determined the contours of the general scheme of the ancient irrigation networks and was able to estimate the date and conditions of abandonment of these "lands of ancient irrigation."

After a detailed study of the sites of the Amirabad Culture, Tolstov (1948) and Itina (1959) concluded that the Khorezmian transition to a farming economy took place around the end of the second millennium BC and the beginning of the first millennium BC, when the first separation of farmers and semi-nomadic cattle breeders can be detected. The remnants of the ancient irrigation infrastructure in the Pre-Aral region and the southern Akcha Darya delta, on the so-called Tazabagyab sites of the second half of the second millennium BC, also date from this period. There are bunds some one metre wide that make up the flood checks across the flats in squares of 2.6–3 m and 3.5–4.8 m located along the extended riverbeds in the delta (Kokcha I). Traces of irrigation canals, 2 m wide and 50–70 cm deep are found along these little bunds. According to Andrianov (1969), instead of the Egyptian basin method of irrigation which is based on the sequence "river – deltaic floodplain – field" here the water supply and irrigation were established according to the sequence "river – canal – field" (see Figure 2.7).

Ancient irrigation canals diverted water from the distributaries[22] at a sharp angle. When they ceased operation, water flowed back towards the former riverbeds. It was typical for these canals to frequently change course and to block each other because of siltation. Irrigation canals of around 1 km long

[21] Abu Raihon Mohammad Ibn Ahmad Al-Beruni.

[22] A branch of a river that does not return to the main stream after leaving it (as in a delta).

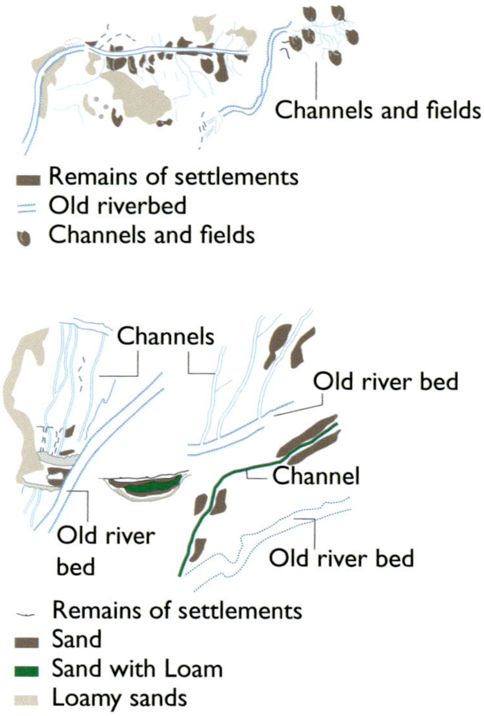

Figure 2.7 The Amudarya in different stages of development (B. Adrianov 1969).

are typical of the so-called Amirabad Culture. These canals had command areas of about 200 hectares. It was typical practice to turn former riverbeds into manmade canals.

The "tribal stage" of life in ancient Khorezm came to an end in this period, as a new class-based society developed around the first quarter of the first millennium BC.

The pre-Achaemenid period in the Khorezm oasis was notable for its considerable natural and human resources. Centripetal forces prevailed and a powerful military and democratic tribal confederation of Saks and Massagets was formed. It gradually became a unified state headed by the Siyavush tribe. This integrated internal structure was a historical prerequisite for the progressive development of the economy in the Khorezm Oasis that followed. At first, according to Engels, when traditions of ancient democracy were still in force and wars were "the constant practice", the strong unified tribes of Saks and Turs (or Saks and Khorezmians) aimed all their forces to conquering the more developed territories to the south.

The history of ancient Khorezm and neighboring states in the eighth and seventh centuries BC clearly shows that Khorezmians were acquainted with the more ancient agricultural practice in the Margian and Bactrian oases located in the south, and this enabled a rapid development of irrigation in the Khorezm oasis. It is most likely that an influx of experienced farmers (irrigators) from the south played a key role in this progress. Close relations with the cultures of southern Central Asia, especially with Margiana, were established following the Amirabad culture, and these characterize the archaic culture of Khorezm in the sixth and fifth centuries BC.

In the sixth and fifth centuries BC, the construction of the early irrigation systems coincided with the creation the ancient Khorezmian state. Here, irrigation systems were built according to the layout sequence "main canal – branch canal – irrigated field". Some of the main branch canals were as long as 8–9 km, with a width of 40 m and an elevation above the land surface of 1–1.5 m.

The Kaltaminar irrigation system (Figure 2.8) is an example of the irrigation systems that existed from the sixth and fifth centuries BC until the fourth century AD. The canals in these irrigation systems quickly become silted, and after rehabilitation (cleaning), they became so large that the ancient farmers

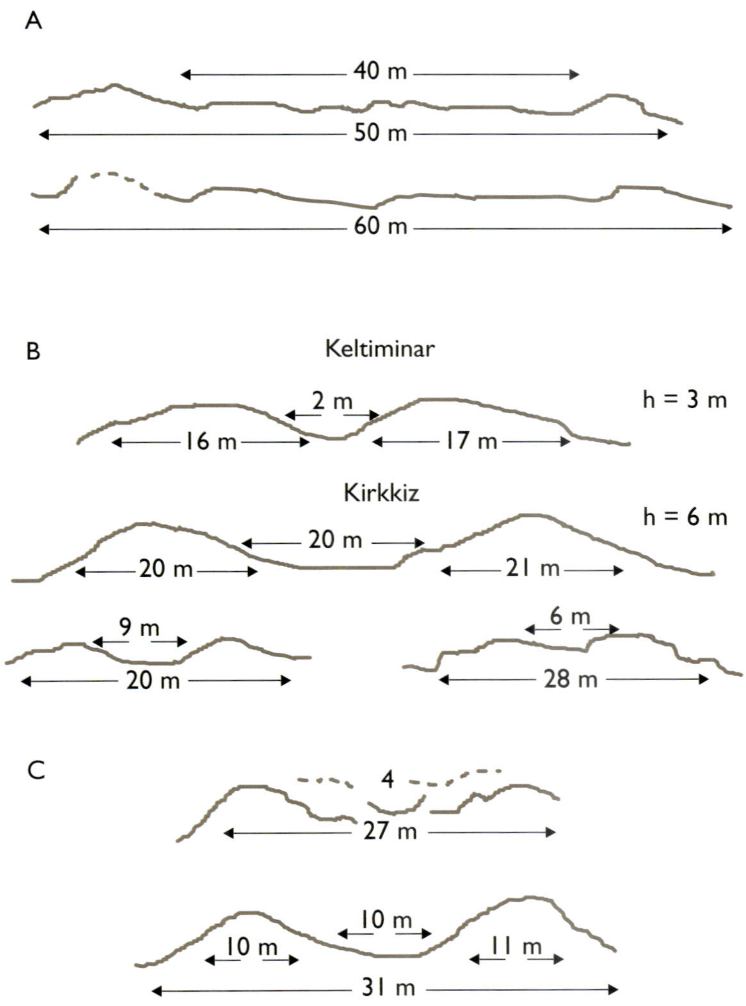

Figure 2.8 Cross-sections of the Kaltaminar system of canals (Andrianov 1969).

preferred to construct parallel canals. Four parallel canals have been found in Kaltaminar irrigation system separated by 25–100 m.

Laterals located at right angles to the main canal were typical for the later Kaltaminar systems. The command area of each of these lateral canals amounted to approximately 2,000 hectares. It took no less than 2,000 people to clean such a canal system. It is clear that the construction and maintenance of these systems required a very highly developed state. A similar system, the Dingeldze irrigation system, has been found close to the Kaltaminar irrigation system.

Canals within the Sarykamysh delta near the Daudan River bend are also from this period. The route of this canal exactly followed the Akcha Darya channel in the first millennium BC on the right bank of the Amudarya. There were quite advanced water intake structures from an engineering point of view. The canal that diverted water from the Daudan riverbed had three water intakes, the third of which acted as a spillway during the spring flood season. The foundations for this structure rested upon hard bedrock, and the width of the spillway reached 40 m. The Chermen-Yab Canal dates from the same period. It irrigated the environs of the famous Shakh-Senem. The Kuni-Jaz canal is impressive because of its 50 kilometre length and 70–75 m width. This canal existed for several centuries, and it had an intake dam that reached a height of 3–5 m.

The large 90 km long Kirk-Kiz Canal (with its headworks on the Amudarya) dates back to the Kanguish and Kushāna periods. Some 6,000–7,000 people were needed to maintain this canal. Gavhore, Tash-Kirman and other canals were not so long at that time. The Kanguish period in general is characterized by an extensive network of so-called arterial[23] canals, which were up to 120–150 km long. These resulted in a total canal network of 250–300 km. Headworks were constructed with multi-headed idle sections (*saḳas*). Water was diverted from far upstream and conveyed in canals parallel to the channels of tributaries and the Amudarya itself.

Canals of shorter length were also built. The Bazaar-Kali Canal, which was only 20 km long, was constructed to replace a system of meandering canals that formed a tree-like shape. The new canal was straight throughout its length. The wide canals, which characterized archaic Khorezm, were gradually replaced with deeper but narrower distributaries. For the first time, farmers started to undertake some land reclamation and to improve the soil by applying manure (in the form of a material that under certain conditions is called *dual*) to increase soil fertility, adding sand and using other methods.

During the period between the fourth and sixth centuries AD, the situation in the Khorezm oasis regressed considerably in both general development and in terms of irrigation, especially on the right bank of the Amudarya. Irrigated lands in Kaltaminar and Tash-Kirman were abandoned. A revival of irrigation in the Khorezm oasis on both banks of the Amudarya did not take place until the period between the ninth and eleventh centuries AD. Tolstov (1948) provides a good description of

[23]An arterial canal is a main canal within a complex canal system.

this time: "*The Ayaz-kala system irrigated large family manors that were still incorporated into the overall framework of settlements. These settlements were scattered along canals within an area of over 30 square kilometers. Manors were located at several hundred meters between each other and were surrounded by fields and orchards.*"

This period is characterized by a very high level of agricultural production, and the population density increased by a factor of four. The Arabic geographer and traveler Jakub, who visited the Khorezm oasis in 1219, wrote: "*I have never seen such a densely populated region with such numerous trees and gardens. You wouldn't meet uncultivated plots.*" The irrigation system consisted of a network of canals running parallel to the Amudarya River, incorporating arterial canals (*saka*), head canals (*arna*), distributaries (*yab*), small distributaries (*badak*) and laterals (*salma*). Distributaries (*badaks)* were equipped with structures (*doldorga*). There were also drainage canals (*mufriga* or *bedray*) and, in this period, water wheels (*chigirs*) were widespread.

Obligatory duties (*hashar* and *bigar*) were set for cleaning irrigation canals. *Bigar* was a kind of forced work given as a penalty, while *hashar* was a form of voluntary public work. The primogenitors of the canals which irrigate the modern left-bank area date back to the Middle Ages. At the end of the ninth century and the beginning of the tenth century, canals such as the Heikanik (today the Palvan-Ata Canal) and Madra (today the Gazavat Canal) were built. The "genealogy" of the Shavat Canal, the largest modern irrigation canal in Khorezm Province, is more complicated. At the beginning of the ninth century, there were two canals: the Vadok and Buve; later they were combined into the Shah-Abad Canal, which then became the Shavat Canal in the seventeenth century.

2.5 Water is the Great Educator

Water is the great educator in Central Asia. It has asserted a supremacy over laws, norms and social moral – and is a force greater than human individualism and egoism. The moral codes governing relations in the triangle "humans – land – water" are rooted in the attitude towards water. Water established unified moral laws that accord to its purity and sacred status. In Central Asia, water not only became a material but also a moral value. Stealing water, for example, was completely out of the question, with every person knowing from childhood the saying "who steals water once, remains a thief for life". A community could forgive theft of property or cattle but never theft of water; the label of "water thief" was attributed not only to a thief himself but also to his family and descendants.

Religion played a leading role in forming the water cult by presenting water as God's creation and turning the worship of water into an ideology. Religious dogmas were transformed into a general public value system, and religious canons underpinned the principles and rules of water use. The scripture Avesta marked the very beginning of deification of water in Central Asia.

2.5.1 Zoroastrianism

The Avesta (the Right Way) is the sacred text of Zoroastrianism, and comprises a religious code and legal regulations, prayers, psalms and hymns devoted to deities. This ancient religion emerged five centuries before Christianity. In 1857, the German historian H. Weber expressed the view that Zoroastrianism arose in the middle of the fifth century BC in lands located between the Caspian Sea and the Amudarya River. According to Weber (1892): "He (Zoroaster) planted a cypress near the Caspian Sea and engraved an inscription in its rind about the adopting of his teachings by Hystaspes[24], a Persian Shah" [79]. Weber's book was first published in Heidelberg in 1857, and republished in 1864 and 1892 (its first Russian edition).

Weber's findings caused a real sensation, as previously Balkh was considered the cradle of the Zoroastrian religion. At the end of the nineteenth century, the monograph *Media, Babylon, and Persia* was published in London. Its author, Mrs. Z. Rogozina, categorically declared that the Avesta originated from the eastern coast of the Caspian Sea (Tzinzerling 1927). It is believed that 21 devotional poems known as the Gathas were written by Zoroaster, and these form the core of the Avesta or Zoroastrian scriptures, although only two of them – *Vendidad* and *Yasna* – have been preserved. According to the order of Hystaspes, these poems were rewritten in golden letters on ox hides and kept in Persepolis (the capital of ancient Persia). In 331 BC, after the defeat of Darius III, Alexander the Great nd incinerated the Khan's palace, destroying invaluable Zoroastrian scriptures together with carpets, furniture and curtains (Herzfeld 1947).

Weber noted that the Avesta represented a religious doctrine that assimilated exceedingly ancient conceptions and customs, which it then revised according to notions and requirements of a later time. In the Avesta, a doctrine with exact regulations for religious life and rites was combined with writings establishing codes for ordinary life. For many centuries, the Avesta was taken as the official code for governance and for social and family life in Central Asia, Iranian, Afghani and Indian societies. Within the cultural arena, the Avesta established the norms and laws of social life. Priests (called *mobeds* or *magi*), who were quite influential in political and social life and were the hierarchs of spiritual life, played a special role in the Zoroastrian religion. Even now, the expression "magic word" which means that the *magi*'s word is beyond any doubt or disobedience because it proceeds from Ahura Mazda (Wise Lord of the World) occurs in many languages. Priests had the exclusive right to manage the rite of sacrificial offerings and officiate and sing theurgic hymns to the Gods. They always operated side by side with the king, interpreting the meaning of his dreams and any unusual natural events, influencing his decisions with their advice and specifying a place for his interment.

Until the dissemination of Islam in Central Asia, people worshipped this sacred doctrine and rigorously followed all its spiritual instructions. Zoroastrianism laid the foundation for Christianity and Islam. According to the Avesta, the universe was divided (ethically and morally) between two spirits: Ahura Mazda, the perfect, rational and all-knowing entity, and Angra Mainyu, who creates sin, disease,

[24]Darius I, called The Great (558–486 BC), king of Persia (522-486 BC), son of the Persian noble Hystaspes, and a member of a royal Persian family, the Achaemenids.

death and similar evils. They correspond to God and Satan in Christianity, and Allah and Eblis in Islam. An implacable cosmic struggle takes place between them until the time when Ahura Mazda will finally conquer Angra Mainyu. Followers of Ahura Mazda are obliged to participate in this struggle, and confront the deeds of Angra Mainyu, and when their dying day comes, the eternal bliss of heaven at the throne of Ahura Mazda will be their reward.

Earth, heaven, fire and water are considered to be the sacred creations of Ahura Mazda. The earth and water are also considered to be living creatures that produce the cereals and trees that provide food for people and the grasses that provide forage of cows and horses. Water was declared holy and to be protected against pollution by sewage. Water in creeks and rivers could only be used for drinking and irrigation, bathing was forbidden and even wading in the river was forbidden (a river could only be crossed over a bridge). *Magi*s were vigilant guardians and tireless monitors of these sacred instructions. Water that falls on the Earth's surface in the form of rain and snow was also considered holy. According to the Avesta, rivers are holy and divine entities. They were glorified in hymns and sacred myths. In line with the Zoroastrian cosmography, the rivers flow not only over the Earth but also in Heaven.

The Avesta laid the foundation for the deification of water and for public attitudes that regard water as a divine creature, which is remains even today a key component of social awareness in the nations of Central Asia. Byzantine ambassadors Gordian, Paul Cilitsya, and Valentinus, who visited Central Asia in 591 AD, reported to Imperator Tiberius II that people living in this region "worship air and water". Grousset wrote that reverence for water as a sacramental creature became the basis for spiritual awareness. Even inhabitants of Genghis Khan's empire, who followed Islam, established rules for the use of running water for Islamic ablution and washing in line with the prescriptions of the Avesta (Grousset 1996). The Avesta was "the gold bar" of thoughts and words that united a human being with nature (with earth and water). According to the Avesta, fire is also a sacramental creation. Even now, lanterns for a sacred fire are lit every evening in Khorezm Province. An obligatory part of wedding ceremonies in Kashkadarya and Surkhandarya provinces requires the bride to jump over a fire for her absolution from evil spirits.

Concern for water has united states and nations. In 630 AD, during the travels of the Chinese Buddhist monk and pilgrim Hsuan Tsang (600–664), the Maverannakhr was politically disintegrated, split into many small and independent domains, but from an economic point of view, it was an integrated unit with market towns and fields irrigated by canals. One part of the community was busy trading and another was involved in farming. However, everbody regardless of their economic activity, supervised water quality and monitored the condition of irrigation canals, jointly solving any water problems as necessary. The leaders of these domains knew one firm truth: "if you would like peace and prosperity, you should take proper care of joint water use" (Bartold 1966). They also knew that it is impossible to maintain irrigation systems properly without the participation of all inhabitants. In the ninth and tenth centuries AD, the maintenance of dams on the Zeravshan River and different parts of the irrigation system was entrusted to the inhabitants of separate blocks of houses, settlements or city districts. In return for looking after the irrigation system, people were exempted from land tax or, for adherents of other faiths, from capitation tax (*jiziya*). In Central Asia, the word *paykal*, which means the daily sequence of water use, dates from this period.

2.5.2 Islam

In the seventh century AD, Arab tribes conquered Central Asia and forcibly spread the Moslem religion. They met persistent and fierce resistance from the local population, who did not wish to submit to conquerors nor adopt a new religion. However, Islam was not only a religion. It created one of the great civilizations of the world, promoting the development of scientific research centres and universities, and the study of literature. In 774, the first university was established in Cordoba attracting students from all over Europe. In 815, the caliph Abdullah al-Mamun established the first academy of sciences in history, which was soon followed by a similar academy in Khorezm. Europeans should remember that the system of Arabic numerals used by all mankind is the fruit of this civilization, and many words such as *mile, admiral, gas, alchemy, algebra, mirage* and *algorithm* are of Arabic origin. Islamic civilization merits as much recognition as Greek or Roman civilizations.

Islamic rulers adapted to local conditions and used the best cultural traditions of the conquered nations rationally to further strengthen their position. This approach helped Islam to become one of the world's leading religions and spread its influence all over the world. Islam supplanted Zoroastrianism, but Zoroastrian ethical fundamentals enriched Islamic concepts. The conquerors did not destroy the temples of earlier religions but used them as mosques.

Islam has played a considerable role in the development of the irrigation culture in Central Asia. Arabs well understood the particular natural conditions of Central Asia and they made energetic efforts to transform this region, so that it might become a base for the supply of different goods to the Caliphate and a source of enrichment. They held to the idea that regional development relied on expanding irrigation areas and constructing water infrastructure. The Caliphate allocated considerable funds for irrigation works and used local knowledge as well as their own great experience of constructing irrigation canals in the valleys of the Tigris and Euphrates rivers. The Caliph Mutasid (833–842) personally contributed a large amount of money for constructing the large canal in Tashkent, which remained in operation for more than 400 years until the eighteenth century.

The Quran is the holy book of Islam. It provides the fundamental principles of the universe and ethical norms of Islamic life. The main doctrines laid down in the Quran are that only one God and one true religion exist. The Quran summons humans to acknowledge God's sovereignty over their lives and invites them to submit to his will. The word Islam means "obedience". For Muslims, it is the very word of Allah, the absolute God of Islamic faith as revealed to the prophet Muhammad. The angel Gabriel is said to have spoken Allah's words into the prophet's ear. All Muslim events start with the words: "*There is no God but Allah, and Muhammad is his prophet.*" The prophet Muhammad did not appreciate questions about God. He said: "What are your questions for, if I myself pass his words to you." The prophet was illiterate. He could not read and write, and the Quran was written after his death by the caliph Uthman (611–656). The first version consisted of reminiscences provided by Muhammad's worshippers, by those who heard his sermons in Mecca, and by his friends and relatives who recalled his words by heart believing that these words represented the true words of God.

The prophet Muhammad was not a theologian. He could be considered a practitioner and he had to keep in mind the traditions, culture and customs of the people he wished to convert to Islam. Most rules suggested by Muhammad were slightly revisions to legislative acts that already existed in Mecca and Medina. There are no doctrines or instructions which prescribe concrete rules or codes in the Quran itself. Therefore, Ali, the prophet's son-in-law and the last of the orthodox caliphs, founder of the Shiah sect and its spiritual leader (566–653), and his follower (*Ibn-Abbas*[25]) began to formulate the norms of Muslim life. After their death, people who attended their lessons opened schools of Muslim jurisprudence. Islamic jurisprudence was named *usul al fiqh,* and jurists who were specialists in Islam jurisprudence were called faqihs. Faqihs rapidly became quite influential and the main Islamic lawmakers.

With time and complication of social relations, there was a need to establish a code of social behaviour for true believers. Norms and rules of social life, and unwritten laws that regulated economic relations, as well as ethical instructions for settling family problems started to be formulated. These were included into the Sharia[26] (the Right Way). The Sharia is the most important part of Muslim jurisprudence and it reflects the general statements of the Quran in instructions for solving specific and vital problems to society. Faqihs were responsible for the development of the system of norms for regulating general Islamic social life. The task of the faqih was to formulate laws, and the Sharia set procedures for putting these laws into practice.

In Central Asia, Arabs faced the challenge of regulating land and water systems. There were quite developed irrigation systems, but few rules governing water use or for dealing with water problems because there was no legal framework that could regulate the use of water resources. Mutual relations were mainly grounded in customs and traditions, and this resulted in frequent disputes and conflicts that were harmful to economic activity. In the reign of *Abdallakh Ibn-Takhir* (830–844), ruler of Khorasan, there was a dispute concerning use of karez and *ariqs* in Bukhara and it was decided to apply to faqihs for a legal solution. There were no existing legal regulations that could resolve this problem. All faqihs of Khorasan were consulted, and with the participation of theologians from Iraq, they were entrusted with the elaboration of regulations for water use. A water code, *The Book on Ariqs*, was written and this code became the guide for regulating land and water relations for many centuries in Central Asia. Unfortunately, this text has not survived, but its regulations were reflected in the Sharia, showing their importance in the development of irrigation in the Muslim world.

The Quran sets out general rules governing human relations with nature. Water is mentioned as the first of the generous gifts of Allah and as the basis for all life on Earth. Nobody but Allah has the right to be the master of water. Plants are growing, animals are breeding, and life on Earth is developing only thanks to this creation of Allah. The *surah* Al-Nahl states that: "It is *He (Allah) who sends down water from the sky, whence ye have drink, and whence the trees grow whereby ye feed your flocks. He makes the corn to grow, and the olives, and the palms, and the grapes, and some of every fruit – verily, in that is a sign unto*

[25]Al-Abbas ibn al-Muttalib.

[26]The Sharia is the philosophy of the Islamic jurisprudence with a typical syncretism – that is, it is integral to the many spheres of spiritual, secular and practical activities of a Muslim.

a people who reflect" (Ayah 11). The *surah* Al-Jaziya states: "*And God created the heavens and the earth in truth; and every soul shall be recompensed for that which it has earned, and they shall not be wronged.*" (Ayah 23). And the *surah* Ash-Shuaura states: "*He said, 'This she-camel shall have her drinḳ and you your drinḳ on a certain day; but touch her not with evil, or there will seize you the torment of a mighty day!*" (Ayah 156) (Khidoyatov 1990).[27]

Faqihs have interpreted these sentences, and these interpretations became the basis for the Sharia instructions. According to these interpretations, water is a gift of God that replenishes nature and cannot be private property. It belongs to all and each in equal parts. Each person can only use water for drinking and irrigation, unless it is collected in a jar. Water cannot be bought and sold without a plot of land. If there is a shortage of water for irrigating fields, water should be divided equally and the sequence of use established by drawing lots. Diverting more water than set out in established agreements is considered a crime, and should be subject to punishment. These rules do not apply to karez (groundwater wells). This water can be for the exclusive use of the person who built the karez.

Arabic faqihs came to the conclusion that agricultural production on the vast areas of land that were suitable for farming in Maverannakhr required skillful management of water resources. They noted that the water reserves in the region (in absolute terms) were quite negligible in comparison to the area of so-called "dead earth" that awaited irrigation and reclamation. Arabic scientists based established a proper system of water distribution based on exact mathematical calculations (which were translated from Arabic into Russian by the Russian colonial administration and later by the Soviet authorities). The works of eminent scholars from Baghdad such as Fath Al-Qodir, Ibn-Abidin and Ash-Sheikh Muhammad Ilyas were also translated into Russian. These works have been carefully studied by modern scientists when developing new irrigated farming systems. In 1924 a special issue of the journal *Bulletin of Irrigation* was devoted to water legislation. It was issued under the title "The Collection of Muslim Regulations (*Sharia*) concerning Water and Land Use", and quoted 205 instructions which covered basic problems of water and land use. Readers might be surprised by the fundamental nature and universality of the Shariah's instructions concerning the regulation of water use and the resolution of water problems.

A version of these regulations became the basis of special water laws that sought to preserve the continuity of age-old traditions and stimulate further agricultural development in Maverannakhr. The Sharia's regulations cover five key problems:

- Dead earth
- Regulation of water use
- *Shif'at* (the preferential right of a neighbor or land co-tenant to acquire land)
- *Muzaraat* (an agreement between two persons on the cultivation of a plot belonging to one of them for a share of the crop yield)
- *Musaḳat* (an agreement on cultivating orchards or vineyards using payments in kind).

The first two issues that directly refer to water problems and are of interest here.

[27]These citations were adapted from the Quran translated by E.H. Palmer (Sacred Books of the East).

2.5.3 Dead Earth

1. The dead earth is land which cannot be used due to a lack of water or flooding problems, or for other reasons that impede tilling.[28]
2. It is also land which remains without owners and lies fallow; that it is located at such a distance from the nearest settlement that a loud human voice from that settlement cannot be heard on the land.
3. Only the Imam (the leader responsible for the local mosque) can allocate dead earth.
4. A man who cultivates dead earth gets the right of ownership by the authority of the Imam or Qadi.[29]
5. The title is paid from the yield grown on the dead earth under cultivation.
6. If the Imam or Qadi gives the dead earth for tilling to somebody with only right of use, the designated person does not have the right of ownership.
7. A non-Muslim has equal rights to a Muslim with regard to the possession of dead earth tilled by him.
8. Any former river channels which may over time become rivers again should not be cultivated, as they will be needed when they revert to being rivers.
9. If a person who has set an abuttal around a plot of dead earth does not cultivate this plot within three years, he loses his right of possession and the Imam can give it to another person.

Over time, some new articles and regulations were introduced that covered matters relating to private property, including even seeing water as a deity. One regulation concerning private property was: "If any outsider wishes to irrigate the land cultivated by him using water which is in private use, owners of the river have the right to prohibit this action, regardless of whether harm is caused by this prohibition or not, because owners have the exclusive right to the river." Article 36 of the code (Sharia regulations) states "If a draw-well, water source, pond or a river is the property of somebody, this person has the right to forbid others entering his plot to drink or water cattle if within a distance of one mile it is possible to find water [that is not on private land]. If it is impossible to find water nearby, an owner of a river should give water for drinking and for another person's cattle or permit to an outsider to scoop up water without spoiling riverbanks."

There are some other articles of special interest:

Article 37: "If a draw-well, water source, pond or a river is on dead earth, an owner cannot forbid to somebody to drink water from this water source or to water his cattle."

Article 38: "If an owner of water wants to prohibit somebody to use water, and the person who needs water fears for himself or for his cattle, he has the right to struggle with the owner of the water by weapon, because those who refuse him water cause his death."

[28]See Fath al-Qadir Vol. IX, page 2; Ibn Abidin Vol. V, pages 306 and 307; Mukhtasar al-Quduri, page 75; Sharh Ilyas Vol. III, pages 258 and 259.

[29]A judge.

Article 39: "A person, who is not the owner of water, has no right to irrigate the earth, date trees and other trees using water from another's draw-well, *ariq* or river without the permission of the owner of the water source."

Article 40: "A water source or pond in which water collects by natural means and which is linked to a river that is in private use is property, and others do not have any right to use it. A person who has stolen the water collected in a pond in a district where it is impossible to get natural water is not to be punished."

According to the Sharia, there are three types of rivers:

- Rivers, which belong to nobody, and use of their waters is not subject to distribution
- Rivers that are in private use, but anybody has the right to use their water resources for drinking and watering cattle
- Rivers that are in private use, and their waters are completely withdrawn from common use.

Dredging and cleaning of any rivers and canals in private use was implemented at the expense of their owners. Maintenance tasks in rivers and canals in public use were allocated to water users in proportion to their share of water use. Special regulations concerned the right to use river water, and "in case of disputes concerning the volume of water which co-owners [or co-users] can extract, amounts are estimated in proportion to the size of each user's plot". This regulation was introduced by Fath Al-Qadir in the tenth century (Collection of Works. Volume IX, Page 13). These general regulations were revised over time taking into consideration local conditions. In some Muslim countries, they acquired the force of of international law. In countries where the indigenous langauge was not Arabic, it was difficult for ordinary people – and even non-Arabic faqihs (lawyers) – to study these laws. Therefore, the most important and frequently used laws and norms of the Sharia were translated into the local language and orally circulated along with the rules and norms of the Odat. The following Sharia regulations were included into the Odat[30]:

1. Waters of rivers and lakes are property of the community.
2. A ban on the sale of land without water.
3. In the event of water scarcity, there should be fair and equitable allocation of water (proportional to land area).
4. Water allocation into *ariqs* (irrigation ditches) to be proportional to their command areas, or water is distributed by turns.
5. Participation (in the form of personal labour or supply of building materials) by each water user in implementing socially necessary works related to the construction, repairing and cleaning of canals and water regulators.

[30]The word "odat" literally means habit, custom or tradition. Rules and provisions of the Odat in the legal sense are regulations of the customary law.

6. The cultivation of "water-loving" crops (for example, paddy) should be coordinated within a community, and consideration should be given to limiting the crop area or obtaining consent of all downstream water users of the canal.
7. When constructing an *ariq* through the land of another person, any damage should be paid for.

Conscience, a certain order of doing things, and rules and words of eternal relevance became the fundamentals of Odat. The instructions were to be impeccably executed. *Mirabs* (water managers) were appointed in the field to ensure proper implemention of the Sharia's regulations. General supervision was entrusted to the Imam (a priest from the mosque). The *mirab* monitored the water system to ensure that water was distributed correctly and helped to settle of disputes. He also took responsibility for water supply in the territory entrusted to him. It was a very responsible post and mirabs continued to be appointed for many centuries right up to the Soviet period, when the construction of water systems that provided a continuous supply to consumers under state support made the positions unnecessary. *Mirabs* were elected and were paid. All members of the local community (*makhalla*) participated in the election of a *mirab*, but there was no set wage (each member of the community could contribute according to their means). The most respected, influential, honest and fair people were elected to this post. Some people served as a *mirab* throughout their working life, and sometimes the position was inherited from an elder relative.

The *mirab* stayed in his box-room near the regulator each day and each night. Many people accorded the *mirab* special respect, even after his departure, as G. Khidoyatov, Professor at the Tashkent University of World Economics and Diplomacy and former adviser to the Ministry of Foreign Affairs of the Republic of Uzbekistan, recalls:

"Prior to 1937, our family lived in the old district of Tashkent. We had a small traditional kitchen garden with fruit trees that was a great help to our household economy. We received water two times a week. At that time, there was no running water, and water was supplied through the public main canal. The regulator, which distributed water into three small ariqs, *was located half a kilometre from our house. The* mirab's *box-room was located near the regulator, where he lived practically all the time. Inhabitants of our residential neighbourhood brought meals for him, as he could only boil water for tea. Special gates turned water in the required directions. Water supply to each household was scheduled. We received water during the day on Tuesdays and at night on Saturdays. Our* mirab *was an honest citizen and a man of principles. He was rather pious and, as a true Muslim, prayed three times a day and rigorously observed the fasts and all instructions of the Sharia. It is over 70 years since that time, but I have remembered his image all my life as an example of honesty, nobility and incorruptibility. He worked as a* mirab *more than forty years and left fond memories with all inhabitants of our* makhalla.

"An inviolable procedure was employed to sequence the water delivery. Poor widows with children received water first, and then single women, handicapped and old people, and only thereafter the rest of the community. Water allocation was implemented precisely according to the established schedule. Inhabitants waited patiently for water. When water was distributed at night, we took torches or lamps and all members of our family, young and old, helped to distribute the water, as well as to fill water into buckets and other tanks for drinking, cooking, washing the dishes and bathing. Our mirab *came two hours later and asked whether there*

was sufficient water and, in cases of insufficient supply, he prolonged water delivery for an additional hour. As far as I know, practically all mirabs *were similar, as if they were formed according to a single standard, because each of them knew that he protected and distributed the great gift of Allah – water.*"

The Sharia set out specific instructions concerning maintenance, cleaning and dredging of rivers and the water infrastructure. Works were usually implemented at the expense of the state treasury, however if there was a lack of money the Imam had the right to force inhabitants to dredge and clean a riverbed at their own expense. The rivers in private use were cleaned at the expense of their owners, but this was rare practice in Central Asia. Special instructions existed that governed any problems or cases of abuse by owners. For example one stated: "An upstream owner can open the gates of the regulator only according to his turn and by approval of co-owners." Today, this regulation is of special interest: "If one of the owners wishes to dam the river and the co-owners agree, then the sequence of irrigation provisions should be started from the downstream section of the river and proceed upstream. When the sequence reaches the head stretch, damming of the river is authorized (that is, nobody has the right to dam the river if this action harms the interests of downstream water users). If co-owners cannot come to an agreement, but at the same time they cannot draw water without damming the river, the Imam distributes water among them according to the days allocated with each co-owner at his turn."

Islam established the legal base for the development of irrigation systems in Central Asia, and introduced social and economic mechanisms for joint water use. As the supremacy of Islam in Central Asia strengthened, historical development in the East and in West diverged. Private land ownership, legally established under Roman law, became the basis of economics and social life in the West. In contrast, state ownership of land, established under the Sharia and expressed in collective land and water use, emerged in the East. This was an important step in establishing a mode of production called the "Asian mode of production" in semi-forgotten Marxist literature (Marx 1969).

2.6 Water is the Basis for Central Asian Economies

Each nation has distinctive traits that are formed by the environment in which it exists. Humans in turn impact on their environment. These relationships between human beings and nature create a social passionarity[31] that was first studied by L. Gumilev (1990), a Russian researcher of ethnic processes. It is impossible to understand the history of Central Asia without clarifying the mechanism behind this phenomenon. The scientific concept of a biosphere was first developed by the Russian scientist V.I. Vernadsky (1863–1945). The biosphere is the earth's relatively thin zone of air, water, living organisms and their vital reproduction functions, as well as the soils capable of supporting life. In practical terms, this is the historical and

[31]Passionarity (from the Latin passio, passion), a term introduced by Russian ethnographer and historian Lev Gumilev to signify the ability to and urge for changing the environment, both social and natural, or, in physical terms the disturbance of the inertia of the aggregate state of an environment (Gumilev 1990).

geographical environment in which a nation lives and develops. The ethnos[32], like Antaeus, is tied to the biosphere from which it extracts its force and creative energy. The landscape is "a melting pot" that forms nations. They arise and develop as the landscape changes and with the increase in amount of land under irrigation.

In Gumilev's opinion, the biosphere gives rise to passionarity – the impulses that incite human development. These impulses have formed ethnic groups and given rise to world religions, civilizations, cultures and natural cataclysms. Each change in climatic or natural conditions has led to shifts in food production methods and resulted in the rise or decline of cultures or victories and defeats. These events cause the sudden emergence of new nations at the scene of history. Genghis Khan can serve as an example. His devastating military campaigns in the thirteenth century changed the natural and historical development of Central Asia nations and China. However, in accordance with Gumilev's concept, Genghis Khan himself was the product of the kind of passionarity that arose in Mongolia and initiated the movement of enormous population masses.

In spite of all its uniqueness and atypicality, the example of Central Asia should, to some extent, provoke a more attentive attitude towards this idea. Consider, for example, the two Turkic nations Kazakhs and Kyrgyzes which formed in this region, and whose people led a nomadic way of life until the 1930s when the Soviet authorities forced them to settle, unlike the Uzbeks and Tadjiks who had been settled farmers since ancient times. Central Asian nomads were quite satisfied with their way of life, and they did not experience many ups and downs in this region because there were apparently not many impulses to awake the vital energy of the Mongols. An absolutely different situation existed in Maverannahr[33] located between the Amudarya and Syrdarya rivers.

Many ancient Asian civilizations emerged in territories located between two great rivers. The Chinese civilization was formed on the terrain situated between the Hwang Ho and Chang Jiang rivers; Indian civilization between the Indus and Ganges rivers; Assyrian civilization between the Tigris and Euphrates rivers, and the Central Asian civilization between the Amudarya and Syrdarya rivers. The terrain between two rivers abounding in water seems to create most favourable conditions for new civilizations to emerge. The struggle for the preservation of natural resources, the effort needed to extend irrigated areas and develop and construct irrigation infrastructure requires a special spirit which gives rise to an energy for improving life. This energy inspires poets, thinkers, philosophers and politicians to great creativity. It provokes the need and aspiration for richness and prosperity, as well as for creative work and beauty.

The physical properties of nature affect the physical properties of human beings and their intellectual development. Scientists have proved that water deficiency leads to cretinism, the abuse of narcotics leads to the degeneration of entire nations and deficiency in phosphorus may lead to mental retardation. The struggle for water resources and the construction of water management

[32]People of the same race or nationality who share a distinctive culture, from the Greek word ethnos, meaning "nation" or "race."

[33]In Arabic, *Ma Wara' un-Nahr* means the territory beyond a river; in Latin *Transoxiana*.

infrastructure give rise to a passionarity of nations, and activity and creativity in all spheres of life. The outstanding Uzbek poet Alisher Navoi[34] (1441–1501) has written about farmers who irrigated their fields as "the fervour of passion is seethed in their job" (in the poem "Farkhad and Shirin"). Here we put special emphasis on this splendid poem and legend that glorifies the productive work of irrigators.

Water in Central Asia has the highest moral value. The attitude to water, based on the recognition of its importance for life and its holy blessing, has been reflected in the culture of everyday life, literary works, philosophical treatises, myths and legends. Water does not allow amorality, material egoism, mercenary motives and malice. The outstanding Eastern poets always appreciated heroic deeds for the sake of water. This idea (the heroic deed for the sake of supplying water to people) was incarnated in Alisher Navoi's poem "Farkhad and Shirin". Alisher Navoi was the Grand Vizier at the court of Husayn Bayqarah, the representative of the Timurid Dynasty in Herat. His poem was based on a love story, but this is only a backdrop to emphasize the heroic deeds of Farkhad who hacked a pass for water through rock to supply water to the population.

Farkhad was the son of a ruler and, from his childhood, his mentor taught him to make iron tools for hacking rocks; tools which could crush even the hardest kind of granite. When seeking his sweetheart Shirin, he arrived in Armenia and witnessed the imminent danger threatening the inhabitants of one city. He saw thousands of people crowded near a rock making unavailing efforts to hack a pass for water that was trapped behind this rock. People were dying of thirst on one side of this rock, while there was a spring on the other side where the "*water was so clean and sweet, that even a deadman would resurrect after one sip*". Farkhad decided to help the people. He collected their iron tools and melted them down in accordance with the instructions of his mentor. Then he made a huge pick and hacked a pass for the water through the rock and made a channel in the granite of 70 km long (7 *yagaches*). People said: "*He made us happy with water.*" Farkhad learned that Shirin's cherished dream was to have an *ariq* around her palace and so he hacked an *ariq* towards her palace, which was called "*the ariq of life*" by the people. There are words in this poem, which indicate the deep respect paid by the great poet to farmers' work and which show his knowledge of manmade irrigation, as evidenced in the saying that "*on solonchaks, only weeds grow*".

It is amazing how the poet knew even the insignificant details of irrigation works. According to the poem, firstly, Farkhad drew two parallel lines, each one thousands ells[35] long (one ell equals about 80 cm). This means that this *ariq* was 800 m long. Its width was 3 ells (almost 2.5 m) and its depth was 2 ells (1.6 m). Using an adze, Farkhad trimmed the ariq's walls and then finished them so skillfully that it was as if he worked in wax rather than in granite. The *ariq's* walls were like a mirror in which each sand particle was reflected. According to the poem "*sands are an inappropriate bed for stream*", so when a section of the *ariq* with a stony bed ended and exposed sands risked collapsing the channel, Farkhad lined the bottom and walls of this *ariq* with hundreds of granitic slabs so that "*sands could not drink all water*". He cut these slabs with special teeth along each edge for a leak-free connection and

[34]Nizām al-Din Alī Shīr Herawī was a Central Asian politician, mystic, linguist, painter, and poet of Uighur origin.
[35]Ell is a measure of length locally variable.

polished them after they had been laid. He did this job following strictly established standards and even the most captious specialist could not find the joins. In the event of droughts, the city could use water from the lake, which Farkhad created according to the regulations of water use established at that time. Alisher Navoi wrote that the water from this lake "was as the elixir of life, so fresh, cool and transparent..."

Farkhad then built the distributary canals, which took water from this lake to the residential quarters of the city. This irrigation system was modelled on those that were built in the fertile valley of Herat under the supervision of Alisher Navoi, the Grand Vizier himself. Even today, the Herat oasis remains the breadbasket of Afghanistan and the irrigation canals that were built in the times of Shah Rukh (the son of Tamerlane), Husayn Bayqarah and Alisher Navoi irrigate crops in this valley. The life of Farkhad had a tragic end in the dungeon of Shah Khosrau II[36]. However, his name, glorified by Alisher Navoi, is remembered in many legends. His name lives on: mothers call their sons after Farkhad and want these children to resemble their namesake. In 1948, the hydropower plant built on the Syrdarya at the site where it turns sharply to the north was called the Farkhad Hydropower Plant. His heroic deeds are indeed immortal.

This way, the unity of humanity, water and earth underpinned the norms of behaviour, lifestyle and ethics that became the basis for Central Asian civilization. Remnants of ancient groves, forests and pastures have been discovered in the steppes of Mongolia and Kazakhstan, and it is unknown whether they perished because of the rivers dried up or whether they were cut down by people. A lack of passionarity leads to victory of spontaneous forces over people. However, on the territory between the Amudarya and Syrdarya rivers, passionarity provided for a victory of people over spontaneous and the often destructive forces that threathened their region.

Water resources and manmade irrigation systems were the basis for life of the people in Central Asia and the source of their well-being. The supremacy of water established the base for distinctive Asian methods of production and patterns of development. This was in historical contrast to European development, which went through four successive stages of a primitive communal system, slavery, feudalism and then capitalism based on private ownership of land and market economies. In Central Asia, water scarcity did not allow a social system based on the private ownership of land to develop. The need to construct irrigation systems that would allow the continuous expandion of agricultural land to meet the needs of a rapidly growing population predetermined collective initiatives organised by a state. The first state in Central Asia was established because of the need to establish a proper system of water resources management. The governance of irrigated farming and the upkeep of a regular army were the key functions of these early states. The state played the role of chief national irrigator. The Quran states that the Earth is the property of Allah, and the sultan is his shade on the Earth and the executor of his will. Therefore, the sultan is the owner and custodian of water and land. Private ownership of land and water was unallowable. While in the West, private ownership of land and fixed assets formed the basis

[36]Khosrau II, also spelled Khosrow II, called Parvez ("the victorious"), was a Persian king (590–628) of the Sassanid dynasty and the grandson of Khosrau I.

for legal and productive relations, and property was considered an inviolable right, in the East, water and land resources were the sovereign's property.

After the Battle of Dandanaqan in 1040, where they defeated the ruler of Ghazny Masud I, Seljuk Turks became the dominating force in Central Asia for a short period. Their administrative centre was located in Nurata (in the present-day Navoi Province of Uzbekistan). Seljuk leaders decided to divide the land into feudal domains, called *iktas*, and the rulers of these domains had to collect taxes, store up forage for their army and maintain the irrigation systems in proper order. This system was kept until the fourteenth century AD (prior to Timur's reign).

A lack of laws governing private land ownership did not prevent the purchase and sale of all land by private persons. However, this had to be in accordance with Sharia regulations. Karez were allowed to be privately owned. They could be sold "fair and square", and after due registration they became "sacred and inviolable" private property. Even a sovereign ruler couldn't withdraw these rights, he could only purchase them using his treasury funds. One ancient document provides evidence of the sale of "one dry karez and one karez in good order". Each of these had command areas with soils suitable for farming (Ivanov 1954).

Developing statehood is inseparably linked with developing civilization. In our view, the most accurate explanation of the term *civilization* in modern historical literature comes from Bongard-Levin (1989). The distinction between "civilization" and "primitive society" is seen in the level of economics, formation of classes, the establishment of nations, towns and civil society, and finally, written language, that serves as the basic functional tool in a complicated social life. To this, we also need to add the money-based economy[37], which is the most important feature of a civil system.

Large-scale irrigation requires well-organized management of considerable water volumes and involves the use huge manpower resources both for the operation and maintenance of irrigation systems for agricultural production. A well-developed irrigation network was created between the seventh century BC and the third century AD. At that point, many historians (Marcwart, Bartold, Ranov (1965) and others) consider that this system could exist only because there was a powerful state, which would have the necessary resources to organize its operation and maintenance.

It is assumed that two large states existed in the period of astounding growth of large-scale irrigation from the end of the second millennium BC to the beginning of the first millennium BC. These were Khorezm (identical with Qanghoy) centered in the Southern Pre-Aral region and Bactria centered in the upper Amudarya (historians now identify this state with ancient cities in the Surkhandarya Province in Uzbekistan). Referring to Al-Beruni, Tolstov considered that the transition towards large-scale farming was assisted by the consolidation of military confederations of tribes under

[37]The functions of money as a medium of exchange and a measure of value greatly facilitate the exchange of goods and services and specialization of production. In a money economy, the owner of a commodity can sell that commodity for money.

the leadership of powerful chieftains personified by Siyavush (the thirteenth century BC). One culture succeeded another through a period of intense irrigation development, from the Achaemenid period (the sixth to the fourth century BC), when a system of large irrigation canals was built up, to the Kanguy (Kanquish) period (the fourth to the first century BC). This is known as the brilliant period of the powerful Kushāna Dynasty.

Margiana (in the valleys of the Herirud and Murgab rivers) and Irkania (Eastern Turkmenistan) were becoming established in the first millennium BC in Central Asia, where large urban settlements such as Maracanda (Samarkand), Shash (Tashkent) and Urusshana (Ura-Tepe) came into existence (Figure 2.9).

The development of urban areas started prior to the Zoroastrian period, when fortresses were built among rural settlements. Ancient settlements of the Chust culture were discovered in the Fergana Valley (Alabaman), in Khorezm (Kusaly-Kyr), in Shash area (Chirchik-Rabat) and near Samarkand (Afrosiab). These populated areas were developed as the result of the transition from farming on naturally irrigated plots to an intensive agriculture based on advanced irrigation systems.

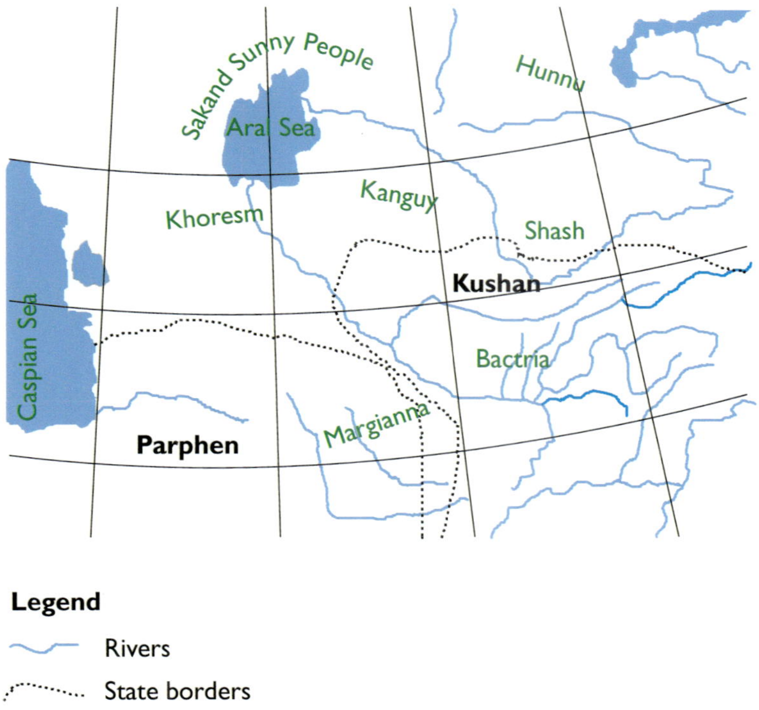

Figure 2.9 Central Asian states during the first millenium AD.

There was intensive exchange of goods between agricultural oases and towns at this time. Quintus Curtius Rufus [1948][38] noted that "the nature of Bactria is quite diverse; abundant waters irrigate lands; crops are cultivated on fertile fields; vineyards yield sweet grapes and irrigated farming is the base of the state economy and tax regulations. There are three kinds of taxes: production tax, land tax and water tax."

There are also the interesting historical monuments in the Syrdarya Valley (Alexandria Eskhata[39], near nowadays Khodjent) dating to the time of Alexander the Great's invasion, which later grew during the Greecian and Bactrian cultures (300–100 BC). About 2,000 documents related to the administrative regulation of the Parthia tax system and other aspects of economic activities were discovered during archaeological excavations in ancient Parthia (near modern Ashkhabad) (Sarianidi 1965).

The military aristocracy and priests of this state gradually ceased relying on the unstable spoils of war for their wealth and found a more stable and reliable income from the irrigated land in common use. (One of the monuments of this epoch is the Dargom Canal near Samarkand, which is still operable today (Figure 2.10) and is the main water artery of the Zeravshan oasis with a water consumption of

Figure 2.10 The Dargom Canal (2010).

[38]Quintus Curtius Rufus was a Roman historian during the reign of Emperor Claudius (41–54 AD). His only surviving work, *Historiae Alexandri Magni*, is a biography of Alexander the Great in Latin in ten books, of which the first two are lost, and the remaining eight are incomplete.

[39]Alexandria Eskhata means Outermost Alexandria.

more than 100 cubic metres per second.) There was a strengthening of the aristocracy but, by the end of this period, the slave state had collapsed and sharecroppers, who rented land belonging to a feudal lord, became the main producers. Feudal lords, who possessed most of the agricultural land, lived in towns. They had mercenary troops (*chakirs*) at their disposal and were quartered in fortresses that were usually located near the headworks of irrigation canals. This dislocation of functions was typical for the period between the sixth and eighth centuries AD. The feudal lords understood the importance of the water management infrastructure and placed their troops near the water sources to protect the system.

The ninth and tenth centuries AD in Maverannahr is characterized by strong economic growth and a golden age of culture. There was further development of the feudal states. Irrigation was the basis of the agriculture sector. In irrigated fields in Soghd, Fergana, Shash and Ustrushany, farmers cultivated wheat, barley, millet, vegetables, oil-bearing plants and other crops, including even cotton. Geographers of the ninth and tenth centuries AD, such as Qudama bin Ja'afar and Al-Muqaddasi[40], enthusiastically described the skills of farmers in the Central Asian oases. For example, Al-Muqaddasi wrote about ancient Merv and its complicated system of irrigation: "*Four main canals run through the town from the Murgab River. The first one is the Al-Zork Canal that flows through the fields and approaches the town gates from the side of the suburbs. This canal enters into the town, and its water fills a few deep tanks. The second one is the Al-Adi Canal, water of which is used by people who live in the settlement near Sin-Djan and the Miremakhan gates. The third one is the Hurmus-Kharre Canal, which flows from the side of Serax and carries its waters to the outskirts and farmsteads, and the fourth one is the Al-Madjan Canal (with bridges towards the main street), which crosses the town and supplies water to the markets. Inhabitants have open and closed tanks with ladders and gates which can be opened allowing the canal water to fill the tanks as necessary*." Al-Muqaddasi noted that the irrigation system of ancient Merv was maintained by 12,000 people, because even a small oversight could lead to terrible consequences for the local population. At that time, the Murgab River flowed away from the town in a north-western direction and it was only thanks to the construction skills of irrigators that waters were directed into the right riverbed. During the destructive wars in the period between the ninth and eleventh centuries AD the irrigation network was damaged and, thereafter, the town could not be rehabilitated to its former economic power, and it gradually fell into dilapidation and desolation. Today, only ruins of some ancient buildings and the remnants of the Sultan Sandjar mosque are reminders of the greatness of the ancient oasis which covered 10,000 hectares ancient Merv.

Hojamuradov, Mamedov and Taganov (2007) have attempted to chart the chronology of the key structure (the Sultanbent Dam) that regulated the Murgab near ancient Merv: "*In historical sources, this most ancient and largest hydroengineering structure that regulated the main stream of the Murgab River is mentioned under different names: the Mary Bent, Serybent, Bash Bent, Bendi-Soltan, and Bendi-Mubarek. The dam was built in the seventh century AD. During its 1000-year history, it went through different tragic events. The dam was repeatedly destroyed by floods and by foreign conquerors. In 1162, the dam was breached*

[40]Muhammad ibn Ahmad Shams al-Din Al-Muqaddasi, also transliterated Al-Maqdisi and el-Mukaddasi, (c. 945/946–1000) was a notable medieval Arab geographer, author of *Ahsan at-Taqasim fi Ma`rifat il-Aqalim* (*The Best Divisions for Knowledge of the Regions*).

by high waters. In 1219, it was destroyed by the Mongols. In 1409, the Sultanbent Dam was reconstructed by Shah Rukh (the son of Timur) and in 1566 it was again destroyed by Abdullakhan Sheybanid. Nadir, the Shah of Iran, destroyed this dam in the winter of 1728 and later, in 1734, a new dam was built on his order. The Sultanbent Dam was destroyed for the last time by Shamurat Veliamy, the Emir of Bukhara." At present, the Sultanbent Dam, with a reservoir capacity of 30 million cubic metres, serves as the regulator for the Khan-Yap and Han-Yap canals (Figure 2.11).

Thus, the history of Central Asia reaffirms the key role of the state in supporting irrigation and water management in these arid regions. Marx (1969) wrote: "*Climatic conditions and the peculiarities of the landscape, especially the presence of large desert areas stretching from the Sahara Desert through Arabia, Persia and India to the mountain valleys of Asia, made the manmade irrigation systems consisting of canals and hydraulic structures the basis of oriental agriculture. This system of artificially improved soil fertility depended on a central government, and was ruined due to a lack of state attention to irrigation and drainage works. As a result the present existence of deserts in former oases such as Palmyra, Petra, Yemen and large provinces of Egypt, Persia and Hindustan can be explained.*"

Figure 2.11 The Sultanbent Dam (Hojamuradov, Mamedov and Taganov 2007).

2.7 Irrigation in the Time of Timur and Timurids

In Asia, economic activity – the method of production – was based on mass collective activity to the detriment of private initiative. This production method was quite sufficient for developing irrigation systems and waging military campaigns under primitive conditions, but it impeded socioeconomic progress. It was conservative and its political superstructure, in the form of oriental despotism, provided a brake to keep things unchanged. However, it suited the feudal military elite that came into existence within this framework.

The despotism of oriental empires was exemplified by the absolute and implicit obedience of citizens to their ruler. A ruler was considered not only as a sovereign but also as a commander of the faithful. The Asian production method was at its apogee under the reign of Timur (1336–1405) and his heirs (Timurids). This period can be compared with the Renaissance in Europe. It is a remarkable historical period in Central Asia because it saw real breakthroughs in science, culture and economics. It is interesting to note that the Renaissance period in Europe also coincided with a period of domination by absolute monarchies.

Timur introduced a new system of land ownership (*soyurgal*), according to which his close relatives, commanders, emirs and the highest-ranking priests could possess vast plots. Sometimes they were awarded entire cities and provinces, but under the condition that they had to ensure the development of commerce and handicrafts, oversee the rise of culture, and expand the amount of arable land through irrigation. *Soyurgal* was not real private ownership: land was granted for lifelong possession and land tenure could be inherited, but sale of land was forbidden. In Europe, dukes and princes could sell or exchange their land, but this kind of action was banned in Central Asia (Vyatkin 1902).

Timur achieved absolute power, and his personal traits and intellect raised his power to the point that it was incontestable. He created a class of feudal military aristocracy that also had absolute power of their land (*soyurgal*). After his military campaigns, Timur's main passion became developing the irrigation systems. His concern for water was like part moral imperative and part charitable work, which each Muslim should execute as his sacred duty. This passion for irrigation, water management, planting of orchards and construction of canals was found in all the Timurids: each aspired to follow the example set by their famous forefather.

Considerable progress was made in developing the irrigation systems during the reigns of Timur and the Timurids. Ruy Gonzalez de Clavijo was a member of the second embassy of King Henry III of Castile to Timur (1403–1404). The Spanish diplomat, who visited the court of Timur at Samarkand, wrote a valuable account of his visit and noted that Timur admired vast blooming gardens in the cities of his Empire,the seething streams of irrigation canals and the large water reservoirs that were built along the way from Tavriz where his son Miran Shah was the ruler to Samarkand, the capital of the Timurid Empire. In his diary, Ruy Gonzalez de Clavijo (1990) wrote that Termez was a large and densely populated city and "there are many canals and orchards all around the city".

Gonzalez de Clavijo was especially impressed by Timur's Palace in Shakhrizyabs (Ak Saray), which towered over an ocean of flowers in the huge garden. He wrote: "*This garden was surrounded by a deep ditch full with water since water continuously discharged into it from a special water conduit. Citrus trees with lemons grew in this garden and a spacious area of fields that was crossed by the river and numerous canals spread in front of the garden.*"

Ruy Gonzalez de Clavijo also mentioned that, following Timur's instructions, all land owners had to show special concern for irrigation and the beautification of gardens. Commenting on one of the land tenures, he wrote that "this terrain belongs to one very important person. It is located in the river valley, and there are many irrigation canals here. And the population that cultivated beautiful orchards and vineyards was quite numerous."

Valuable information on the construction of irrigation infrastructure during Timur's reign is contained in the book *Zafarnama* (the Book of Victory) that was written by the secretary of Timur's grandson who was the governor of Shiraz, Sharafitdin Ali Yazdi. It is based on the original documents that were kept in Timur's archives after his death. The book was written in 1427 and ornamented with miniatures by Behzod, the outstanding Central Asian painter. This book was recently published in Russian (Sharaff Ad-Din Ali Yazdi 2008).

In the *Zafarnama*, it is noted that "Allah the Almighty has granted devoutness and generosity to some servants of God so that they could leave their good heritage to the world such as cities, mosques, gardens and canals. And from all this great blessings will be granted to these servants of God and good deeds will remain as monuments of their memory under the sun." During his Caucasus campaign, Timur had seen the channel of a dry canal that had connected the Aras River with Karabakh. He decided to rehabilitate this canal and fill it with water. By his order, the canal was restored in one month through the work of his huge army of 200,000 people. The 100 km long canal was so deep that it became navigable. The canal was called "Barlas" after Timur's native tribe. Many settlements were improved, orchards planted and fields were irrigated by this canal, with so many buildings and irrigated fields in each province that it was difficult to count them (Sharaff Ad-Din Ali Yazdi 2008).

In the time of Mohammed Shaybani, Khan of the Uzbeks, an aqueduct bridge across the Zeravshan River was constructed at the site where it branched into the Ak Darya and the Kara Darya (Figure 2.12). In the same period, the ancient Eskituyatartar Canal was rehabilitated. This had been built at the beginning of the Christian era for transferring the flow from the Zeravshan River through the mountain valley to the Sanzar River, and the canal is still used today. The influence of the Timurids on water infrastructure extended beyond Central Asia. They encouraged the people of the nations they conquered to build irrigation canals and dams, and to reclaim the wetlands by transforming them through a system of canals and *ariqs*. All Central Asian rulers were great irrigators who considered irrigation and land development as their sacred duty and their service to the people and their God.

Figure 2.12 Remains of the aquaduct accross the Zeravshan Canal

One outstanding descendant of Tamerlane was Babur[41]. His real name was Zāhir ud-Dīn Muhammad (1483–1530). According to *The Oxford History of India* (Smith 1981): "*Babur, Zāhir ud-Dīn Muhammad, King of Kabul, was the most brilliant Asiatic prince of his time and worthy of the highest place among all sovereigns of any century or country*."

Numerous books and articles have been devoted to Babur. In 1825, his autobiographical work *Baburnama* was translated from Uzbek into English, and this monograph of provides a unique history and geography of Central Asia, Afghanistan and India. The book was written in the sixteenth century, but it has not lost any of its scientific and educational value. It has been republished many times in England and in the USA. The best translation of this book, in an edition first published in London in 1895 and then republished many times, was made by Annette Beveridge. Researchers of Babur's life and activities were mainly interested in his military feats (he was a very brave warrior) as well as by his activities as a ruler and statesman (he was as wise as Buddha).

Babur became the ruler of Kabul (in Afghanistan) in 1506. Kabul was a part of the empire established by Timur; and the former ruler welcomed the twenty-two-year-old young Timurid who arrived together with his family and fellow-fighters after their defeat by Mohammed Shaybani. The transition of power was peaceful, due to the authority and glory of the Timurids. Babur ruled Kabul for 20 years and governed India for four years. He died when he was only 47 years old. In 1526, he defeated the huge Indian army of Sultan Ibrahim Lodhi in the Battle of Pānīpat, despite its substantial power and a herd of 100 war elephants that horrified enemy troops. Babur achieved victory through a combination of superior tactics and use of artillery. The roar of the field guns frightened the war elephants and caused them to stampede, trampling Lodhi's own warriors. During a short period, he managed to consolidate

[41]At that time, the Chaghatái were very rude and uncultured (*bázári*), and not refined (*buzurg*) as they are now; they found Zāhir ud-Dīn Muhammad difficult to pronounce, and for this reason gave him the name of Bábar. The name *Babur* is derived from the Persian word *babr*, meaning "leopard" or "tiger".

almost the whole Deccan Peninsula into a single state and, through astute diplomacy and military skill, to establish the Mughal Empire that existed until 1858.

Many authors pay scant attention to the fact that Babur constructed new irrigation systems that vivified India. The Indian historian R. Nath is right to reproach Western historians for their superfluous attention to his military feats, while ignoring the other side of his activities, the construction of water infrastructure all over India. In *The Immortal Taj Mahal*, Nath writes that historians describe, with pleasure, his battles of Pānīpat and Khanwa, but take no notice of Charbagh Garden[42] in Agra. This is now a world-famous park, "where Babur implemented his *charbagh* project and constructed a terraced garden with manmade canals, chutes and lily ponds".

Babur was born on February 14, 1483 in the town of Andijan, in the Fergana Valley, surrounded by irrigated fields, orchards and kitchen gardens. He started his governance in Afghanistan by constructing the water supply system of Kabul, which was a display of his natural talents rather than some kind of political act. He decided to supply clean fresh water to the inhabitants of Kabul, which he loved for the rest of his life, and to plant gardens to beautify his new capital and to fill it with the fragrance of flowers. In mountains close to the city, he found a spring which issued water with an amazing taste. "*I ordered that the spring should be enclosed in mortared stone-work, 10 by 10, and that a symmetrical, right-angled platform should be built on each side, so as to overlook the whole field of Judas trees. If, the world over, there is a place to match this when the arghwans are in full bloom, I do not know it. The yellow arghwan grows plentifully there also, with the red and the yellow blossoming at the same time. In order to bring water to a large round seat which I had built on the hillside and planted around with willows, I had a channel dug across the slope from a half-mill stream, constantly flowing in a valley to the south-west of Sih-yaran. The date of cutting this channel was found in jui-khush (kind channel)*" (from Beveridge's translation, *The Baburnama: Memoirs of Babur,* page 217). Later on, the water supply system that distributed water into the residential areas in the city was constructed. Kabul's running water is a monument to Babur and his invaluable legacy. He wanted to be buried in Kabul, and his last will was implemented. His tomb is in downtown Kabul today. Even the Taliban fundamentalists left it untouched, although they blasted the huge stone statues of Buddha in the Herat Valley.

When Babur became the ruler of India, he decided to train Indians in artificial irrigation. He studied the agricultural sector and came to the conclusion that: "*The greater part of the Hindustan country is situated on level land. Many though its towns and cultivated lands are, it nowhere has running water. Rivers and, in some places, standing waters are its "running water" (aqdr-suldr). Even where, as for some towns, it is possible to convey water by digging channels (ariq), this is not done. For not doing this there may be several reasons, one being that water is not at all a necessity for cultivating crops and orchards*" (*The Baburnama: Memoirs of Babur*, page 486). Babur decided to introduce new irrigation systems to India. He got this idea when he visited Agra (150 km from Delhi), when looking for a site for the capital of his new empire. He wrote: "*One of the great defects of Hindustan is its lack of running water, it kept coming to my mind that waters should be made to flow by means of wheels erected wherever I might settle down, and also that grounds should be laid out in an orderly and symmetrical way. With this objective in mind, we crossed the Jun-water to look*

[42]This garden symbolizes the four rivers of Paradise and reflects the gardens of Paradise derived from the Persian *paridaeza*, meaning 'walled garden'.

at garden-grounds a few days after entering into Agra" (ibid. page 531). On August 28, 1528, Babur visited Agra and surveyed possible locations for his future palace. He found the place disgusting and dirty, with a repulsive boggy smell. The fierce heat persisted day and night, and dust was blown around by the wind all the time. Despite this, Babur chose Agra for his permanent residence, selected a site for his residence on the left bank of the Yumuna River and designed the future irrigation system. At its heart was a large 300-metre square *charbagh*, a Mughal garden with a raised marble water tank at the centre of the garden from which water flows by gravity, and avenues of trees and fountains. The city palace in Agra, along with Taj Mahal, became the most beautiful and expressive embodiments of this project.

Shortly after the promulgation of his project, his relatives and dependents started to build residences on the riverbanks. Beautiful and well-designed gardens with canals, green lawns and water reservoirs were developed. The local inhabitants of India had never seen such gardens, and they called this district Kabul because the gardens were exact copy of the garden complex constructed by Babur in Afghanistan. Gardens were his brainchild and his passion, and he often spoke about them to his new subjects. "*Charbagh fever*" spread over India. People started to look for water sources, to dig wells and to construct canals throughout country. They found waterfalls in the mountains and built canals from these water sources to feed water by gravity to fields, pastures and settlements. If it was not possible to build a canal, Babur advised people to use bamboo trunks. By burning out the inner part of bamboo trunks, and linking them using bulls' guts, they created long pipelines that could be used to supply clean water. The *charbagh gardens* stimulated the development of India, and contributed much to its beautification. During the four years of his rule, Babur revolutionized attitudes to water and raised Indian people's interest in water as the basis for their well-being. The Taj Mahal, built by Babur's grandson (Shah Jahan) in the seventeenth century, is a worthy crown to his creations. The best traditions of Central Asian architecture and brilliant solutions to water supply problems are incorporated in this project. After 400 years, the 85 fountains are still smoothly operating in this beautiful complex. A raised marble water tank is still kept here, and water from mountain springs flows through the pipelines.

This recognition of the importance of irrigation resulted from ethics, laws, intellectual and moral responsibility that emerged from the historical experience accumulated in Central Asia. Respect for water as the basis for life and well-being is a distinctive trait of all rulers of the Timurid Dynasty. The traditions that they established are still observed everywhere where they rule.

In 1539, Mirza Muhammad Haidar Dughlat (1499–1551), another descendant of Timur, assumed the throne in Kashmir with the help of Babur's son Humayun. Mirza Muhammad Haidar was the son of Muhammad, the ruler of Tashkent, and Babur's cousin. Tashkent was famous for its perfect irrigation system its abundance of water, and for its formal and kitchen gardens. So when Mirza Muhammad Haidar became the ruler of Kashmir he also started to build an irrigation system. Like Babur, he was rather inventive. He decided to take advantage of the huge volumes of snow in the mountains near Kashmir. He constructed canals to capture and transport meltwater that provided water for irrigation of newly-developed arable land and a water supply to the city and settlements. The French physician F. Berne, a personal doctor to the Mughal emperor of India Shah Jahan (1592–1666), enthusiastically described the irrigation system in Kashmir: "*Numerous streams and rivulets flow down from the mountains.*

Local inhabitants manage to divert their waters to their paddy fields, and even lift them to low plateaus by means of constructing earthen dams" (Berne 1936). Berne was partiucarly taken by the fact that: "*all these rivulets flowing from mountains make the fields and hills so beautiful and fertile that the entire state looks like a huge green garden, in which here and there amongst trees settlements and villages are met and sometimes for variety's sake small steppes and fields under rice, grains and other different crops exist. All this terrain is crossed by canals full of water that in some places create lakes.*"

From 1550 to 1585, the rulers of the Bukhara Khanate constructed many irrigation canals that diverted water from the Vakhsh, the Murgab and other rivers. These included the Khodja Qaab Canal for the Shah Rukh irrigation system (built 1557), the canal diverting water from the Amudarya to Chardzhou (built 1569) and the canal diverting water from the Vakhsh for the irrigation of six districts located on the territory of present-day Tajikistan (built 1579–1585). At the same time, the Eskiangar Canal, which was constructed for delivering water to the Kashkadarya region, was restored.

The Middle Ages – the period of flourishing feudalism when the large states of the Samanids and Qarakhanids as well as those of the Khorezm Shahs were established – saw the intensification of irrigation all over Central Asia. According to Tolstov, 1,400,000 hectares were being irrigated in Central Asia at that time. Some components of the elementary water infrastructure dating from this period still exist. In Samarkand, there were numerous stone structures, such as bridges, aqueducts and chutes. About ten years ago, one could still see the ruins of the water distributor near the Ak-Karadarya waterworks on the Zeravshan River. Developing intensive and sustainable irrigation over large areas in this region was possible only when people could manage natural floods and train the main watercourses. They built dozens of dams and other hydraulic structures. Traditions of permanent agriculture inherent to the local population and a strong unified state have favoured the implementation of these works. Many historians (see, for example, Ostrovski 1907), have taken on Engels' idea that "*irrigation should be the concern of the community, and regional or central authority*" and have vividly shown that the sequential highs and lows in irrigation are closely correlated to the highs and lows of social development. Each rise in political centralization coincides with a rise in development of large-scale irrigation and land reclamation. Many examples of this relationship between power and water can be found in the history of Turkistan.

Civil war followed the death of Timur in 1405. After the Timurids had completely exhausted themselves fighting each other, they fought the Shaybanids, Uzbek tribes that came from Siberia. Shaybanids and Timurids were ethnically identical, sharing a common Uzbek language and the same Muslim religion and traditions. The power struggle between them in Central Asia bears strong resemblance to the war of the Roses in England (the English civil wars between the Houses of York and Lancaster) and these wars occurred almost concurrently (between 1455 and 1485). The Shaybanids gained victory over the Timurids in Central Asia. Many Timurids were killed, and the remainder headed by Babur escaped into Afghanistan and then to India and Kashmir. However, the Shaybanids were soon defeated by Shah Ismail I of the Safavid Dynasty founded by Sheikh Safi of Ardabīl in Iran. In 1510, the Shaybanids were defeated in the Battle of Mevr and Mohammed Shaybani was killed. Other Uzbek tribes filled this power vacuum. Three new states, the Khiva Khanate (1515–1920), the Bukhara Emirate (1785–1920), and the Kokand Khanate (1700–1875) were consequently created in Central Asia (Gulyamov, Islamov and Askarov 1966).

All three states were based on the Asian production method. Narrow-minded and poorly educated rulers (khans and emirs) were zealous defenders of archaic social relations, and were despotic regimes. For hundreds of years, rulers defended their privileges and became a constraint to technological progress and good labour management relations. All things in the world are imbued with and related to the notion of "necessity". The reality of Central Asia –its geographical position, climate and soils – made water resources the most important political factor, and one which had to be taken into consideration by each ruler in both domestic and foreign policy. All rulers were forced to take into account the necessity of properly maintaining the irrigation systems, constructing new canals and expanding the amount of irrigated arable land in order to meet the needs of growing populations in this region. However, one war could easily destroy what had been created by hundreds years of labour. These moral responsibilities forced the authorities to adopt certain basic rules and meet obligations, which they had to execute in line with the requirements of the Muslim religion. Although there were some minor local differences, the Sharia was applied throughout the region and its carried a uniform set of requirements. A primary duty was to take care of irrigation and developing new irrigated areas. Murat Shirin ibn Amir Shah, the famous historian from Bukhara, and vizier and mentor to the heir to the throne, said to his pupil: "*Allah never grants the power to somebody by mistake, however, if He granted you the kingdom, you have to hurry to do good and if you planted a seed then you should irrigate it and try to defend it from thirst and cold.*" He persistently advised the heir to the throne to take care of water as the guarantee of the sustainability in his state and he taught him that "crops will not ripen without water, and a dry tree will not bear fruits" (Masalsky 1913).

Through a combination of the behavioural psychology of farmers, a widespread understanding of the value of water for life, and the adherence to the Sharia in the Central Asian khanates, irrigation systems developed even under conditions of general backwardness. The Asian production method combined with Eastern despotism suppressed personal initiative, but it did ensure mass participation in public works. The agricultural sector remained generally conservative, maintaining patriarchal relations and employing primitive agricultural implements. The basic tool of tillage was the *omach* (wooden plough), which did not turn over but only loosened the topsoil. The only improvement to this main tool of production came from applying a cast-iron or iron plowshare. Harvesting was done by hand, and threshing was done by walking animals over bundles of harvested cereals.

Nevertheless, some technical progress was observed in the construction of irrigation systems, both in the technologies used and in the scope of works. Most progress was achieved in the Bukhara Emirate. From the earliest times, the traditions of artificial irrigation were well established in the Bukhara region. In the tenth century AD, during the Samanids period, eleven irrigation canals were dug in this city, which transformed it into a blossoming oasis. Bukhara was able to develop and exert its influence mainly thanks to this well-developed manmade irrigation system. There were manmade ponds with flowing water, from which water carriers delivered water in waterskins to markets, mosques, offices and private houses. The profession of water carrier became a respected role in the emirate (Gulyamov et al. 1957). The Labi-Hauz Pond is a brilliant creation and example of the Bukhara irrigators' art (Figure 2.13) Water infrastructure in Samarkand, which until 1873 was a part of the Bukhara Emirate, was very impressive. The Hungarian traveller and specialist in Turkic philology Arminus Vambery (1873) visited this city in 1864 and wrote: "*Samarkand is located much higher than Bukhara, and was always notable for its bracing healthy climate, but at the same time, it gained a reputation as the Paradise of Islam*

Figure 2.13 Lyabi Khauz.

thanks to the wealth of water that was conveyed through numerous canals and streams from the neighbouring mountains onto the plain. According to Al-Belhi, all this water comes from the Sogd River. A reservoir existed not far from the river. From there the water was presumably conveyed to Vargas (also called Bargas or Burgas) and flowed through canals eastward and westward. The key reservoir looked like a small lake with villages along its shoreline and the principal canals such as Barmish and Dekish made fruitful large tracts through this country up to over a six-day journey distance."

Water granted great power to Central Asia. Many monarchs from the region strived for immortality by constructing water infrastructure, hoping to atone for their sins. They understood the value of irrigated land for the prosperity of the state. Wars could only lead to ruin, but water brings development and well-being. In spite of the overall backwardness of Central Asia, there was a high level of irrigation and farming, and as a consequence the rural population did not aspire to migrate to the cities.

2.8 Progress of Science and Historical Development of Irrigation in Central Asia

A reverent attitude to water is an indispensable precondition for life in this arid region. The need to use water and land in conformity with the laws of nature has lead to the development a knowledge base for cultural and scientific progress. To sustain agriculture, especially irrigated farming, there has been a continuous modernizing of the methods of land treatment, and agricultural implements and tools. Some historians have assumed the culture of ancient Central Asian civilizations – in Parthia, Khorezm and Bactria, for example – was based on the cultures of the civilizations of Akkad, Sumer, Babylonia and Assyria in Mesopotamia, but local and Russian scientists have challenged this assumption and come to a different conclusion. They argue that the basis for later Central Asian civilizations, with their outstanding scientists, writers, painters and architects, came from the ancient Central Asian culture that has its origins in the Bronze Age.

The diversity of natural conditions in Central Asia, the contrast between desert and steppe landscapes and oases and mountains, shaped specific individual cultures and civilizations. These were nevertheless closely related to each other, and had links other ancient centres of civilization over a long historical period. However, a lack of understanding of the multiplicity of natural processes impeded improvement of the modes of production and life. Religious teachings tried to fill this knowledge gap.

Prior to the Persian conquest of Central Asia in the first half of the first millennium BC, the religious teachings of Zarathustra were widespread. These reflected the process of the collapse of the classless society and the transition towards widespread farming and cattle breeding. The process of forming the first states in Central Asia, which was accompanied by numerous political and social conflicts, was reflected in religious ideology and in the passionate sermons of Zarathustra, who advocated farming as the first virtue of veritable supporters of Ahura Mazda. Zoroastrianism undoubtedly played a great role in presenting farming as a sacred duty. The third chapter of the Avesta (the Vendidad: Fargard 3. The Earth) sets forth: "*O Maker of the material world, thou Holy one! Which is the third place where the Earth feels most happy? Ahura Mazda answered: It is the place where one of the faithful sows most corn, grass and fruit, O Spitama Zarathushtra! where he waters ground that is dry, or drains ground that is too wet.*"

According to Bongord-Levin (1950) and other researchers, the Avesta (the sacred book of Zoroastrianism) may date back to the pre-Achaemenid period. Scientists investigated some manuscripts belonging to the Kushana period, and these showed that an ancient school of irrigators was formed consisting of scientist-priests, and this existed until the time of Kutaiba's military campaign against the Khorezm kingdom. This school combined knowledge of mathematics, cartography, astronomy, hydraulics and other sciences. The creation of the lunisolar calendar of 12 lunar months of 30 days with five extra days at the end of a year may date back to that same period (the seventh century BC). Studying the natural seasons cycle was needed for the proper application of irrigation and required knowledge of meteorology, astronomy and other sciences. Abu Abdulloh Mohammad Ibn Musa al-Khorezmi in his famous book *Al-Jabr va-al-Mifuqabilah* wrote that this knowledge was necessary for various calculations and also for measuring plots, canals, etc.

By applying traditional methods in irrigation and water use, humans accumulated knowledge, especially in agriculture. Knowledge of the trajectory and periodicity of the movements of the stars, as well as climatic and weather phenomena, was used for assessing land use options. All this lead to the introduction of water management and irrigated farming based on knowledge of astronomy, meteorology, pedology and many other scientific fields. It is not by mere chance that in Central Asia in the eighth century, the basics of algebra *(Al-Jabr)* and the use of algorithms *(al-Khorithm)* were developed by the mathematicians of that period, such as Al-Khorezmi and Abu Ali al-Husayn ibn Abdullah ibn Sina.

Al-Khorezmi was not only a famous mathematician, but also one of first encyclopedists in Central Asia. In his glossary of irrigation terms compiled at the end of the eighth century AD and the beginning of the ninth century AD, the founder of algebra mentioned various types of water-lifting mechanisms, including *duliab, dalia*, *garrafa*, *zurnuk*, *nasura*, *mandjanup* and irrigated highlands *garb*. The terms in this glossary, such as *tiraz* for watershed, *mussanna* for dam and *kazain* for karez, are evidence of the high level of irrigation engineering development in that period. Unfortunately, only a small part of the cultural heritage of the Khorezm scientists (*madgus*) has survived. They wrote numerous manuscripts on astronomy, mathematics, hydraulics and other sciences, but this heritage was destroyed by the Arabic vandals of Kutaiba in 712. Abu Raihon Mohammad Ibn Ahmad Al-Beruni was an outstanding follower of Al-Khorezmi. In 1010, he was the leading member of the famous academy established by the caliph Abdullah al-Mamun, where together with Al-Beruni, such outstanding medieval scholars as Ibn Sino, Abu Nasr Iraqi and the philosopher Abu Sahl Masihi worked.

At this time, the development of natural sciences reached a high level in Khorezm and included established canons covering architectural proportions, irrigation canal design, accurate land levelling, the Khorezmian calendar and detailed astronomical terminology (as described by Al-Beruni). The richness and diversity of their mineral paints can be seen in the decoration of Toprak Kala. All this would not have been possible without the progress made in this period in geometry, trigonometry, astronomy, topography, chemistry and mineralogy.

The scientific works of Al-Beruni mark the culmination of the development of Khorezmian science. Astronomy, geography, mineralogy, ethnography, history and poetry belonged to his spheres of interest. He wrote many treatises in different areas of science. He suggested some ideas on the heliocentric structure of the universe, which were more progressive than Ptolemy's geocentric theory that dominated the thinking of Arab scholars. His historical and geological theories on the formation of the northern Indian lowland landscape and the shifts in the directions of the Amudarya went far beyond the science of the time, even approaching the level of some modern scientific concepts.

It is regrettable that wars and invasions have left only few monuments of this cultural and scientific heritage in Central Asia. At present, these chronicles and scientific works can only be found in libraries in London, Madrid, Cairo and India. Almost no record remains in the cultural centres of the time, including the big cities of Central Asia such as Bukhara, Samarkand, Balkh, Merv, Khodjent and Shash. Scientists, poets and painters who lived in Samarkand, Bukhara and Guragandj amassed considerable knowledge, but the famous and in former times rich library *Savlady Hikmat* (the Storage of Wisdom) only preserved a negligible number of historical documents. This loss of the scientific record and cultural heritage is to be linked to the destruction of many achievements of this age.

For example, monuments of the tenth and thirteenth centuries such as the large dam near Gurgandj[43] and the numerous dams on the Amudarya's tributaries have been completely destroyed. In 985, Al-Muqaddasi wrote that he admired the outstanding creations of Khorezmian irrigators including two main dams, one which was built near Gurgandj and the other near Charjou. Taking into account

[43]Now Kunya-Urgench in Turkmenistan.

the size of the Amudarya at these sites, we share his opinion and raise our hats to the achievements of the irrigators of that epoch. Bartold (1966) wrote that there was a major dam on the Daryalyk with a much ramified network of canals, but after the attacks of Genghis Khan's descendants, many different rulers governed Khorezm including the Timurids (the descendants of Berqe Khan) resulting in the deterioration of this network.

Only in the sixteenth century, during the epoch of Abulgazy Khan, with the strengthening of centralized power, were new irrigation works initiated, promoting some sort of a renaissance in Khorezm. Neglected canals in the north were cleaned and reconstructed. A new water intake structure for the Sarykamysh irrigation system was moved from the Daryalyk River to the Amudarya. Several dams were built on the Kunyadarya River, including Ushak-beit, Salat-beit and Egen-Klitch (the largest one), remnants of which still remain today. This dam was 6.5 m high, with a bottom width of about 40 m, and a crest width of about 13 m. These remains provide evidence of a rather high level of irrigated agriculture practice during the different periods of development of ancient Khorezm.

This experience should be a warning to our time, when new leaders often try to destroy everything that has been achieved in the past. Our heritage, especially the scientific and cultural legacy, is a unique asset. Everything that has been created and accumulated from the past should serve the future. Ruining water infrastructure therefore is a crime. Over the centuries, the systems of irrigation and water supply have, step by step, become more complicated and they have been perfected through the accumulation of knowledge and the growth in technical prowess. Drawing on relevant monographs and documents, we have systematized the available information on the development sequence of these systems (see Table 2.1).

In spite of wars and conflicts, continuous progress can be clearly observed over a long period in history. Though lacking written information about developing irrigation systems and water management skills, the workmanship of ancient water managers *muhadys* demonstrated considerable knowledge and expertise in the construction of the irrigation systems – knowledge that was passed from generation to generation.

Masalsky (1913) distinguished different irrigation patterns depending on the available water source: "Mountainous creeks and small rivers, due to their gradient, were most suitable for diverting water for irrigation at the site of their outlet from a mountain range into the lowland. These rivers divide into a few small streams, creating a cone-like body of alluvial material, and supply their water to fields and then disappear near the border with the steppes. These waters are most applicable for irrigation use. Tributaries of the great rivers such as the Amudarya, Syrdarya, Chu, Ili and Zeravshan are of secondary importance, because diverting water from them requires to special structures such as barrages, water intakes with headgates[44] or water intakes equipped with *sypays* or *ķarabūrs* (prototypes of modern spur dykes or fascine dykes)." Therefore, rivers such as the Amudarya and Syrdarya were regarded as less suitable for irrigation due to the need to regulate these large streams with their considerable flow rates and a high sediment load.

[44]A structure with gates for controlling the water flowing into a canal (an irrigation ditch).

Table 2.1 Chronology of Irrigation Development in Central Asia.

Period	Canal	Location of water intake structure	Type of irrigation	Dimensions of canals	Idle section of a canal	Remarks
Old Stone Age, 6000 to 8000 BC	Jeytun (Turkmenistan) Tutkaul (Tajikistan)	On small rivers in piedmont valleys	Mountain creek type of irrigation (primitive forms of basin irrigation)	0.2 km long and 0.5 m wide	No	
Old Stone Age, 5000 to 3000 BC Bronze Age 3000 to 2000 BC	Geoksur, Annau (Turkmenistan) Dahistan (Turkmenistan) Chust, Dalverzin (Ferghana) Murgab (Turkmenistan)	On small rivers in piedmont valleys Distributary channels in deltas	Seasonal liman flooding;	1 km long and 0.5 m wide	Only near a river	Riverbed flanked by levees
Early archaic irrigation systems Beginning of the first millennium BC	Amirabad (Khorezm) Murgab (Turkmenistan) Katadian (Tajikistan) Otrar (Kazakhstan)	On tributaries Appearance of karezs	Seasonal liman (basin) irrigation	Up to 20 km long and 20 m wide	Available	Construction of small dams or karez
Kangay period: second half of the first millennium BC	Chermen-Yab (Khorezm) Varahsha (Bukhara) Merv (Turkmenistan)	In deltas of large rivers, (with dams, multihead water intake structures in deltas)	Seasonal liman (basin) irrigation with use of water-storage basins in former riverbeds; controlled former riverbeds	Up to 60 km long (20 km on average) and 10–30 m wide	Available	

Kushana period: the first half of the first millennium AD	Gavhore (Khorezm) Zang (Surkhandarya) Dargom (Zarafshan)	Multihead water intake structures in deltas of large rivers, One water intake structure in the upper and lower reaches of rivers	Regular basin irrigation with water delivery by gravity canals	Up to 100 km long and 20–30 m wide	A deep idle section at the beginning	Construction of dams
Second half of the first millennium AD	Kyrkyz (lower reach of Amudarya) Otrar (lower reach of Syrdarya) Soltan Bent (Merv) Sokh (Ferghana) Dargom (Zarafshan)		Regular basin irrigation with use of regulators on distributors	Branched systems	Available	Intensive development of irrigation science, dams on the Daryalyk, canals were lined with stone
From the first half of the ninth century until the fourteenth century	Shamurad (Khorezm) Nurata (Uzbekistan)	Construction of large barrages with water intake facilities (*Han bandy*)	Regular basin irrigation with use of diversion weirs	Branched systems with distributors	Available	Interbasin canals (Eskitya Tartar Eskiangor and others)
From the fifteenth century until the eighteenth century	Palvan-Ata Hanka-Arka (Khorezm)	Multihead systems "*saka*" on large rivers	Wide use of *chigirs* on distributors; regulators		Available	Construction of drainage canals

The technical level of ancient irrigation practice can be analyzed from the numerous descriptions of Russian researchers, who studied the experience of local water managers in Turkistan after colonization in 1870. Dingelshtadt (1895), a well-known researcher of irrigation practice in Central Asia, wrote about the very interesting indigenous structures, such as aqueducts and water distributors. "For transporting water across a valley or a ravine, aqueducts were built with water channels made of wooden planks or hollowed-out logs. Sometimes, water channels of aqueducts were made of sod laid on the planks, creating a gutter like structure." This method was used only for short aqueducts, but for longer structures (sometimes dozens of metres) wooden flumes were built mounted on piles or trestles. For example, Dingelshtadt mentioned the Baldyrbek aqueduct across the Upper Aksu in the Chimkent region. This was 156 metres long, 17 metres above the riverbed had a carrying capacity of 1.2 m^3/sec, and enabled the irrigation of 1,500 hectares.

There were several types of river training structure that diverted water from a river into an irrigation canal. The most commonly used were *karaburs* (fascine dykes) (see Figure 2.14) and *sypays* (spur dykes). Dingelshtadt described these water diversion structures: "*There are many pebbles and much brushwood in the Syrdarya region. Therefore, in the absence of timber at these sites, fascines are made from bundles of brushwood or reed filled with pebbles and sod. These are used for protecting the headworks of canals. Such fascines can be up to 3 metres long and 1 metre in diameter, though their size is not limited to these dimensions. In spite of the large size and weight of these fascines, the local population very easily copes with manufacturing and placing them into water, even into high velocity streams. Construction of karaburs is not expensive. Karaburs are of a very simple construction and are effective in operation. Flexible fascines neatly fit together*

Figure 2.14 Construction of karaburs (picture by Kildushov, from *Irrigation in Uzbekistan*, 61, Volume 1, page 148).

under the pressure of water, and alluvial material fills all the internal space. As a result, they provide a strong and cheap structure that is very difficult to break."

Dingelshtadt also described local methods of water measurement. "*There were various primitive systems of water measurement. The unit flow rate that was used everywhere was the* kulag, *an amount of water per second that freely passes through a large clay pipe (*katta-kaur*) 1.5 feet long (0.45 m), with an inlet diameter of 7 inches (0.175 m) and an outlet diameter of 6 inches (0.15 m). Another unit was the* tigerman *or* tegerma. *One* tigerman *equals 5* kulags. *The* tigerman *meant the flow rate was sufficient for operation of a water mill. The tenth part of a tigerman was called* dagana. *A unit water volume called* ariq *was applied in the Fergana Valley. This is the amount of water sufficient to irrigate several* chireks. *One* chirek *meant both a unit weight equal to 5 poods (one pood equals to 36 lb), and an area that was sown with 5 poods of grain seeds. Local irrigators considered that for irrigation of 15* tapaks *(400* sazhens, *1* sazhen = *2.134 m) of* jenushka *(alfalfa), one* kulag *of water per day was sufficient. Using this information, it is possible to calculate that one* kulag *equals 375 cubic* sazhens *(or 1,488 cubic feet) per second, which was the local water requirement for this crop. There were, for example,* guza-kulagy-su (kulag *for cotton),* shaly-kulagy-su (kulag *for rice), and* budai-kulaky-su (kulag *for wheat), and so each crop had its own* kulag. *Thus, the term* kulag *means a water volume supplied from a canal to a field in an amount sufficient for water application for a specific crop. There was another meaning of the term* kulag, *which is the energy momentum sufficient to operate a local water mill. Such a* kulag *was called* tigerman-kulagy-su, *and was a unit flow rate sufficient for rotating one millstone.*"

Guidance for land levelling was also developed. Primitive surveying instruments, including a bowl filled with water, wooden sticks and ropes, were used for levelling. The process was to use a stick with a tied rope that was stretched in the required direction and fixed next to a water source. A bowl full of water was then placed near the lower end of the rope. The ancient surveyors tried to keep this simple instrument such that water within the bowl was in a strictly horizontal position without spilling over the edges of the bowl. Using this method of levelling, they obtained a perfectly horizontal line. After laying a horizontal section, these surveyors then introduced a step-by-step drop in elevation to provide a gradient in the canal bed.

The use of groundwater provides further evidence of the professional art of ancient irrigators in Central Asia. *Karez* or *qanats* were constructed as a series of well-like vertical shafts connected by gently sloping tunnels. This technique tapped into a subterranean water supply in a manner that efficiently delivered large quantities of water to the surface without the need for pumping. The water drained by gravity to a destination lower than the source, which was typically an upland aquifer. Vertical shafts were used for monitoring water extraction and to store and remove stones and excavation material during construction of the connecting tunnels between the shafts. For a long time, *karez* were the sole source of water supply to Krasnovodsk (Balkan was founded as a town in 1869), located in the area between Jebel and Kazanjik in south-western Turkmenistan. *Karez* were still being used up the the colonization of the region by Russia (Khlopin 1964).

Irrigation, using water from *karez,* was widespread in Turkistan, Kyrgyzstan (Osh region) and near the foothills of the Gissar Ridge. The Ahaltekin and Aytek oases in Turkmenistan are specific examples of *karez* developments. However, one of the oldest *karez* is located near the settlement of

Ulugdepe in the Kopet Dagh. It consists of 20 vertical shafts and dates back to the eighth century BC. This *karez* was based on a tunnel of 1.8–1.4 m wide and 1.6 m high. Sometimes, *karez* (see Figure 2.15) were several kilometres long and could yield up to 100 l/sec.

From 1893 to 1845, the Land Reclamation Department carried out special field surveys under the leadership of the engineer L. Tsymbalenko[45], which described in detail the system of *karez* water abstraction and provided information on the chemical composition of the water, *karez* structures, and repair and maintenance procedures.

A *karez* water supply system is based on the interception of a subterranean water flow. It comprises a system of underground tunnels, with a tapping section placed within an upland aquifer and a water-carrying channel for delivering water to the surface. Apart from the tunnels, vertical well-like shafts are also an integral part of *karez*. They are needed for tunnelling and for removing of excavated material, and they can be used for monitoring purposes when the *karez* is in operation. The vertical shafts are spaced at intervals of 10 *sazhens* (about 20 m) on average, though they can be between 2 and 20 *sazhens* apart depending on the hardness of the rock and the skill of the workers (*karezchi*). The total length of a *karez* system may reach 10 *versts* (*1 versta* = 1.0668 km). The cross-section of a *karez* is rectangular (0.2 × 0.5 *sazhens*) and gradients of the tunnels typically vary from 0.002 to 0.005, with some even having gradients of 0.1. Tunnels are lined with stonework, and vertical shafts are oval-shaped ceramic tubes with wooden chamfered edges. Shafts may be to 40 m deep. During a field survey, 42 operable *karez* were found in Ashkhabad Uyezd (district) with discharges of up to 50 l/sec.

The ruins of some ancient water structures that were found near Afrosiab in Samarkand at the end of the last century are rather interesting. There is an underground water storage area accessed by several shafts going down at an angle of 35° from the land surface. These shafts were big enough for a man to walk through with jugs in hand. The shafts went down to an arched hall at a depth of 42 m, which had a round pit in the middle and was filled with spring water.

Figure 2.15 Karez in South Turkmenia.

[45]Tsymbalenko, L.I. 1896. *Karezes in the Trans-Caspian Region*. Printing house of Kirshbaum. St. Petersburg, p. 71.

According to ancient manuscripts (Istahari) on the history of Samarkand in the tenth and twelfth centuries AD, water was delivered into the town through a conduit called a lead stream. "Flowing water entered the town through the lead conduit. This conduit had a headwork and started from a rockfill dam. Water flowed from "medini" through this conduit and entered the town through the Southern Kesh gates. The internal surface of the conduit (*vadj*) was lined using lead plates." The remaining section of the conduit was lined by burnt brick cemented with lime-ash mortar. Already by the eighth century AD, a new canal had been built lined with burnt bricks with many outlets and made of ceramic thick-walled pipes (*kuburok*) (Khanykov 1849).

One can say that the development of water-lifting wheels (*chigirs)* led to the transition from irrigation by gravity to so-called "pumping irrigation". *Chigirs* can arguably be seen as forerunners to the Karshi pumping cascade and the Bukhara pumping cascade in Uzbekistan that raise water by 150–180 m. According to Tsinzerling (1927), the use of *chigirs* in Central Asia can be dated back to the ninth and tenth centuries AD. A *chigir* is a biaxial system of wheels placed in horizontal and vertical positions. Special teeth or gears were used for connecting the wheels. The driving force of a *chigir* is supplied by running water or by an animal, which walked around in a circle pulling a lever fixed to a horizontal wheel. Clay pots were mounted to the vertical wheel for withdrawing water. There were 24–26 or 32–38 pots depending on the size of the wheel.

There were several types of *chigirs* (Figure 2.16) that were powered by working animals rotating the geared horizontal wheel, including *chigir-tulegen* (where the water-lifting wheel was outside the circle made by the animals) and *chigir-tartma* or *chigir-keltyk* (where the water-lifting wheel was

Figure 2.16 A chigir.

inside this circle). It was calculated that the capacity of a *chigir* could be as much as 350–400 liters/minute.

Remnants of ancient structures inspire respect for the knowledge of our ancestors. One of the first interbasin canals (the Eskytujatartar Canal built in the fourteenth century) was more than 60 km long and had a carrying capacity of 30–35 m^3/sec. It was constructed for diverting water from the Zeravshan River to the Sanzar, and it had the same bottom gradient all along its length.

In the Farish region, a dam up to 25 m high was built at the foot of the Nurata mountains in the tenth century AD. This stone masonry dam (see Figure 2.17), which still exists today, made a storage reservoir with a capacity of 1.5 million m^3. Local saw-cut rocks and gypsum mortar were used in its construction. The dam had two semi-conic buttresses and nine water outlets built at different levels. Today, the storage reservoir is completely silted. A similar dam, the Abdul-Khan Dam, was discovered near Akchai Village in Samarkand Province (Dingelshtadt 1993). Built later in the fourteenth century AD, the body of the dam was made of large shale stones fixed with *ganch* (special water-resisting alabaster). The length along the base of the dam reached 75 m, with a crest length of 85 m and height of 14.5 m. The width at the base of the dam was 15 m and the crest width was 4.5 m. The dam's design was based on rigorous calculations and had a terrace-like shape. The dam site was quite well selected, closing a narrow gorge through which the river flowed. The reservoir thus created was 1.5 km long with a width reaching 125 m. A few gated structures and conduits were built to control water releases.

Constructing water intake facilities was a very special area of engineering. In ancient times several types of water intake facilities were developed. *Sypays* were built on mountain rivers with high gradients and rapid flows (in the Ferghana Valley and on the Chirchik, Vakhsh, Surkhandarya and

Figure 2.17 The Abdullahandandi dam.

Amudarya rivers). A *sypay* was a pyramidal structure made of logs and filled with stones and brushwood (see Figure 2.18). In the lower reaches of the rivers, fascine types of water intake facilities were used, called *vardy*, *uluky* and *naval* locally. These looked like cylinders (1–1.5 m in diameter and up to 8 m long) made of brushwood and were filled with excavated material. These structures were used to dam a river. Khiva's chronicles of the nineteenth century describe the story of Khan Muhammad Rahim who wanted to punish the inhabitants of the Aral Khanate and ordered his army consisting of 10,000 warriors to dam three branches of the Amudarya completely. These were the Shumanay, Changly-Barry and Ters-Akar. The dams completely stopped the water supply to the town of Kungrad at that time.

Rockfill dams and brushwood training dykes were built in the Zeravshan Valley (for example the water intake structure of the Dargom Canal). There was the knowledge to construct dams with a rockfill body and an impervious screen made of logs. (According to the description bu Al-Muqaddasi, such a water intake facilitated water supply to Bukhara from the Zeravshan River and was built in the tenth century AD.) In the Khorezm region, multihead water intake structures with an overflow spillway disposing extra water through a special canal (*bederak*) were built on canals with high flow rates. There were separate water intake systems for the different seasons, with systems suitable for low and high water levels in the river.

V. A. Latynin (1953) has clearly shown the gradual development of irrigation systems in the piedmont valleys using the transformation of irrigation technologies in the Ferghana valley as an example (as shown in the Figure 2.19).

Figure 2.18 A Sypay.

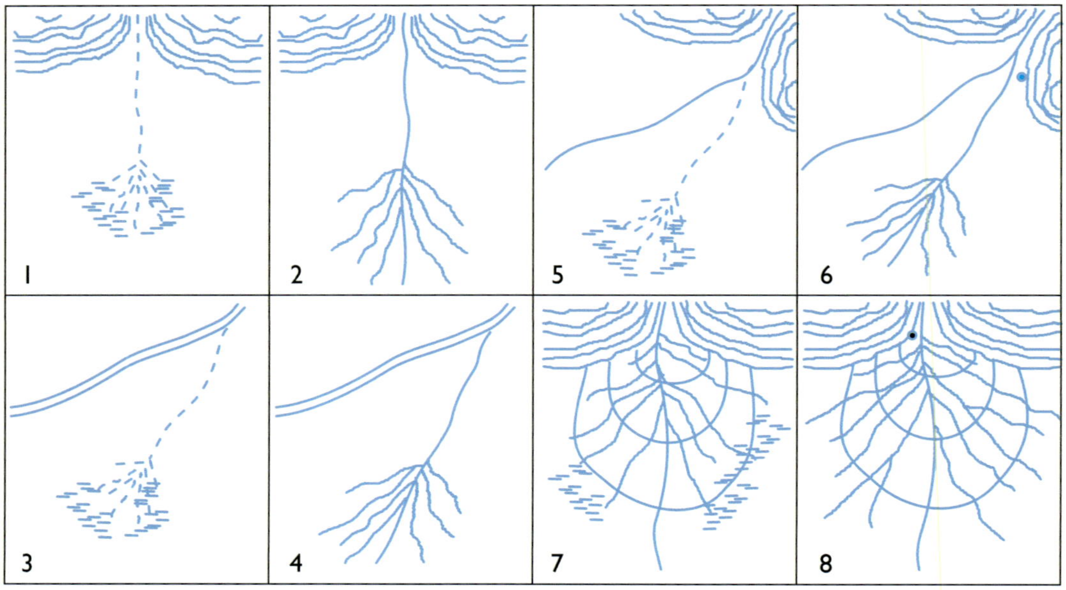

Legend

1. Frequently flooded area in the lower reaches of a mountain stream at the outlet from the mountain range into the lowland
2. Regular transformation of a flooded area into a small ancient irrigation system by means of cleaning up of a channel of the stream and by arranging lateral canals
3. Flooded area in the lower reaches of a dammed river branch
4. Transformation of a flooded area into a manmade irrigation system by means of cleaning of the channel and arranging lateral canals
5. Flooded area created by a side stream overflow of a mountain river at the point of entry from a mountain range into the lowland
6. Transformation of a flooded area into a fan-shaped irrigation system by means of cleaning of the channel and construction of a water intake structure (site of the water intake structure is shown as a black dot in the schemes 6 and 8)
7. Fan-shaped hydrographic network formed by a mountain river at the point of entry from the mountain range into the lowlands
8. Transformation of the natural fan-shaped hydrographic network into a fan-shaped irrigation system by means of cleaning of channel branches and arranging for laterals, which is typical for the Fergana Valley.

Figure 2.19 Scheme of Development of Irrigation of the Sokh basin in the Fergana Valley (according to V. Latynin 1953).

2.9 The Origin and Dynamics of the Aral Sea

The geographical and hydrological past of the Aral Sea and its basin is not clear even today, although there has been a tremendous amount of geographical, geological and historical research, including field expeditions and desk studies undertaken by numerous Russian and foreign scientists since the

start of colonization of Turkistan by Russia. During the Soviet time, there was considerable research including archaeological, paleontological and anthropological studies and fieldwork. The findings from this work were mostly inaccessible to foreign researchers for a long period. After independence, the research became more accessible with the rapid intensification of scientific exchange with European and American institutions – an exchange which also introduced researchers to some advanced and new methods for geomorphological research.

Many researchers are confused by the contradictions about the actual existence of this sea in ancient records. For example, Herodotus (1968), a most competent ancient author, wrote: "A plain, wide and endless for the eye, stretches east of the Caspian Sea." It is doubtful whether Herodotus with his meticulous and inquisitive mind would have forgotten to mention the Aral Sea if it existed as a separate body of water at that time. However, Polybius wrote about the Tanaid River (Syrdarya) that discharged into "Lake Mestid". Ancient chinese authors also do not give direct evidence of the existence of the Aral Sea. The Chinese pilgrim Zhang Qian (147 BC) informed his emperor about "a western sea", but it is not clear if he was referring to the Caspian Sea or the Aral Sea. Lake Daligan was also mentioned in some Chinese sources, but without a proper definition of its location.

The Aral Sea was mentioned for the first time by Arabs in the tenth century AD. All Arab authors refer to this body of water as a sea. Al-Muqaddasi (in the first half of the tenth century) called this body of water Lake Jurdjany (see *Facts related to the history of Turkmenistan and Turkman*, Volume 1, Leningrad, 1939, p. 166), Abu Ishaq Al-Istahri called it the Khorezm Sea (ibid p. 179) and Muhammad ibn Najibi referred to it as the Djend Sea (ibid p. 349).

Some researchers make much of the lack of historical records about this sea, but European travellers only visited the area where the waves of the Aral Sea still wash much later. A famous Flemish Franciscan missionary and explorer Wilhelmus de Rubruquis journeyed two times across the area but he did not catch sight of this of water body. When he travelled to Karakarum, he went no more than 10 km from the northern coastline of the Aral Sea, and he returned by a track road that was about 20 km from the sea. However Rubruquis never mentioned the sea. The outstanding Russian orientalist P. Chikhachev (1982) considered that Rubruquis kept silent about the Aral Sea because he thought that the water he saw was an extension of the Caspian Sea about which he had already written. This may be true, but then why did the Venetian traveller Marco Polo, who travelled along the southern coast of the sea, also kept silent about it? This suggests that the surface area of the Aral Sea changed from time to time. During the wettest periods, it extended from the north to the south-east over a distance of about 450 km, with a width 250 km and water surface area of 67,000 km^2. It is worth noting that the Aral Sea's water surface has an elevation of up to 74 m above the Caspian Sea and 48 m above the Black Sea.

Many aspects of the genesis of the Aral Sea basin and the sea itself remain at a level of supposition and hypothesis. Some ideas are in accord, but others are diametrically opposed and there are different interpretations of the same data. Recent insights obtained through new approaches in geology and geomorphology should probably shed more light on the origin and history of the Aral Sea.

Estimates for the geological age of the Aral Sea depression vary between 2 million years (Pinkhasov and Pryadchenko 2008) and 3 million years (Aladin and Plotnikov (1995) and Rubanov et al. (1987)). This vast depression 150–200 m deep was formed during the Pleistocene period and was originally named as the Palaeogene Sea (Obruchev 1951), Aral-Caspian Sea (Kovda and Egorov) and Turkistan Basin (M. Mushketov 1886).

According to data from German and Tajik glaciologists (Oberhansli and Boroffka 2007–2008) who studied the characteristics of the Pamir glaciers and their impact on Central Asian hydrology, after repeated periods of extreme glacier growth (from 13 to 9 million years ago and from 8 to 6 million years ago), the last period of extreme glacier growth was 2 million years ago. One may assume that maximum water abundance in the middle and lower reaches of the pre-Amudarya and in the lower reaches of the pre-Syrdarya coincided with the last extreme period of glacier growth. According to Kaulbars, with reference to Obruchev, the Amudarya initially discharged into the ancient sea near the present gauging station at Kelif. It formed a delta, which was similar to the later Khorezm delta near present day Charjou. In Kaulbars' opinion, at that time the Zeravshan River also discharged into this ancient sea, and formed the Karakul delta.

In the map of Ptolemy (*Claudius Ptolemaeus*), reproduced on page 92 of *The Aral* (Letolle and Mainquet 1993), the Zeravshan River (*Polytimetus*) is shown as discharging into the Caspian Sea further north than the Amudarya. However, it is impossible to decipher the word that is indicated on the map between these two rivers, and that would probably shed more light on this situation.

As the glaciers retreated, we can assume that there was less water in the rivers and, simultaneously, a reduction in the area covered by the ancient sea. This would intensify sediment deposition and change the riverbed direction towards Unguz, where an emerging chain of shors[46] became the silent witness of this process. Support for this hypothesis comes from the ancient parable about Hazret Ali, who crashed a rock cliff on the right bank of the Amudarya and turned the course of this river along the Kopet Dagh Ridge.

An interesting point was made by Aladin and Plotnikov (1995). They did not assume that the ancient Zeravshan delta was connected to the ancient sea. In their opinion, the Amudarya with its main tributaries (Zeravshan, Murgab and Tejen) flowed towards the Caspian Sea parallel to the Kopet Dagh Ridge. This oldest channel, south of the Uzboy, flowed directly towards the Caspian Sea and formed its Caspian Delta (Figure 2.20) They argue that "*in the Late Pleistocene, the Amudarya flowed through the Kara Kum Desert towards the Aral-Sarykamysh depression. This shift in direction of took place during the so-called Lyavlyakan pluvial period*[47]*... Eventually, according to the assumption of A. Kes, the joint flow of the Zeravshan and Amudarya rivers broke through the second barrier near Tyuamuyun and most likely filled the Khorezm Lake. This lake existed during the Khvalyn period (the Eneolithic period). Later on, the*

[46]Brackish lakes or saline desert areas.

[47]In the Lyavlyakan pluvial period (7,000-4,000 years ago) the climate in the Kyzyl Kum Desert was like the climate in the European steppe zone, with an average temperature in July equal to 21–23°C, and about 400–450 mm/year percipitation.

Figure 2.20 Redrawn situation of the last geophysical transformations of the region on the basis of new GIS data.

Khorezm Lake extend into a northern direction and finally connected with the Aral Sea through the Akchadarya Corridor, forming a delta of the same name."

The system consisting of the Aral Sea, Amudarya and Syrdarya rivers, and the Caspian Sea was then subject to many metamorphoses, which explains to some extent the contradictory historical information on the existence of the Aral Sea. Some researchers have attempted to explain this phenomenon by assuming that the Aral Sea drained into the Caspian Sea. We attempted to generalize the present understanding of the hydrological cycles of the Aral Sea Basin in our previous work, *South Prearalye – New Perspectives* and some quotations from this work are given below (Dukhovny e.a. 2003).

In the Pleiocene period, the Great Aral Sea flooded the part of the Kara Kum Desert located between the Usturt Plateau in the north, the mouths of the Murgab and Tedjen rivers in the south and the foothills of the Kopet Dagh Ridge in the west. In the opinion of some scientists, the Unguz cliffs were the border of the former Karakum Bay on the eastern half of the Aral-Caspian Sea. This united sea would have covered a wide strip of the present day Trans-Caspian area to the foothills of the Kopet Dagh Ridge, and was linked with the Karakum and Chilmetkum bays through the two sea straits Big Balkh and Small Balkh. At that time, the Aral part of this sea filled the whole Sarykamysh depression and formed the Pytnayk Bay that is now the Amudarya delta and the Khiva oasis (this also explains the origin of shor deposits near the Pytnayk). The Uzboy was the strait that linked both these water-filled areas, but its present-day channel with its large gradients was obviously formed during the gradual separation of the Caspian Sea from the Aral Sea and the resulting increasing difference in their respective water surface elevation.

The process that saw the division of the united Aral-Caspian basin into two separate parts – and its gradual reduction in size – continued during the next geological period and practically remained in progress until today. First, the watershed ridge between the Aral-Sarykamysh depression and the Caspian Sea near the Balla-Ishem arose on the Usturt Plateau, and then the Uzboy channel was gradually formed The sequence of the drying up of this territory is confirmed by transitional deposits, starting from the most recent deposits of Caspian mollusks (along the Uzboy channel in the sands of Chilmetkul and the south-eastern coast of the Caspian Sea) that were overlaid by unfixed sands with rare vegetation, up to the ancient formations in the central Kara Kum Desert. These have eventually transformed into shors, takyrs and compacted sandy hillocks fixed by arboreous plants. Shors, the lowest sites on the seabed that were fed by artesian brackish waters, have preserved the shape of the ancient coastal lakes.

Since ancient times, explorers and historians have attributed the transformation of the Aral and Caspian seas to the changes in the volumes of water in the rivers in their joint basins – changes that are in part the result of the development of irrigation systems. Historians noted the complete disappearance of Sarykamysh Lake at the end of the sixteenth century, when the Amudarya stopped flowing into the Sarykamysh depression through the Kunya Darya and Daudan and further onwards through the Uzboy channel. The Uzboy channel from the Caspian Sea to the watershed at Balli Item rose 40 m over a distance of more than 200 km. According to Obruchev, Sarykamysh Lake existed from the seventh century BC until sixteenth century AD. Obruchev cites Antonine Jenkinson who, when travelling to Khiva in 1559, wrote about Sarykamysh Lake, which he took for the estuary of the Oksus River leading to the Caspian Sea. He was also guided by the testimony of Abdulgazi-khan, Gaydula and other Khorezm chroniclers (Obruchev 1951).

Based on geological and historical surveys, most researchers (B.V. Andrianov, A.S. Kes, P.V. Fedorov, V.A. Fedorovich, Y.G. Maev, I.V. Rubanov, and others) have come to the conclusion that there is a correlation between the prehistoric increase in humidification of the region and the formation of vast surface bodies of water. During the humid climatic phase, the Syrdarya and the Amudarya rivers were abound with water and the sea reached a maximum level of some 72–73 m +BSL (Baltic Sea Level). In contrast, during the arid climatic phase, both rivers became very shallow, and the Aral Sea level also dropped and salinity levels increased. In historical times, during the period of ancient Khorezm, the changes in the water level of the sea depended to some extent on climate changes but mainly on the irrigation activities in the basins of the two rivers. In periods of intensive development in the Aral Sea basin, the increase in land area under irrigation resulted in much more water being diverted from the rivers leading to a further lowering of the sea level. When irrigation systems collapsed or fell into disuse, through periods of war or revolution, the rivers discharged more water again.

The largest drop in the Aral Sea level was observed at the beginning of the nineteenth century. By 1846, parts of the former sea had turned into two vast desert territories, some 2,200 km^2 and 2,800 km^2 in surface area respectively. At the time, many explorers affirmed that this natural cataclysm was related to an overall increased aridity in Central Asia. However, at the beginning of the 1880s, the sea water level suddenly started to rise again. According to Masalsky, in 1908 it had risen by 3 m and reached again the level it had at the beginning of the nineteenth century. This means that in the course of one hundred years, a drop and a rise of almost similar magnitude were observed. In 1913, Masalsky (1913) wrote: "*During recent years, the Aral Sea level has continued to rise at a rate of more than 20 cm per year, and peninsulas have been turned into islands, and vast lowlands and depressions near the coast are again covered with water. This is a very important and interesting phenomenon, and it is obviously closely related to the increase in precipitation, which has been experienced all over Central Asia during recent years.*"

Most researchers came to the same conclusion. Natural changes in the level of humidification in the atmosphere adjusts long-term precipitation and evaporation rates. The Russian administration monitored the status of the Aral Sea closely and carefully studied the works of Arab geographers about the sea. This work also found periodic rises and drops in water level. This is from a report on Syrdarya Province prepared in 1910: "There was the assumption that the sea dries up, meaning that the evaporation from its surface exceeds inflow of the Syrdarya and Amudarya flowing from the ridges of Turkistan Tien Shan." The report emphasized that the sea water area had considerably reduced since 1856, based on findings from a field survey carried out by the Admiral Butakov. However, it noted that since 1896 this trend had been reversed and water started to cover the dried-up coast again, and filled previously empty bays and gulfs, and the islands got smaller.

The authors of this report noted that although the study confirmed "fluctuations in the water levels of the Aral Sea, there is still not sufficient [evidence to make] understanding of the phenomena for a categorical prediction of its volume [in the future]. We are most likely seeing periodic fluctuations of its water level that depend on meteorological phenomena in the upper watersheds of the Syrdarya and Amudarya rivers." At present, although human factors are likely to play a more significant role in the decrease in the sea water level, there is no basis on which to disagree with this report's conclusions. It is still most likely that the real reasons for the fluctuating water levels will be found in the upper watersheds of the two rivers. The formation of rivers, seas and lakes result from natural processs. Humans can make use of these resources but cannot predict the impact of natural cataclysms, such as earthquakes, hurricanes, landslides and flooding, on water courses. Undoubtedly, in the future, global warming will also play a significant role, causing intensive evaporation of the rivers and water bodies in region which may lead to further desertification and drying up of the Aral Sea.

The Amudarya and Syrdarya have regularly changed direction and shifted location throughout their long history. They did often not reach the Aral Sea and, as a result, the sea dried up with desert areas forming on the dried out land. As water levels in the sea fell, salinity sharply increased, producing the salt depositions that have been found by geologists among other deposits on the Aral Sea bed. The thick layers of sedimentary mirabilite found by Rubanov (1987) are especially interesting. The shifting river deltas of the Amudarya and Syrdarya have created a very peculiar terrain in the lower reaches, where depressions that were filled with marsh sediments alternate with significant desert sandy-loam deposits.

Drawing on numerous historical sources, Table 2.2 shows the interaction between the rivers, the Aral Sea and the Uzboy channel, through which a part of the Amudarya waters discharged into the

Table 2.2 Historical sources with information on the water systems in Central Asia.

Period	Source	Status of the Aral Sea	Status of the Uzboy Channel	Level of the Caspian Sea with respect to that of 1990, m + BSL	Note
Fifteenth century BC	The Avesta	A dry area			Wetlands
Fifth century BC	Herodotus	The sea exists	Uzboy = Amudarya		
Third century BC	Patroclus	The sea is filled up with water	A dry channel		The Amudarya and Syrdarya flow into the Aral Sea
First century BC	Strabo	The Amudarya and Syrdarya rivers flow into the Aral Sea, but the latter not completely	The Amudarya parallel to Uzboy	+25	
891 AD	Al-Balhi	The sea exists	Along the Uzboy Channel to the Caspian Sea	+9.28	
960	Al-Isthari	The sea exists		–4.2	
1211	Jivendi Murkhand	Almost dry	With a flow		Descendants of Genhgis Khan diverted the Amudarya from Gurganj I in 1221
1325	Marino Sanuto	At mean level	There is a flow through the Uzboy channel from the Sarykamysh Lake into which the Amudarya empties, but one branch discharged into the Aral Sea		The Small Aral Sea is identical to the small lake Sarykamysh

Table 2.2 (Continued)

Period	Source	Status of the Aral Sea	Status of the Uzboy Channel	Level of the Caspian Sea with respect to that of 1990, m + BSL	Note
1375	Catalan Atlas	The sea exists	With a flow	+5.64	The Syrdarya flows into the Aral Sea and Amudarya flows into the Sarykamysh Lake
1400	Zair Al-Merashi	A low level	The main channel		The same
1575	Abdul Ghazi	A high level	A dry channel		All water flows into the Aral Sea
1638	Olirey	A low level	With a flow	+5.34	The Amudarya and Syrdarya flow into the Aral Sea
1680	Abdul Ghazi Baghadur	The sea exists			The Amudarya empties into the Caspian Sea since 1220 and finally they were separated in 1575
1734	Kirilov	Not mentioned	They alternated	+4.03	
1826	Kolodkin	At high level	Not shown	+3.12	
1858	Ivanichev	At high level	A dried channel	+0.99	

Caspian Sea in the past. In late Stone Age, about 20% of the Amudarya water ran into the Caspian Sea, through two linked lakes (Sarykamysh and Assake-Daudan) and the Uzboy channel which created a unique but irregular connection between the Aral Sea and the Caspian Sea. Today, there is an opportunity to compare our assumptions with data produced by German researchers, who researched climatic changes during the Holocene period. Based on investigations of gypsum deposits using carbon tracing methods, Oberhansli and Boroffka have determined that the lowest water level in the Aral Sea occurred in the third century AD, this coincides with account of Patroclus, and then in the sixth century AD. The subsequent two minima in 1220 and 1400 coincide with the evidence of Jivendi Murkhand and Zair Al-Merashi. This last finding is supported by the presence today of the ruins of settlement of Kerderi that are underwater on Aral Sea bed. This settlement dates back to the fourteenth or early fifteenth century.

We can compare these findings with historical record given by Herzfeld (1947). He cites Al-Muqaddasi (about 985 AD) who confirms the account given by Al-Balhi that a stream from the Amudarya flows into the Uzboy channel and who mentions the low Aral Sea level. In the period 1304–1316, during the reign of Oldjaytu, according to Hamdalah Al-Mustafa, a great waterfall existed on the Uzboy, while only a small inflow into the Aral Sea was observed. The inflow into the Caspian Sea was so great that the Abaskun peninsula was formed in the Caspian estuary at that time. This is also confirmed by Al-Bakuvi at the beginning of the fifteenth century AD. There is also record of the English ambassador Jenkinson (1558–1559), who provided evidence that the Uzboy existed but did however not reach the Caspian Sea.

Most researchers consider that any changes in the direction and volume of rivers and inflows into large water bodies in these periods were the result of natural processes, such as increasing humidity or lower percipitation. This was the case at least until the third century BC. However, some believe, including Tolstov, that during the new era, changes in the scale and extent of irrigation had a major impact on the formation (ane volume) of inflows into the Aral Sea. There is some historical evidence of the construction and subsequent destruction of dams on the Amudarya, which affected runoff into the Aral Sea or through the Uzboy channel into Caspian Sea. In 985 AD, Al-Muqaddasi described a huge dam near Gurgange, an impressing hydraulic engineering structure that was built using trees and braided twigs. Ibn Al-Athir later described the destruction of a dam on the Amudarya near Gurgange by Genghis Khan (it is not known if this is the dam described by Al-Muqaddasi or another one).

Numerous hydrological and geological changes have resulted in a somewhat confusing history of the Aral Sea. Nevertheless we share the conclusions made in a most rich and versatile generalization by Nikolay Aladin and his colleague: "*During its history, the salinity and level of the Aral Sea were initially subject to impacts of local climatic factors, which caused changes in river runoff. However, later on, anthropogenic impacts (mainly irrigation, wars, economic and political decisions) became a primary factor influencing these fluctuations as well. It is incorrect to think that the environmental problems existing today are completely new and unique for this region. Similar events repeatedly happened before.*"

References (Chapters 1 and 2)

1. Aladin, N.S., Plotnikov, I.S. 1995. Changes in the Aral Sea level: paleolimnological and archaeological proofs. Collected papers: «Biological and environmental problems of the Aral Sea and adjacent areas», Proceedings of the Russian Academy of Sciences, Volume 26, E. 22. St. Petersburg, pp. 17–43. (in Russian).
2. Al-Istahri, IGA UzSSR, f. S-4, ok 2, d. 414. (in Russian).
3. Al-Muqaddasi, IGA UzSSR, f. H-1, d. 324. (in Russian).
4. Andrianov, B.V. 1969. The ancient irrigation systems of the Trans-Aral region. Moscow. (in Russian).
5. Arrianus, L.F. 1908. Anabasis of Alexander. St. Petersburg. (in Russian).
6. Barker 1932. The Mumie, Leningrad. (in Russian).
7. Bartold, B.V. 1966. Full collection of works. Volumes 1, 2, and 4. (in Russian).
8. Bekchurin, N.Y. 1950. Collecting the information on people inhabiting Central Asia in ancient times. Moscow & Leningrad, p. 223. (in Russian).
9. Berne, F. 1936. The history of last political upheavals in the Grand Mogul's state. Moscow & Leningrad, p. 318. (in Russian).
10. Bongard-Levin, G.M. Editor-in-Chief. 1989. The ancient civilizations. Publisher: Mysl. Moscow. (in Russian).
11. Bulletin of Irrigation, June 1924, No. 9. (in Russian).
12. Bukinich, D.D. 1924. The history of primitive irrigation farming in the Trans-Caspian region. Journal "The Cotton Business" No. 3 & 4. (in Russian).
13. Chikhachev, P.A. 1982. The page about the East. Moscow, p. 182. (in Russian).
14. Dingelshtadt, N. 1895. The experience of studying irrigation in Turkistan, Syrdarya Province. Volumes 1 & 2, St. Petersburg. (in Russian).
15. Dingelshtadt, N.G. 1993. The experience of studying irrigation in Turkistan. St. Petersburg, pp. 1–2. (in Russian).
16. Dukhovny, V.A. 2007. Globalization and water in Central Asia. (in Russian)
17. Dukhovny V.A., De Schutter J.L.G. 2003. South Priaralye – New Perspectives NATO SfP. Tashkent (in English)
18. Engels, F. 1956. Full collection of works. Volume 18. (in Russian).
19. Gafurov, B.G. 1989. Tajiks, Book 1 & 2, Dushanbe, «Irphon», p. 377 and p. 480.
20. Grousset, R. 1996. The Empire of the Steppes: A History of Central Asia, New York.
21. Gulyamov, Y.G. et al. 1957. The history of Samarkand. Tashkent, Volume 1. 1969. p. 468, "The history of Khorezm irrigation from ancient times", Tashkent. Publishing House of the Academy of Science of the UzSSR. (in Russian).
22. Gulyamov, Y.G., Islamov, U., Askarov, A. 1966. Primitive culture and impacts of irrigated farming in the lower reach of the Zeravshan River. Book 1, Tashkent. (in Russian).
23. Gumelev, L. 1990. Ethnogenesis and the Biosphere of Earth. Leningrad. (in Russian).
24. Herodotus, 1968. The History. Publishing House of the USSR Academy of Sciences. (in Russian).
25. Herzfeld, E. 1947. Zoroaster and his world. Volume 2.
26. Hojamuradov G., Mammedov A., Taganov C. Explanatory dictionary on the water management in Turkmenistan, Ashgabat, 2007, p. 173. (in Russian).

27. Itina, N.I. 1959. The memorials of primitive culture in Upper Uzboi. TKHAEE, Volume 4. (in Russian).
28. Ivanov, P. 1954. The economy of Juybar Sheikhs. Moscow, p. 319. (in Russian).
29. Kaulbars, A.V. 1903. Ancient Amudarya Riverbeds. (in Russian).
30. Khanykov, N.V. 1849. The descriptions of Bukhara khanate. St. Petersburg, p. 279. (in Russian).
31. Khidoyatov, G.A. 1990. My native history. Tashkent, p. 32. (in Russian).
32. Khlopin, I.N. 1964. Geoksur settlement groups of Eneolithic epoch. Moscow-Leningrad. (in Russian).
33. Konshin, A.M. 1883. About ancient Amudarya stream. Proceedings of KO IRGO, St Petersburg, Volume 15. (in Russian).
34. Kuftin, B.A. 1954. The studying of Annau culture. Publishing House of the Academy of Science of Turkmenistan. (in Russian).
35. Kyugelgen, A. 2004. Legitimization of Central Asian dynasty of Mangits in the works of historians of the 18th and 19th centuries. Almaty, p. 254. (in Russian).
36. Latynin, V.A. 1953. Some aspects of the method for studying the history of irrigation in Central Asia. Publisher: "Fan". Tashkent. (in Russian).
37. Letolle, R., Mainquet, M. 1993. The Aral. Publisher: Springler-Verlag. Paris.
38. Levi, D.N. 1955. The historical types of economic cultures in geographical regions. (in Russian).
39. Lisitsina, G.N. 1965. Irrigated farming of the Neolithic Period in the south of Turkistan. (in Russian).
40. Livshits, V.A. 1963. The ancient state formation. ITN, Volume 1. (in Russian).
41. Masalsky, V.I. 1913. Turkistan. Editor: Semenova-Tesompasona, St. Petersburg, p. 861. (in Russian).
42. Masson, M.E. 1935. Problems of studying cisterns – sardobas. Tashkent, p. 43. (in Russian).
43. Masson, V.M. 1970. Central Asia and Ancient East. Moscow-Leningrad. Science. (in Russian).
44. Minashina, N.G. 1962. Ancient irrigated soils in the Murgab oasis. Journal "Soil science", No. 8. (in Russian).
45. Markvart, I. 1951. Is Central Asia drying up? VG, p. 24. (in Russian).
46. Marx, K. 1969. The British empire in India. Full collection of works, Volume 31. (in Russian).
47. Mukhamedjanov, A.R. 1969. Samarkand water-supply. Volume 17. (in Russian).
48. Mukhamedjanov, A.R. 1978 & 1984. History of irrigation in the Bukhara Oasis since earliest times until the beginning of the 20th century. Tashkent & Moscow, p. 175. (in Russian).
49. Mushketov, I.V. 1886. Turkistan, the geological description. St. Petersburg. (in Russian).
50. Nemtsova, N. 1976. Ancient source of water supply, Tashkent, Fan, p. 8.
51. Oberhansli, H., Boroffka, N. 2007–2008. Holocene climate variability during the past 2000 years and past economic and irrigation activities in the Aral Sea Basin, Documents of the INTAS DFG project "Climate".
52. Obruchev, V.A. 1951. Selected works on Geography of Asia. Volume 1, Moscow. (in Russian).
53. Okladnikov, A.I. 1962. The Kara Kum antiquities. Dushanbe. (in Russian).
54. Okladnikov, A.P. 1966. Paleolith and Mesolithic in Central Asia. Collected papers: «Central Asia in the Epoch of Stone and Bronze», Moscow & Leningrad. (in Russian).
55. Ostrovski. 1907. The irrigation and farming experience of colonial India, St. Petersburg, p. 106. (in Russian).
56. Pinhasov, B., Pryadnenko, T. 2008. The past, present and future of the Aral Sea. Proceedings of the German-Central Asian Symposium. Tashkent. (in Russian).
57. Ranov, V.A. 1965. The Stone Age in Tajikistan. Dushanbe. (in Russian).
58. Rauner, S.N. Editor. 1887. The artificial irrigation of lands, St. Petersburg. (in Russian).
59. Rizenkampf, G.K. 1930. Survey data for the Golodnaya Steppe Irrigation Project. Leningrad. (in Russian).
60. Rubanov, I.V. et al. 1987. Geology of the Aral Sea. Tashkent. p. 248. (in Russian).

61. Rufus, Quintius Curtius. 1948. The travel comments, Moscow-Leningrad. (in Russian).
62. Ruy Gonzalez de Clavijo. 1990. A diary of travelling to the court of Timur at Samarkand in 1403–1406. Moscow. 100 pages. (in Russian).
63. Sadykov, A.S. Editor-in-Chief. 1975. Encyclopedia: Irrigation of Uzbekistan. Volume I, Publisher: "Fan." Tashkent. (in Russian).
64. Sachau, E.D. 1873. Zur Geschichte und Chronologie von Khwarizm, SBWAW, PLHSI, B73.
65. Sarianidi, V.I. 1965. The monuments of late neolith of the South-East Turkmenistan. Publishing House of the Academy of Sciences. (in Russian).
66. Shahermayer, F. 1948. Alexander the Great. Minsk, p. 175. (in Russian).
67. Sharaff Ad-Din Ali Yazdi. 2008. Zafarname. Tashkent, p. 290. (in Russian).
68. Shishkin, V.A. 1969. To the history of Marakand archaeology. p. 152. (in Russian).
69. Smith, V. 1981. The Oxford History of India. Oxford University Press. p. 964. (in Russian).
70. Solovyev, S.M. 1989. Readings and stories about Russia's history. Publisher: "Pravda." Moscow, p. 768. (in Russian).
71. Suleymanov, R. 2000. The ancient Nahshab. Uzbekistan civilization problems between the 7th century BC and the 7th century AD. Samarkand & Tashkent, p. 77. (in Russian).
72. Tchaikovski, A.I. 1904. Turkistan and its rivers. (in Russian).
73. Tolstov, S.P. 2004. Following the Tracks of Ancient Khorezmian Civilization. Publishing House of the USSR Academy of Sciences, Moscow-Leningrad.
74. Tolstov, S.P. 1948. Ancient Khorezm. Publisher: Moscow State University, p. 440. (in Russian).
75. Tsimbalenko, L.I. 1896. Karezes in the Trans-Caspian region, St. Petersburg, p. 71. (in Russian).
76. Tzinzerling, V.V. 1927. Irrigation in the Amudarya Basin. Publisher: Central Asian Water Management Administration, Moscow, p. 712. (in Russian).
77. UNESCO 2006. UN World Water Development Report 2.
78. Vambery, A. 1873. The history of Bukhara. London p. XXYII. (in Russian).
79. Vavilov, N.I. 1967. The agriculture. Selected works in two volumes, Leningrad. (in Russian).
80. Victor, A. Dukhovny, 2003. South Prearalye - "New Perspective", Joop de Schutter, Tashkent, p. 149.
81. Vyatkin, V.V. 1902. The materials to historical geography in Samarkand Viloyat. SKSO, Samarkand, Paper 7, p. 83. (in Russian).
82. Weber, H. 1892. The history of the East. St. Petersburg, p. 374. (in Russian).
83. Yakubovsky A.Y. 1947. "The findings of the Soghd-Tajik Expedition" Proceedings No. 15, Moscow & Leningrad. (in Russian).
84. Zhukovsky, V.A. 1894. Antiquities of Trans-Caspian region. Ruins of ancient Merv. Proceedings: "Materials on Russian Archaeology." St. Petersburg. p. 24. (in Russian).

Social obligations were used for construction of irrigation works during the period of Russian colonization

3

Russian Colonization and the Soviet Era in Central Asia

– *One way Towards Economic Growth and Meeting Future Challenges*

3

Russian Colonization and the Soviet Era in Central Asia – One way Towards Economic Growth and Meeting Future Challenges

Close ties between Russia and Central Asia, based on trade relations mainly, existed even prior to the emerging Russian state. After their liberation from the Tartars, Russia expanded towards the southeast through military campaigns of various princes, independent atamans and Ivan the Terrible (in the sixteenth century). In the second half of the nineteenth century, tsarist Russia completely dominated Central Asia. Very soon thereafter Russia drew up plans to transform Central Asia into a huge cotton plantation. After the "emancipation manifesto" of 1861 which ended serfdom, the Russian textile industry developed rapidly and dozens of new textile factories and enterprises were established and demand for cotton grew rapidly. Russian textile manufacturers wanted to speed up the development of cotton production in Turkistan and used the construction of railways for developing the region's rich natural resources. The tsarist administration began to expand the farming areas under cotton and develop many new irrigation infrastructure projects. The extreme 1891drought in the whole of Russia enhanced this process.

For development and management the Russians tried to incorporate Central Asia traditions and there were special regulations for the operation and maintenance of waterworks and irrigation canals. Each water user for example had to take part in public work related to the cleaning of irrigation canals, maintaining of water intake structures, and repairing waterworks. This work of fulfilling societal obligations for maintenance of irrigation networks in the nineteenth century was known as *kazu*. Many such initiatives were taken throughout the region, which greatly enhanced the productivity of the systems.

The chapter further describes how as a result of research and field surveys during 1895 to 1905, the extended irrigation infrastructure was developed as we know it today. The best know projects development of the the Golodnaya Steppe Irrigation Project (1910–1915), the 1908 project to transfer Amudarya water into the Merv and Tedjen oases and construction of the Kayrakkum Canal are explained in detail. The chapter closes with the enormous technical, scientific and administrative infrastructure (for example de setting up of the IRTUR and the SANIIRI research center) that was required to develop and operate these activities on a scale not known to the world before.

3.1 Water Management during Russian Colonization

Close ties between Russia and Central Asia, based on commercial relations, existed even prior to the emerging Russian state. An interesting, though rather disputable, approach to the mutual influence on the development of the states in ancient Turkistan and Russia is described in Gumilev's (1984) unconventional work *Russia and the Ancient Steppe.* This book, despite the doubts one may have about many of its propositions and conclusions, contains valuable documents concerning the mutual relations between these two ancient geopolitical regions. Masalsky (1913) mentions that there was intensive trade between the regions from the eighth century until the thirteenth century, mainly through caravans trading with Khorezm and Maverannahr. After liberation from the Tartar yoke, Russia expanded towards the southeast through the military campaigns of various princes, independent atamans and Ivan the Terrible (in the sixteenth century) and thoroughly fortified its position in Yaik on the Ural. The memoirs of Abul-Gazi Khan are evidence of the fact that Cossacks from Yaik were visiting the Khiva Khanate in the sixteenth century. Starting in 1565, Shaybanids began regular ambassadorial relations with Russia. These events are known through the meetings of Tsar Boris Godunov and Tsar Michael I with ambassadors from Khiva and Samarkand, as well as the sending of an embassy of Tsar Aleksey I to Bukhara in 1675.

According to Solovyev (1989, Chapter 2), in 1714, Peter the Great sent an expedition along the Irtysh under the command of Bukhgolts, and in 1716 sent two big armies to the Caspian Sea and Khiva Khanate under the command of Prince Alexander Bekovich-Cherkasky. His task was "*to find the mouth of the Amudarya at the Caspian Sea coast, and to study its channel and the possibilities for constructing a dam in order to return the river into its former channel running towards the Aral Sea*". The troops that went to the Khiva Khanate failed in their mission and the prince was killed, but the Caspian troops managed to find out that the Amudarya did not flow into the Caspian Sea.

Despite some failures, Russia expanded and fortified its positions in the Ural region and in 1847 the first Russian fort (Fort Raim) was established in the lower reaches of the Syrdarya. From here, V. Obruchev started to explore the Aral Sea. Two vessels were supplied from Orenburg, and these were used in 1849 to mount an expedition under the command of Captain Butakov which carried out a cartographical survey and made a description of the Aral Sea. This was the first well-surveyed geographic description of this sea. By the mid nineteenth century, a Russian protectorate was accepted by the Kazakhs of the Big Hordes and Verny Town was established on the Almaatinka River. Soon after in 1856–1857, the well-known Russian geographer P. Semyonov-Tjanshansky made his journey through the region. The advancement of Russian armies and the establishment of Russian protectorates (voluntarily or by force) over vast territories in Turkistan resulted in the creation of Turkistan Province (renamed Syrdarya Province in 1865) and, two years later, Semirechye Province.

In the second half of the nineteenth century, tsarist Russia established complete domination over Central Asia. In 1865, Tashkent was taken by brute force and creation of the Turkistan Governorship-General[1] was declared. In 1873, armies of the Bukhara Emirate and Khiva Khanate

[1]A large administrative and territorial unit governed by a Governor-General (in tsarist Russia until 1917).

were defeated and both states accepted the Russian protectorate. They practically became colonies, with considerable loss of territory. In 1874, the Kokand Khanate was defeated and in 1875 was annexed by the Russian Empire (under the name of Fergana Province). The remarkable organizer and reformer General Skobelev later headed this administrative unit. In 1881, the colonization of Turkistan was completed with the establishment of the Trans-Caspian Province and the start of the construction of a railway line from Mihaylov Bay on the Caspian Sea to Kyzyl-Arvat (Figure 3.1).

Very soon Russia made plans to transform Central Asia into a huge cotton plantation. After the "emancipation manifesto" of 1861 which ended serfdom, the Russian textile industry developed at rapid rates. Dozens of new textile factories and enterprises were established every year, and demand for cotton grew rapidly. Imports of cotton from the USA sharply decreased as a result of the American Civil War (1861–1864) and Russia was compelled to create its own cotton base to meet the growing demand for cotton. In Central Asia tsarist Russia tried to modernize the system of land and water relations by subordinating it to Russian laws. However, the colonial authorities encountered the insuperable

Figure 3.1 Old map of Central Asia[2].

[2]Map from Masalsky (1913). Turkistan Krai. St. Petersburg, Publisher: A.F. Devren.

obstacle that it is impossible to quickly change centuries-old traditions of water use and land tenure. These traditions were specified by existing Muslim legislation, which subdivided all land into state ownership (*amlyak*), manor ownership (*myulk*) and religious institutions' ownership (*vakf*).

After long discussions, the tsarist government decided to legally institutionalize the pre-colonial land and water relations in Turkistan, as the Russian domain in Central Asia was called. On June 12, 1886, Tsar Alexander II approved the so-called Regulations for Governance of Turkistan Krai[3]. These defined the fundamentals of the agrarian policy of tsarist Russia in its new colonial domains. The third chapter of this document was devoted to land and water relations, under the heading "Land Ownership of Settled Population". All land in Turkistan was declared state property and plots of arable land could be worked by private persons as a "hereditary property". The land tenure of the military and feudal aristocracy was abolished. These landowners had possessed 90% of the plots and leased them to peasants, but under the new regulations they lost their land and plots were distributed among peasants as a hereditary possession. In this way the tsarist authorities killed two birds with one stone: they undermined the material basis of the local aristocracy, who were potential opponents, and they created a labor mass which begun to pay rent directly into the tsarist treasury. Some 90% of the arable land resources were granted to peasants and only 10% remained with the ancestral aristocracy. The average size of plot was 5–7 *dessiatinas*[4]. In Bukhara, Fergana and Khiva the average plot was smaller at 1.5–2.5 *dessiatinas*.

The Code of Laws adopted for governing the Turkistan Krai practically changed nothing in respect of water use rules. It stated that "*water in ariqs, canals and small rivers that can be used for irrigation of fields belongs to the state treasury (i.e. to the State), but the population has the right to use the water for those needs which are established according to local customs*" and further that "*waters in the main ariqs, streams, rivers and lakes are granted to the people in accordance with the established custom*". The main *ariqs* were managed by *ariq-aksakals*, and the branches by *mirabs* (Savitsky 1963).

Russian textile manufacturers wanted to speed up the development of cotton production in Turkistan and use the construction of railways for developing the region's rich natural resources. The tsarist administration began to expand the farming areas under cotton and develop many new irrigation infrastructure projects. There was one further circumstance that played a key role in arousing the tsarist authorities' interest in developing the virgin and long-fallow lands in Turkistan. In 1891, the severest drought in the history of Russia affected the country. Hunger, dysentery and cholera were rife and rampant in the central provinces of the country, forcing peasants to leave their villages, and some of them migrated to Siberia and some to Turkistan. In 1891, there were 19 Russian settlements in the Syrdarya Province. After that time, Russian peasants began to settle in the Fergana valley and in the Murgab oasis. The number of settlers grew rapidly after Stolypin's agrarian reform of 1910–1911, which turned resettlement into state policy for the purpose of reducing the tensions that had arisen in Russian rural areas as a result of overpopulation.

[3]An administrative region of Russia: in pre-revolutionary times krais were each made up of a number of provinces, which became large administrative units in the Soviet territorial system in 1924.

[4]A measure of land (1 *dessiatina* = 10,900 sq. metres or 2.7 acres).

Forward-thinking people in Russia clearly promoted what was necessary to develop the agricultural sector in Central Asia. According to a non-scientific report prepared by A. Popov in 1885 for the Russian Academy of Agricultural Sciences: "*Irrigation is a key factor for developing agriculture in this region. There are two large rivers, the Syrdarya and the Amudarya, which flow with their tributaries from upland sources to the flat countries of Turkistan. These rivers and their tributaries originate from a mountainous region of Turkistan, and many of them have glacial sources and therefore have high water discharges in summer. The water resources of these rivers are quite sufficient for developing irrigation in the flat countries. Here, irrigation can be implemented based on large-scale projects with main canals that divert water directly from the rivers. In the past, regular wars have made the existence of such canals almost impossible. However, in times of strong state power these canals were constructed and the remnants of these large canals, which conveyed water at a distance of over tens of kilometers from the river, speak much in favor of this idea. During periods of weak state power, when irrigation canals were out of operation, the prosperous countries turned into desert areas. The total area of the Turkistan region is about 100 million* dessiatinas *and only one-fiftieth of this area is arable. Other land resources are represented by either barren steppe or low productive grazing areas used by nomads. Most of the two million* dessiatinas *of agricultural land were consequently irrigated. Land treatment and application of fertilizers were at a high level and therefore productivity was also quite high. Irrigation infrastructure can be divided into two categories, large-scale and small-scale. The large main canals with appropriate structures and a vast command area belong to the first category. Irrigation based on small rivers and reservoirs with necessary water infrastructure can be referred to the second category. This kind of water infrastructure could be built by private individuals, and it serves small irrigated areas only and is undoubtedly the business of private farmers*" (*Bulletin of Irrigation* 1923 and 1924).

In the opinion of Russian water professionals, the main shortcomings at that time consisted of a lack of knowledge about potential land productivity and crop water requirements, insufficient use of agricultural machinery and fertilizers as well as a lack of specialists for the design, construction and operation of large-scale irrigation systems, a lack of funds and an underdeveloped credit system. These specialists eventually developed a master plan setting out governmental measures for the rehabilitation and modernization of the irrigation systems. Some of these systems remain in use even today because of present-day cuts in financing and operation and maintenance. (Many of today's irrigation systems require enhancing, as they were designed for the water sector as it existed more than 100 years ago.)

These were the main points of the master plan proposed by Russian water professionals.

1. Pilot farms of 250–500 hectares should be designed and constructed, taking into consideration economic, climatic, soil and hydrogeological conditions. On these farms, there should be studies of the productivity of irrigated lands under different crops and under various conditions as well as different irrigation methods and schedules depending on crop water requirements, and the most effective farming practices must be then introduced within the large-scale irrigation schemes.
2. Agriculture businesses must be organized in a way that allows a record to be made of the financial results of each operation. Advanced experience should be widespread among farmers. Agricultural schools for farmers have to be established in all regions.
3. A special system of preferential credits for irrigation and drainage projects has to be created.
4. A normative and legislative basis for water use and irrigation practice should be developed.

5. Financing for large-scale irrigation works must be provided. The construction period of large-scale water infrastructure is rather too long for private business to meet the cost. The government must take this fact into consideration, and benefits from irrigation should be measured not only by direct financial revenues as is the case in private business. Socioeconomic feasibility analysis should substantiate project decisions, with the understanding that each irrigation canal increases the total productivity of a country and may help turn desert areas into oases.
6. Public interests can often conflict with private interests as, for example, is the case with the high taxes that may be used to increase the profitability of the irrigation system as a whole but which may slow down the overall growth of a country.

The tsarist government kept in mind Great Britain's experience in India and combined colonial administration with economic management. The former governor K. Kaufman noted that "*large-scale irrigation works financed by the government not only have economic but also great political significance. Power must be maintained not only through weapons but also through enhancing the economic potential of the country by developing industry and raising welfare standards. Our actions aimed at economic growth should be similar to England's actions in India.*"

The Director of the Department of Land Reclamation of the Russian Empire, Prince V. Masalsky, admired the status of the irrigation systems in Turkistan. "*Almost all irrigation canals were built by the local population and most of them since times immemorial. Some canals are enormous, and when I looked at these powerful streams that transport the life-bringing water over many tens of versts*[5]*, involuntarily I had a profound respect for the nation which, while having so scarce engineering resources and in the broiling Turkistan sun, by means of incredible efforts has covered all the land surface with a network of irrigation arteries*" (Masalsky 1913). The canals that most impressed Masalsky included the Sharikhansay Canal, 101 versts long with a carrying capacity of 7 cubic sazhens[6] per second, the similar Andijansay Canal in the Fergana Valley and the irrigation canals Zakh and Bozsu with the same carrying capacity in the Tashkent oasis. Some 84 canals diverted water from the Zeravshan River, and the two largest canals, Dargom and Narpay, had existed for a millennium already. There were similar canals in the Khiva Khanate, namely the Palvan-Ata, Gazavat, Shahabad, 135 versts long, and the Yarmish.

To the credit of the Russian State, it should be noted that after having discovered the very advanced water sector in the region, with "procedures of which were time-honored", the colonial authorities decided not to interfere with this new and poorly known sphere of activity. They granted all responsibilities for water use and management to the local population (Dukhovny et al. 1976).

According to the key rules of the Sharia and Adat, water cannot be private property since it belongs to all and everyone who creates the facilities for irrigation or other needs for water. Water cannot be an object of purchase or sale, except water gathered in a pond or reservoir. The Sharia states: "*The water which is gathered in ponds is property and other people have no right to use it, but even a proprietor of water cannot refuse water to the thirsty or those who should water cattle, caravans or horses. An owner of a "khauz (pond)" should*

[5]Verst is an old Russian measure of distance equal to 3,500 feet or 1.6 km.
[6]Sazhen is an old Russian measure of length equal to 2.34 metres.

provide a special watering place for livestoḳ." If there is a shortage of water for irrigation, water should be shared and divided into equal quantities. The sequence of water-use priorities is established either by starting from the tail section of an irrigation canal or by the drawing of lots. The withdrawal of more water than the established share or the withdrawal of water outside of an established sequence is considered a crime and is punished. *Ariq-aḳsaḳals* appointed by the authorities and *mirabs* elected by the population execute all functions of water monitoring and management. The *mirabs* establish the sequence and norms of water use, as well as supervise execution of the traditions and customs established in their districts.

The well-known irrigator and researcher of ancient Asia S.N. Rauner wrote that: "*Ancient customs of water use served as a 'code of law'. These customs had existed for thousands of years and were called 'worldly irrigation'. This unwritten code contained many important regulations including: (i) procedures for water measurement and accounting; (ii) duties of land owners in respect of transporting water through their plots to neighboring plots; (iii) the rights to participate in distributing private and common revenues and the use of water or land belonging to a third party in certain cases; (iv) rules for specifying the spacing between canals diverting water from the same water source in order to prevent filtration flows from an upstream canal to a downstream canal; (v) procedures for determining the rates and schedule of water use for private persons, communities, settlements, cities, etc.; (vi) procedures for flood control; (vii) procedures for specifying crop patterns and areas under crops depending on available water resources; (viii) the order and sequence of water use in the case of water shortage; (ix) procedures for resolution of any disputes and conflicts arising among water users; (x) procedures for defending common interests of downstream water users against the actions of upstream water users diverting water from the same canal or river, etc*." (Rauner 1887).

There were special regulations for the operation and maintenance of waterworks and irrigation canals. Each water user had to take part in public work related to the cleaning of irrigation canals, maintaining of water intake structures, and repairing waterworks.

There were two kinds of public work, one related to work on the section of a canal designated for operation and maintenance by villages or groups of water users and the other comprising annual public work (*ḳazu*) under the leadership of *mirabs* and *ariq-aḳsaḳals. External ḳazu* related to the cleaning and repairing of canals and waterworks for general use, and *internal ḳazu* related to the cleaning of a person's own *ariqs*. Peasants and inhabitants participated in public work every year at sites of their responsibility. Masalsky (1913) noted that the cleaning of all canals in the Khiva Khanate required about 700,000 man-days annually or two man-days per hectare under irrigation. In Tashkent Uyezd[7], more than 10 man-days per hectare were required to clean and repair the Zakh-Arik Irrigation System (Figure 3.2).

By giving credit to prosperous local farmers and the aristocracy, the tsarist government of Turkistan sought to improve water intake facilities, construct new irrigation canals and systems, and expand the irrigation area in order to encourage development of irrigation within oases. The Director of the Department of Land Reclamation in Turkistan wrote in his report that thanks to such activities, and by establishing a proper order for record-keeping of irrigated land, the area under irrigation significantly increased. In 1880 the area of irrigated lands was about 1.5 million *dessiatinas,* but by 1910 the total

[7]*Uyezd* is a district (an administrative and territorial unit in Russia).

Figure 3.2 People engaged in *kazu* – fulfilling societal obligations for maintenance of irrigation networks in the nineteenth century.

irrigated area reached 2.5 million *dessiatinas*, including 620,000 *dessiatinas* in Fergana Province, 526,000 *dessiatinas* in Syrdarya Province, 329,000 *dessiatinas* in Samarkand Province, and 500,000 d*essiatinas* in Khorezm and Trans-Caspian provinces. Over 40 km of canals in Kazalinsk Province, and 2,500 *dessiatinas* of the Chilysk Irrigation System in Kyzyl-Orda Province were rehabilitated in the lower reaches of the Syrdarya.

In 1907, the Governor-General decided to develop a water code for Turkistan. Eminent jurists, irrigation engineers, historians and specialists in local customs were invited to help prepare this legal document. The Turkistan Administration of Irrigation Works (IRTUR), which was established by that time, initiated the collection and translation of those chapters of the Sharia that related to use of water and land resources. These were later published in the Soviet period in the journal *Bulletin of Irrigation* (Issue 9, 1924) in an article titled "The Code of Muslim Legal Regulations on Land and Water Use".

Maintaining such a vast area of irrigated land in an operable status required clear regulations and rules for water diversion and distribution, as well as for operation and maintenance of the dense irrigation networks. Therefore, the intention of the tsarist administration to develop appropriate water laws is quite understandable. Although the Water Code was not approved by the Senate, it was introduced at the level of provincial governors and all subordinated local administrations. As the researcher S. Pokrovsky (1926) noted, "the most interesting provision in the water law is the establishment (or rather preservation) of elected self-government institutions, which are responsible for water management at the level of separate water districts". The requirement to set up specially elected bodies (councils) was based on the existing system of water self-management (the whole community of *ariq-aksakals*, *mirabs* and *tuganche*), as well as on the generic character of water interests and overall labor duties required for maintenance of irrigation systems.

Water self-management was maintained at all levels of the water hierarchy, from the water council of the water district, to the elected *ariq-aksakals* (managers of irrigation canals) and the *mirabs*, who allocated the water available in their command areas. One should keep in mind that in 1888 the office of the Turkistan Governor-General Mr Rozenbah had issued *Instruction on the Rights and Duties of Officials, District Chiefs, Ariq-aksakals and Mirabs responsible for irrigation management in Turkistan Krai*.

Professor Pokrovsky argued that some of the interesting features of the system of public water management that existed prior to Russian colonization should be retained: "*The rather complicated irrigation system created in Central Asia has facilitated the perception of water as a public property. Clear rules of equal water allocation according to the principle* 'bir adam – bir su' *(literally translated as 'one person – one water') should be kept with the established distribution priority sequence and the norms of water use over specific periods of time. The clan system should get new public features, where the clan is transformed into a water-sharing community that has to adapt to further economic evolution.*" This transformation of a clan should be directed towards the development of a peaceful communal environment, within which all water disputes can be solved on the basis of the established order of water use in order to guarantee overall equity. "*The practice of artificial irrigation holds a key feature that predetermines peaceful resolution of all water conflicts. In order to prevent severe conflicts both inside a clan and between clans resulting from water use or the organization of the water supply, it is necessary to establish step by step a special water administration using the ariq-aksakals and mirabs who are elected by the population themselves.*" This reliance on *mirabs* and *aksakals,* men of authority with a deep knowledge of customs and the water supply system who were elected by the population itself and capable of acting as mediators in the settling of water disputes, resulted from the wish to prevent severe intra- and inter-clan tensions arising over water use.

Serkara-bashi and *keleme-bashi* were elected to supervise water distribution in the primary social units of *serkara* and *keleme* (a subdivision of *serkara*) respectively. Their duties included supervision of the correct sequence of water supply and also the organization of public works related to construction or repair of the irrigation network at village or inter-village level. *Mirab*s were elected for water management in each village as a whole. They were elected annually in autumn, after termination of water applications to the crops in the current growing season. Their duties consisted of organizing the annual cleaning of canals with the local population and the safekeeping of waterworks at the site under their responsibility; the organization of the sequence of water supply and supervision over its implementation; supervising crop cultivation in the command area of *ariqs* under their responsibility and assistance in water supply to peasants.

Ariq-aksakals were elected for managing the main irrigation canals. They had the same functions as *mirabs* but at a higher level in the water management hierarchy. While a *mirab* executed water management functions at the intra-clan or community level, an *ariq-aksakal* was a water manager at the inter-clan or inter-community level. The *mirab* was responsible for an *ariq* having links with other *ariq*

systems, and he executed his duties within the irrigated area belonging to the community that elected him. An *ariq-aḳsaḳal* was responsible for the main irrigation canal. His duties consisted of maintaining the water intake facilities, including barrages, water intake structures, check structures, regulators and canal banks, as well as maintaining water levels and flow rates in laterals established according to the local customs.

It is interesting that in the Fergana Valley where there were Uzbek and Kirghiz communities, two *ḳaḳ-bashis* were elected rather than one *ḳaḳ-bashi*. One of them represented the Uzbek population and the other the Kirghiz population. The *ḳaḳ-bashi* was responsible for the operation and maintenance of small distributary *ariqs*, and the *mirab-bashi* was responsible for the lateral canals. These water managers received payments in kind, a part of which was a voluntarily contribution by the people rather than a salary in cash. In this way, the assessment of their work was in the hands of the water users themselves. When a dispute could not be settled, the problem was considered and discussed at a general meeting of the community.

From the outside, it appeared that the draft Water Code had kept all the provisions of customary law in principle, but in fact it provided opportunities for reflecting certain tendencies. It would be expected that the development of capitalist relations in Central Asia should be reflected in changes to the legal mechanism of water relations. There was pressure from Russian and local authorities who wanted to take over responsibility for the production of raw cotton and cotton fiber but faced problems given the tradition of water use according to local customs. The authorities took over the overall governance of water resources, preserving on the one hand the customary water use rights of the population (and fixing these rights into legal norms established by the state), while on the other hand redistributing water not belonging to water users according to local customs to the benefit of private owners in the form of water concessions given to new settlers.

On the basis of new provisions in the law, a certain restructuring of the water sector took place in the Turkistan Krai in 1910. Water management at the level of the Governor-General was delegated to a special water department (the IRTUR) coming under the Central Directorate of Land Management. Water districts were established according to hydrological principles, and these were managed by "district and provincial water offices" and the elected water councils that were subordinate to the IRTUR. The water councils were responsible for drawing up and submitting regulations of water use within their districts for approval. Members of a water council were elected for a three year period, but they had to be approved by the Governor-General.

The state, while keeping public participation in water management at the lower levels of the water hierarchy, exercised strict governance and authority over water resources in other fields. The authorities were granting the right of water use, granting permissions for the withdrawal of freely available water resources and on the use of somebody else's plots, and they were granting the right of water management and allocation. These rights were eventually transferred to district-based administrative bodies. In 1910, water self-management at the bottom level became under the responsibility of a special system of water management headed by the Turkistan Administration of Irrigation Works (IRTUR). Local offices of the IRTUR issued special certificates (Irrigation Certificates) with reference to those provisions of the customary law for water use which the population itself considered as obligatory. In passing, we should note that one cannot but agree with A.A. Kadyrov (2005) that by establishing water users associations and councils of canal water users today we, in fact, revert to customs that existed for many centuries.

Throughout this period, the Ministry of Agriculture of Russia kept in mind the potential for a considerable increase in the use of water resources, especially for developing the cotton sector in Turkistan. This sector was of special value and large-scale cotton production would be impossible without specially designed and constructed irrigation systems. To support the planned large-scale water supply and irrigation activity, engineering field surveys were started in 1895 and thence conducted on a regular basis. Around 600,000 *dessiatinas* were surveyed in Syrdarya, Fergana and Samarkand provinces, and some 45,000 *dessiatinas* in the northeast part of the Hunger Steppe were selected for the Irrigation Phase I Project. The choice of these virgin lands in the Hunger Steppe was not accidental. In 1870, this territory drew the attention of the Governor-General due to its very convenient location at the crossroads to the Fergana Valley, Tashkent and Samarkand, while also being near to the Syrdarya. Between 1870 and 1895, more than ten projects were prepared and one small-scale project was even implemented. The project delivered water through a canal named after Nikolay I and irrigated 2,500 hectares.

Meanwhile a new large-scale project for delivering water to Bukhara was being planned. Some 2.25 million Russian rubles were allocated for this project and construction started under the leadership of the Grand Prince Nikolay Konstantinovich. However, progress was unsatisfactory until 1907, when a special government decree pointed out the need to solve some fundamental problems such as improving the overall plan of the irrigation networks, and establishing a plan to develop the virgin lands of the Hunger Steppe including the design of water intakes and other infrastructure. After approving this project document in 1911, work intensified on a contract basis. On October 5, 1913, the Romanov Canal (in the Soviet period this canal was renamed after Kirov and it is now called Dustlik) came into operation (Figure 3.3). The initial irrigation system covered about 69,000 hectares, with a cultivated area of 34,500 hectares, including 19,000 hectares under cotton. This canal initiated the irrigation of the Hunger Steppe based on a specific engineering philosophy. The project

Figure 3.3 Opening of the Romanov Canal headworks [45, p. 210].

encompassed the construction of a main irrigation canal with a carrying capacity that allowed about 81,000 *dessiatinas* of virgin lands (85,700 hectares) to be irrigated as well as the construction of the Shuruzyak Drainage Canal. It also included a desert sink for drainage water disposal, and two thousand kilometers of lateral canals, bridges, roads and telephone communication lines. This project was the major pilot of that period.

However, the first phase of developing the Hunger Steppe had some important shortcomings, which were not completely addressed during the next phase. The main issue was the unsettled problem regarding who should actually develop the lands. The engineers A.F. Ostrovsky and A.I. Kursish suggested building the irrigation network and delivering water up to an irrigated plot of 8–10 *dessiatinas*, however the conveyor canals were built at the expense of the state only up to the borders of the irrigated areas of 150 *dessiatinas*. The distribution network had to be built at the expense of the settlers themselves.

Only Russian peasants were assigned as future settlers of the Hunger Steppe. In 1913, Mr Krivoshein, chief of the Department of Land Management and Agriculture, clearly expressed this aspect of colonial policy in his report to the State Duma: "*The needs of colonizing this territory by Russian people should come in first. Along with this approach, it is important that land settlement promotes expansion of the areas under cotton, which our cotton industry especially needs. At the same time, there should be a sustainable operation, which should regain the capital investment in irrigation that was allocated by our treasury*" (Dukhovny 1976). As a result, a law was adopted that only allowed resettlement of Russian citizens of Christian confession who had a property of no less than 1000 rubles.

The construction of the so-called Monarchic Manor in the Murgab Oasis was another large project implemented through state investment into the water sector of Turkistan. The works started in 1890 with an unsuccessful attempt to rehabilitate the ancient dam of the Sultan-Bend (Figure 3.4) on the

Figure 3.4 Dam in the Sultan-Bend (Hojamyradov 2007).

Murghab River. Then, it was decided to construct the new Gindukush Dam, and later an additional two dams, the Lolotan and Sultan-Bend. The so-called Tsarist Canal, some 26 versts long, diverted water from the river by means of a water intake with Poiret type gates, allowing the irrigation of some 14,000 *dessiatinas*. Subsequently, the irrigated area here was increased up to 25,000 *dessiatinas*.

These two state projects were very significant, since they were examples of rectangular irrigational networks rather than the earlier and existing fan-type irrigation systems, a typical representative of which was the Sokh Irrigation Scheme (Figure 3.5). The Great Ferghana canal was built here during the Soviet time in order to create a more stable water regime. The training of numerous local and Russian hydraulic engineers and land reclamation specialists was based on the experience gained from these projects.

With the exception of these projects, other large-scale irrigation schemes were not undertaken in this period. However, the following smaller schemes were constructed: the Iskander-Aryk Canal, with a command area of 4,140 *dessiatinas* in the Tashkent oasis; the division structure in the head of the Shaarihansay Canal on the Karadarya River in the Fergana Valley; the Palman division structure on the Isfairamsay River (160 m^3/sec); the Wuadyl regulator on the Shahimardansay River (84 m^3/sec); and the Akkaradarya division structure and Ravathodji hydro scheme on the Zeravshan River with several division structures on the Dargom River.

More than ten pump stations were financed and built through private investment in Fergana Province. These were sited on the banks of the Syrdarya and on the Termez Main Canal. This work is evidence of the fact that insufficient state funds were allocated for irrigation in Turkistan. This state

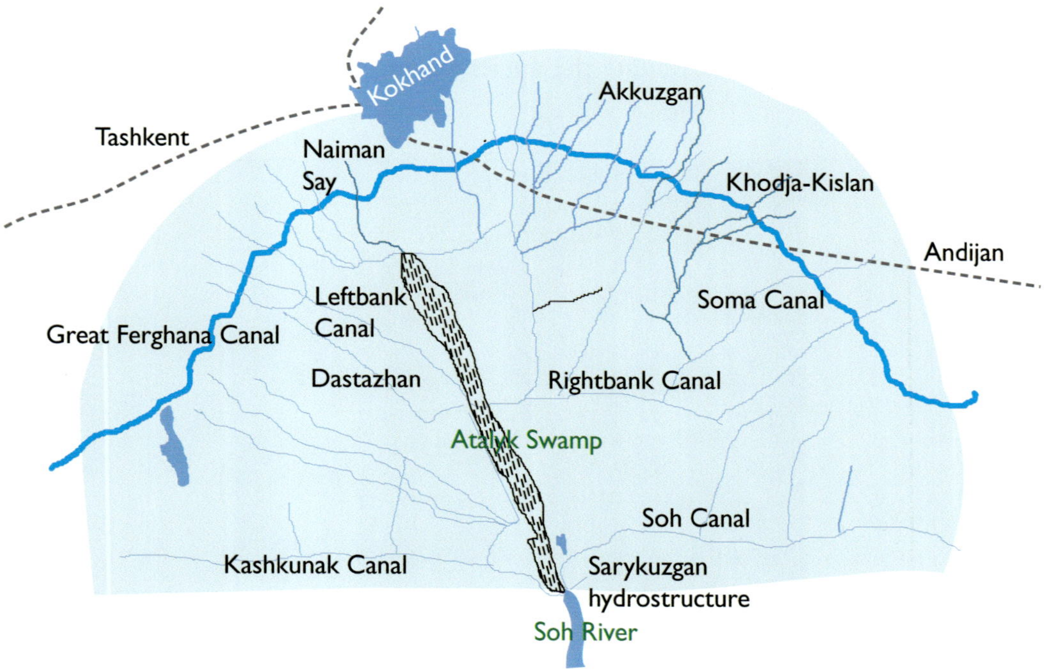

Figure 3.5 The Sokh Irrigation Scheme and the Great Fergana Canal [47].

funding was of the order of 36.4 million rubles over 20 years. Information presented by Krasnopolsky regarding the sums spent on the operation and maintenance of irrigation networks covering an area of 654,300 *tanaps*[8] in the Khiva Khanate is quite interesting: "According to local customary law, since mid February, annual cleaning of *ariqs* has been carried out by local workers at a rate of 15 to 25 man-days for each 10 *tanaps* of irrigated land. In addition, there was an in-kind duty in the form of cleaning dams, training streams and repairing the head structures of *ariqs* during the summer period." Each peasant spent 40 to 54 man-days annually on this work. In money terms, this equates to a contribution of 3,600,595 rubles for the whole Khanate. The total area under crops was around 654,300 *tanaps*, including an area of 97,595 *tanaps* under cotton, so about 6 rubles per *tanap* were annually spent on operation and maintenance (Krasnopolsky 1915).

Thus in one province the input of water users into operation and maintenance of the irrigation network was two times greater than the public investment through state funds! The key problem of this period was the extremely unsatisfactory irrigation water supply which needed to be maintained through a huge in-kind input by local people.

A description of water availability in the Bukhara oasis at the beginning of the twentieth century is found in the work of A. Gubarevich-Razdolsky (1905): "*Available water resources are quite insufficient for such an irrigation system that exists in the Bukhara oasis. Water is also diverted from the Zeravshan River along the upstream river stretch, and therefore the agricultural land in the Bukhara oasis is irrigated only by that water which remains after irrigation in Samarkand Province. Due to the lack of proper control over crop patterns within the command areas of* ariqs *in this province, since 1870 the local population has been cultivating rice, which has a considerably higher water requirement. As a result, the downstream flow of the Zeravshan River has become less and less. Only not long ago, His Highness ordered that one third of the Zeravshan River's flow must be diverted to the Bukhara Khanate. However, due to the lack of a proper dam, reservoirs and water-measuring posts, sharing of water is by use of primitive dykes made of fascines and earth that often break. At present, the actually established procedure for sharing water is not observed.*"

Another important feature of Russian rule in Central Asia was the attempt to establish a bureau for keeping records of land under irrigation. S. Tolstov considered that at the beginning of the first millennium AD and in the first half of the second millennium AD – that is, during the golden age of irrigation in Turkistan – the area of irrigated land had already reached 3.5 million hectares. In 1913, according to V. Masalsky, this was the area of irrigated land in each province:

If these figures are correct, more than 5.1 million hectares were irrigated in Turkistan at that time! However, Gubarevich-Razdolsky (1905) provides a different perspective: "Out of almost two million dessiatinas of land with an irrigation network no more than two thirds were actually under crops."

Table 3.1 presents data from the detailed report submitted by the Senator K. Palen (1911), which was based on statistical data on sown in areas (areas under cultivation) in four of the seven provinces

[8]1 *tanap* = 900 square *sazhens*.

Semirechye Province	703,000 *dessiatinas*
Fergana Province	840,000 *dessiatinas*
Syrdarya Province	635,000 *dessiatinas*
Samarkand Province	480,000 *dessiatinas*
Trans-Caspian Province	150,000 *dessiatinas*
Bukhara Emirate (including part of present-day Tajikistan)	1,600,000 *dessiatinas*
Khiva Khanate	350,000 *dessiatinas*
Total in Turkistan	**4,758,000 dessiatinas**

Table 3.1 Comparison of irrigated and sown in areas in Turkistan (000' *dessiatinas*).

Province	**Irrigated area according to V. Masalsky (as of 1912)**	**Sown in area according to Palen (as of 1907)**		
		Total	**Dry farming**	**Irrigated**
Semirechye	703	556.3	68.3	488.0
Syrdarya	635	653.5	42.8	610.7
Fergana	840	676.1	167.8	508.3
Samarkand	480	409.5	56.5	353.0
Total	2658	2295.4	720.6	1960.0
% irrigated (with *ariqs*)				73.7

mentioned by Masalsky. Thus one can presume that Masalsky presented data for an area with existing irrigation networks of which only between 3,360,000 to 3,760,000 hectares were actually irrigated in Central Asia. In his memorandum "Irrigation in the Fergana Valley" presented in *Irrigation of Uzbekistan* (Sadykov 1975), F. Fedodeev, one of the very experienced irrigators in Turkistan, gave a very interesting description of the complicated system of irrigation according to customary laws that existed in the Fergana Valley until 1928. The Isfayram-Shakhimardan Irrigation Scheme that served

Margelan and then also Fergana was a typical example of the complicated system of water use which existed at that time.

An extreme shortage of water, which during critical periods could mean there was only about 40% to 50% of the required water volume, promoted the establishment of well-developed and firm procedures for water-sharing based on customs that were preserved by the local population. These procedures of water distribution were quite unfair in respect to water users in the tail (or end) of the irrigation networks as these were mainly disadvantaged by the system. The idea of water ownership was mainly practiced in the so-called "everyday" water distribution. Many amendments were made to the basic procedures for water-sharing and the sequence of water supply among *ariqs*, changing the strictly established sequence and amounts of water supply within the irrigation system. For example, Margelan could use an additional volume of water (the so-called *beyliq-su*) supplied through the Shakhimardan system as the gift of Chimion Bey, whose wife came from Margelan. Various additional volumes of irrigation water – *chakob-su, sizov-su* and *khurova-su* – were available as the result of purchase and sale, exchange and gift, violence or deception dating back to the far-off days of the khans' rule.

These unplanned adjustments to the established order of water use considerably reduced the efficiency of irrigation systems. There were significant water losses as a result of delivering small amounts of water through dry channels often kilometers long to a specific and limited circle of water users with the right of sale and inheritance.

The right of water ownership was especially critical within the Shakhimardan irrigation system, where water availability was particularly limited. Starting from the head division structure at Wuadil and almost down to the small division structures, self-operating wooden division devices (later stone and concrete structures) were installed that automatically divided the water in the system into shares, which, of course, did not correspond to either the irrigated areas nor to the natural conditions that determined crop water requirements. Many landless people who had their own water share (*dambyr*) only lived at the expense of selling their water shares or their turn for supply.

The situation in Yazyavan *kishlak* (village) located in the tail part of the Isfayram irrigation system is an example of the unfair water distribution that could pertain under conditions of water shortage. Until June, this *kishlak* received water from the Isfayram irrigation system only once each six-day period, when the lateral canals of the Isfayram Main Canal were closed and water was transported to Yazyavan.

3.2 Surveys, Research and Design in the Pre-Soviet Time

The period of Russian colonization played a specific role in implementing new irrigation projects and putting irrigated agriculture in order. Some important decisions were taken to attract private capital

and to encourage mixed investments for implementing irrigation projects on a contract basis. Some additional incentives, such as granting concessions, were provided in order to attract private capital. For example, one half plot on which an irrigation network was built by a farmer could become his private property while the second half remained state treasury property. Or, after completing construction of the irrigation network, a farmer was exempted from the state and *zemstvo*[9] taxes for the next five years and in the following years was only taxed at half the standard rate. There was also a scheme for farmers to obtain low-interest credit (1 to 2% per year). Another significant privilege for businessmen constructing irrigation canals and other water infrastructure was the option to lease public land without any compensation to the state (for a period of 99 years).

In the period from 1895 to 1905, the research and field surveys necessary for irrigation projects were financed and organized. These became the foundation for the large-scale development of irrigation in Central Asia, including:

- the Golodnaya Steppe Irrigation Project (1910–1915) developed under the leadership of Professor G.K. Rizenkampf
- the 1908 project to transfer Amudarya water into the Merv and Tedjen oases, developed under the leadership of M. Ermolaev, which mainly consisted of constructing the Kayrakkum Canal with the irrigation of 516,000 *dessiatinas* in its command area.

The project prepared by Rizenkampf was most elaborate and unique. The project included the construction of the irrigation networks up to the last irrigated field, land leveling, the drainage system, and other necessary infrastructure. Rizenkampf managed the project works in the Golodnaya Steppe and made a great contribution to developing irrigation there. His civic-minded responsibility and in-depth understanding of the importance and significance of large-scale irrigation in Turkistan enabled him not only to brilliantly solve many engineering problems but to also elaborate the fundamentals for the development of virgin land and to adopt an integrated approach to this work. This resulted in international fame for the Golodnaya Steppe Project and put in place the basics for integrated water resources management and irrigation development of that time.

In the preface to the Survey Report for the Golodnaya Steppe Irrigation Project (published in the Proceedings of the IRTUR 1921), Rizenkampf wrote: "*Restoring life in this desert burned by the sun and reviving dead lands are the task of irrigation specialists. An engineer-irrigator can consider his mission as completed when a water intake structure for diverting water from the river has been built, the network of canals with life-giving water has covered the whole area, and regulators for delivering a sustainable water supply to each settler have been constructed. The task of irrigation systems makers is very difficult. The irrigation network is a canvas on which "life" will be embroidered; and in the process of its creation it is necessary to see all the aspects of future life very clearly. Developing an irrigation system should not be the end in itself; it is part of an integrated whole – the revival of the desert – from which the basic assignments arise and with which the irrigation system should be inherently linked. A key requirement is to provide the most rational arrangement*

[9]One of a system of elected councils established in tsarist Russia to administer local affairs after the abolition of serfdom.

of all life rather than only focus on the construction of irrigation networks, as well as to achieve the maximum effect as a whole rather than in any details.

It is necessary not only to design the irrigation system but also to draw up the plan for developing the area under consideration, including the scheme of roads, sites for industrial and market centers and the most rational sources of energy in order to supply future factories and workshops, as well as to prove that the designed irrigation system is inherently linked with future life arrangements and is a well designed part of the integrated whole."

In contrast to a project proposed by F. Morgunenkov, Ermolaev planned to build a water intake structure on the Amudarya near Kyzyl-Ayak with the layout of a canal to Murgab along the former floodplain of the Kelif-Uzboy and with water delivery through one of Murgab's channels parallel to the railway to Tedjen. Morgunenkov had developed Tsinzerling's (1927) idea regarding water diversion from the Amudarya at a site upstream from Nukus. Thus, M. Ermolaev preempted the idea of the future Karakum Canal project, rejecting the first option of a Turkmen Main Canal with a water intake structure near Takhiatash.

The Central Fergana Irrigation Project prepared by K. Sinayavsky proposed a water diversion from the Naryn River and a canal that crossed some streams flowing from the mountains (Aravansay, Isfayramsay, Shakhimardan and Sokh); a similar idea to the future Big Fergana Canal although this was designed at higher elevations. However, at that time, this project was rejected due to its "inefficiency". Another project for the future (the South-Eastern Fergana Irrigation Project) was prepared by the outstanding hydraulic engineer I. Aleksandrov. This proposal was for the construction of a multiple-purpose dam on the Karadarya River creating a reservoir with a storage capacity of 1.38 km^3 and a main canal of 232 *versts*[10] long (all this water infrastructure was later built in the Soviet period).

As a result of their work in this region, Russian scientists not only created the scientific basis for irrigation development in Central Asia but also became leaders in the design of water infrastructure and land reclamation for the whole Soviet Union. The outstanding role of pilot farm developers such as M. Bushuev, who established the Golodnaya Steppe Pilot Farm in 1905, needs to be mentioned here. They developed methods of irrigated farming on salt-affected lands increasing the harvest to 3.5 tons of raw cotton per *dessiatina* from the 1.4 tons per *dessiatina* on average over the region that was achieved in 1912. For the first time in Central Asia, pilot plots were constructed with subsurface field drains that made sustainable soil desalinization possible. Experimental work on the Golodnaya Steppe Pilot Farm included the zoning of irrigated lands according to crop water requirements and the scheduling of water applications, an approach also followed at the Andijan and Murgab pilot stations. From 1911, these experimental works were headed by A.N. Kostyakov who became a main founder of Soviet land reclamation science. As the chief of the Hydro Module Division of the Russian Department of Land Reclamation, Kostyakov personally organized the experimental work and published his findings on irrigation scheduling (the terms and norms of water applications for different crops), the impact of irrigation on various soils, and irrigation methods. This work provided the basis for his key publication

[10]A Russian measure of length equal to approximately 1.1 km (0.66 mile).

The Fundamentals of Land Reclamation, which became the textbook for the training of thousands of water professionals.

The theory developed by A.N. Kostyakov was published in 1913 in "Taking into account the crop water requirements of a hydro module under operation of an irrigation system" (see Kostyakov 1919). The theory provided guidelines for designing irrigation networks. For the first time, this scientist was able to combine the design of an irrigation system with an analysis of factors affecting its composition and efficiency. Kostyakov wrote that "*when designing the irrigation system it is necessary to specify the parameters and mutual arrangement of its components, but it is much more important to know the actual conditions for its operation and for water use. At the same time, both aspects of irrigation system's operation statistics (parameters and mutual arrangement of components) and dynamics (the process of water use) are greatly interrelated. On the one hand, irrigation system's operation depends on its design, but, on the other hand, the design should allow keeping a close interrelation of all components*." This was a really innovative approach. Based on his five-year field observations in different regions of Central Asia, Kostyakov (1956) tried to establish an harmonious framework for the interaction of all components of an irrigation system.

The establishment of the Hydrometric Division of the Turkistan Administration of Agriculture and State Affairs under the leadership of V. Glushkov in 1892 played a significant role in the development of the water and agricultural sector. Its activities created the basis for almost all hydrological and meteorological observations and for the development of an appropriate database in Central Asia. This division was responsible for establishing the gauging-stations on all rivers, the hydrometric posts and weather observation stations, as well as workshops for calibrating and repairing hydrometric equipment. Research was also carried out at these stations and posts according to well-established methodologies. These included taking water level measurements in irrigation water sources, measuring and recording flow rates in rivers, analyzing the suspended load and content of dissolved substances, making meteorological observations, keeping records of atmospheric precipitation and evaporation from the water surface of rivers and lakes, and studying the hydraulic parameters of river streams.

By 1912, six gauging-stations and 50 hydrometric posts (including eight hydrometric posts with automated recorders of water levels), three weather observation stations (Class I, Category 2), 34 rain-gauging stations (Category 3), seven river evaporation measuring posts, a chemical laboratory and a calibration station with a workshop for repairing hydrometric devices and tools were operable in Turkistan Krai.

In summarizing the period of Russian colonization in Turkistan, it is fair to say that the tsarist government lost its competition with Great Britain's colonial administration in India, which managed to build head structures for irrigation systems over an area of 6,000,000 hectares inherited from the former authorities of India and to extend the irrigated area up to 15,000,000 hectares. Recalling this period, Rizenkampf (1921) emphasized that "due to the lack of understanding of the economic value of reviving irrigation in Turkistan by some high-ranking officials, the works were implemented at a snail's pace, restricting the irrigation plans to some small areas selectively taken from the large whole" . Nevertheless, many Russian specialists who, according to Rizenkampf, "were full of enthusiasm" were attracted by this grandiose task of reviving vast areas and developing the water and irrigation sector in this region. These specialists made a great contribution, creating the achievements in science and

design in Central Asia that received world-wide recognition in the following years. They also clearly realized that "the vast irrigation systems built by the native population in Central Asia consumed water resources non-rationally and could not provide sustainable water use and efficient use of irrigated land. The problem of improving indigenous irrigation is basically the problem of remodeling it radically" (Proceedings of the IRTUR 1921).

These well-motivated persons generated a huge list of design ideas – creative and absolutely unconventional ideas – which were implemented in the Soviet period through inherited knowledge, scientific methods, and enthusiasm and devotion to the great cause of reviving the desert areas. In the list of irrigation projects recommended for implementation by the IRTUR under the leadership of Rizenkampf set out in the 1923 report *Data on available land resources suitable for irrigation in Turkistan*, we can find almost all the irrigation schemes that were developed by these pioneers as well as many hydraulic structures which are still in use today. This list starts with the Uchkurgan Steppe Irrigation Project with water diversion from the Naryn River, the Ulugnor Irrigation Scheme, the Kampir-Ravat Dam on the Karadarya River and the Golodnaya Steppe Irrigation Scheme. Also on this list are the Big South Fergana Canal, the Begovat Dam, the Dalverzin Irrigation Scheme and practically all the water infrastructure still under operation as well as some not yet constructed on the Syrdarya, including the pumping stations for irrigation of the present-day Tajik part of the Golodnaya Steppe. There were plans for the Boz-Su Cascade and for dams on the Chirchik, Chatkal and Pskem rivers to be built in the Tashkent region. An irrigated area in the Syrdarya Basin was clearly specified at the present-day level (3,200,000 hectares), but irrigation requirements appeared to be set 25–30% lower than used now, even under conditions of a permanent water deficit.

One other feature of these works that should be mentioned. In contrast to the tsarist government's idea to colonize Turkistan with Russian settlers, the developers of these projects supported the participation of local population. "The agricultural practice on irrigated lands in Turkistan is quite specific and requires special knowledge, skill and people accustomed to work under specific climatic conditions. In this case, native inhabitants should be considered as more suitable settlers… It is only necessary to help them to acquire knowledge of modern technologies in irrigated farming further to their experience" (Tsinzerling 1927). And these ideas are alive and actively being researched again today.

3.3 The Water Sector of Turkistan During the First Years of the Soviet Period

Establishing Soviet power prior to national demarcation in Central Asia (in 1924) was difficult due to class struggle, loss of management structures, war, mutinies in different parts of the region, and *basmachi* opposition[11]. Clear goals and intense organization were necessary in order to retain power

[11]*Basmach* were Muslim anti-Bolshevik fighters in Central Asia during 1917–26.

and get economies going under these circumstances. All irrigation infrastructure and main canals were transferred to the jurisdiction of the People's Commissariat of Agriculture[12] by the decree issued by the Council of the People's Commissars[13] of the Turkistan Republic three months after its election on March 13, 1918. Taking into account that only two irrigation systems had been designed and built in an irrigation area of many millions of hectares (the Romanov Canal in the Golodnaya Steppe and the Monarchic Manor in the Murgab oasis), the People's Commissariat of Agriculture received a rather difficult legacy. In particular, the ramified indigenous irrigation systems were equipped with primitive water intake facilities and these required enormous energy for operation and maintenance and placed constraints on irrigation development.

"*These structures (spur dykes, fascine dykes, etc.) made of stone, brushwood, and pebble are breaking down each year and require continuous repairs or even rebuilding over and again. In addition, over time, sediments are deposited within the riverbed in front of the head intakes of a canal and reduce water inflow. Therefore it is necessary to make longer spur dykes upstream or to clean silt deposits constantly. Moreover, a river stream changing its channel can cut into an opposite riverbank, again requiring new river training works*" (Tsinzerling 1927).

The government of Russia well understood the importance of creating conditions for the rehabilitation of irrigation in Turkistan in order to provide a satisfactory level of well-being to most of the population in this region. At the beginning of 1918, by order of Lenin, the Supreme People's Economic Council reviewed the plan for the top-priority irrigation works, which was prepared by Rizenkampf, Vasilyev and the economist B. Lodygin, and submitted this plan to the Council of the People's Commissars. Based on this document, on May 17, 1918, Lenin signed the famous decree "Allocation of 50 million ruble for irrigation works in Turkistan". In accordance with this decree the following irrigation works had to be implemented:

i) irrigation of 500,000 *dessiatinas* of the Hunger Steppe in Khojent district of Samarkand Province
ii) construction of head structures of the irrigation system covering 40,000 *dessiatinas* in the Dalverzin Steppe located on the opposite bank of the Syrdarya River
iii) irrigation of 10,000 *dessiatinas* in the Uchkurgan Steppe in the Fergana Province and improving water use over an area of 20,000 *dessiatinas* in the same region
iv) construction of the dam on the Zarafshan River (near Dupulin bridge) for regulating the river flow and irrigating about 100,000 *dessiatinas* for cotton production
v) completing the construction works for irrigation systems in the Chu Valley over an area of 94,000 *dessiatinas*.

This decree could not be implemented during the following years, although immediately after it had been issued three special trains with equipment for irrigation works went towards the south.

[12] A government department of the USSR before 1946.
[13] Soviet government from 1917–46.

A terrible civil war had already engulfed practically the whole country and all IRTUR engineering personnel had to work in Moscow. They were not only developing the projects provided for by the decree but also contributed to the plan of the State Commission for the Electrification of Russia. Rizenkampf and Alexsandrov prepared a report on the electrification of Turkistan, which became an integral part of the plan. Some hydropower plants included in the 1923 IRTUR report, Data on Land Resources Available for Irrigation in Turkistan, such as the Boz-Su Hydropower Station Cascade and the Uchkurgan, Farkhad and Khishraus hydropower stations were presented as top-priority projects in this report.

Wars and revolutions always result in the complete collapse of the enormous and creative works of preceding decades and generations and the water sector in Turkistan did not escape such ordeals. One year after the beginning of the First World War the irrigated area in Turkistan reduced by 400,000 *dessiatinas*, and by 1922 the total sown in area was only 1,180,000 *dessiatinas*. Table 3.2 shows the changes in the amount of irrigated land under cultivation in the period from 1916 to 1923. The data in Table 3.2 does not include the agricultural land in the Khiva Khanate and Bukhara Emirate where the situation was a slightly better. The presence of strong Khan power here allowed 76% of the previously irrigated area to remain in operation (1,010,000 *dessiatinas*) (Tsinzerling 1927).

This reduction in irrigated land under cultivation resulted from economic problems including loss of clients and the need to develop self-sufficiency in grain seed, but primarily from the collapse of the central and local authority structure. To an even greater degree, this unprecedented drop in agricultural production was caused by damage to the irrigation systems resulting from the civil war, a lack of funds since 1916 for the operation and maintenance of the irrigation systems, and lower revenues from labor duties in kind.

"The indigenous irrigation networks covering 90% of the farming area were half destroyed. The modern irrigation networks that serve the remaining 10% of the irrigated area needed to be thoroughly repaired. The gauging stations and hydrometric posts were closed. Pilot farms stopped their activities as long ago as the beginning of the revolutionary events. A decline in the technical level of irrigation systems happened in parallel to the collapse of the water sector as a whole. The collapse of central and local management structures, with employees leaving their offices due to a lack of activity, and the loss of confidence in the economic sector all lead to this decline."[23]

Nevertheless, the ninth congress of state farm representatives of the Turkistan Republic held in September 1920 discussed the problems of irrigated farming and the action plan for the Turkistan

Table 3.2 Trends of sown areas in Turkistan (000' *dessiatinas*).

Year	**1916**	**1917**	**1918**	**1919**	**1920**	**1921**	**1922**	**1923**
All crops	2000	1620		no data	no data	1070	1180	1180
Cotton	534.0	339.4	80	119	109	90	42.7	65.9

Water Administration was adopted (Aminova 1963). These were the main resolutions adopted at this congress.

1. All water and land resources within the Turkistan Republic, regardless of their current ownership, are national property.
2. The sale, purchase, mortgage and lease of land and water resources are absolutely banned and considered a state crime.
3. Specifying the size of personal plots for working people is the responsibility of the State Land Committee and local authorities with the participation of *deḳhḳan* (peasant) representatives. [This should] take into consideration the conditions in each agricultural district and the family status. The same principles should be used for distributing water resources.
4. For the purpose of preventing aggressive tendencies, all settlements of Russian settlers created in the process of colonization must be equalized with the native population in respect of their rights regarding land and water use.
5. Water rights should be adjusted to the key regulations of land management in accordance with real need.
6. Water rights cease with the discontinuance of land use rights.
7. Maintenance and repair of waterworks and the main irrigation networks are the responsibility of the State, but other water infrastructure should be supported at the expense of water users.

In the spring of 1921 catastrophic high waters seriously damaged the water and agricultural sector throughout Turkistan. There were disasters in the Fergana Valley. The Kampyr-Ravat Dam was damaged on the Karadarya River, which stopped the water diversion into two large irrigation canals, the Andijansay and Shakhrihansay. The Balykchi Canal diverting water from the Ulugnara for irrigation of 15,000 *dessiatinas* was also destroyed. High waters damaged many intake dams on numerous rivers in the Fergana Valley including Palman, Wuadyl and Besharyq on the Isfayram, Shakhimardan and Margelan rivers, as well as on other small rivers and sources of irrigation water supply. The intake sections of the canals of the Chirchik Irrigation System, including the Zakh, Bozsu, Khanym, Shakh, Tal and Bektimir canals, were also heavily damaged.

The Ravatkhodja water division dam on the Zarafshan River (Figure 3.6a) was completely destroyed resulting in big problems with the water diversion into the Dargom, Yangikazan, Ak-aryq and Tuyatartar irrigation canals. Damage to the Akdarya water division dam caused a sudden redistribution of water between the Zarafshan branches, and most of that water streamed into the Akdarya right branch channel. Canals that diverted water from the Karadarya River lost their regular water supply, resulting in the loss of crops.

The floods of 1921 and 1922 hastened and worsened the collapse of the irrigation systems, and at the beginning of the next year, the water sector faced the challenge of the almost complete rehabilitation of the irrigation infrastructure in Turkistan.

The People's Commissariat of Agriculture attempted to reorganize the management system at a regional and provincial level in the Turkistan Republic. At the end of 1922, M. Rykunov, Deputy

Figure 3.6a View in 1914 of the destroyed Ravatkhodja Reinforced Concrete Dam on the Zarafshan River [Sadykov p.352].

Commissar of Agriculture, held a meeting with employees in the water sector to review the budgetary needs for the coming years, which could inform the 1923 Action Plan and cover key provisions regarding duties in kind and irrigation tax. The five-year plan for rehabilitating the water sector to a pre-war level was based on decisions of this meeting. The total budget amounted to 92 million rubles (18.38 million rubles per year), including 7,944 million rubles of duties in kind, 1.441 million rubles from local budgets, and 9.035 million rubles from the federal budget. This amounts to 3.30 rubles, 0.59 rubles and 3.76 rubles respectively per *dessiatina*. It was planned to rehabilitate irrigation over an area of 783,251 *dessiatinas,* to improve irrigation water supply in 2,090,000 *dessiatinas*, and to develop irrigation on up to 231,900 *dessiatinas* of new land. By the end of the five-year period, the irrigated area should have increased to 3,100,000 *dessiatinas* (Trombachev 1924).

In the section on engineering measures, the five-year plan provided for the replacement of some traditional water intake structures with more sustainable engineering structures. The plan included the construction of regulators on main canals and some lateral canals for the purpose of improving water use, rehabilitation of all destroyed dams, reconstruction of some irrigation systems, reclamation of salt-affected and waterlogged lands, construction of drainage infrastructure, river training works for preventing floods, rehabilitation of gauging stations and posts, creation of pilot irrigation farms, and recommencement of the hydrogeological and socioeconomic surveys of irrigation systems in Turkistan. The IRTUR, which maintained its functions after the revolution, was responsible for implementing these planned works.

From the beginning of 1923, about 1,000,000 rubles (in gold) were budgeted monthly for irrigation works (a total of 3,717,000 rubles over a five-month period). It is significant that during the same period the duties in kind comprised about 620,400 man-days and 3,670 horse-days. When compared with the funds set aside for the irrigation works set out in the report of the commission headed by Senator Palen,

the considerable increase in funding for irrigation works in the first years of Soviet power is obvious. IRTUR documents confirm that during this period the following works were implemented: rehabilitation of the water intake structure of the Zakh Canal that was washed away by floods in 1913; construction of the Karaspan Irrigation Canal and a barrage on the Arys River; a soil survey over an area of 20,000 *dessiatinas* in the command area of the right branch of Hunger Steppe Canal under the leadership of Professor Dimo, and similar work in the Zarafshan Valley and in the Turkmen Provinces (covering about 500,000 *dessiatinas*).

The IRTUR implemented a program of staff training for the water sector. This was for *ariq-akṣakạls* and irrigators, who had to be capable of operating and maintaining the irrigation systems established over the centuries as modern techniques were introduced and parts of the original structures were replaced. It is interesting to note that in support of these works in Turkistan a state water tax for all water users, including any industrial and rural enterprises which use water as a driving force, was now enacted. In 1923, this tax amounted to 0.45 ruble per *dessiatina* of irrigated land and one ruble per horsepower of hydropower. The water tax was collected between August 1 and October 1. All money collected was transferred to the Ministry of Finance for the explicit purpose of financing of the water sector.

The government of Turkistan and the IRTUR planned a considerable amount of work. This work included the rehabilitation of 27 gauging stations and 150 hydrometric posts, as well as hydrometric measuring operations on the Murgab, Zarafshan, Chirchik and Talas rivers and on some rivers in the Fergana Valley. It was proposed that the rehabilitation of the existing meteorological station network and construction of new stations, as well as the realization of the large-scale program of research work that had to be implemented should be done with technical assistance from the Turkistan State University and the newly-established branch of the Petrograd Research Institute of Land Reclamation. Top priority was given to studies related to the problems of sedimentation of irrigation canals and riverbeds, seepage losses from canals, and distribution of piezometric pressure along different underground sections of hydraulic structures.

In accordance with the Decree of the Council of the People's Commissars entitled "On Establishing the Meteorological Service", issued in June 1921, a Turkistan Meteorological Institute was created in Tashkent. (This was later renamed the Central Asia Meteorological Institute.) At the same time, the Hydrometeorological Forecast Service was established under the leadership of L. Davydov and with the assistance of E. Oldekop. This service developed a program of meteorological observations in the mountain regions. It also established the first high-altitude meteorological observatory near the head of the Big Naryn river at an elevation of 3,672 m. Studying the melting regimes of the glacier and estimating water availability was started under the leadership of N. Korzhevsky.

It was also planned to create field schools for water foremen, to establish technical secondary schools for training water technicians, and to enhance the land reclamation department of Tashkent State University. An interesting innovation was proposed at the meeting of the Technical and Economic Panel, introducing general economic and standard regulations for all kinds of water and land concessions in accordance with the 1921 Decree. This document provided for three types of concessions in the field of water supply, agriculture and hydropower. The first type of concession granted those that improved the water supply by means of constructing reservoirs, canals, regulators and pumping stations the right to then collect fees from the local population for water supply. In case of construction of a reservoir, payment was taken for volumes of water released. The second type of concession granted rights in respect of irrigating and using virgin land or reconstructing irrigation systems for the purpose of generating additional agricultural output. The third type of concession granted rights for using energy generated by natural or man-made streams.

However, in practice, these concessions generally could not be used by the private capital sector. A few concessions were granted in the Hunger Steppe, where there existed an excess of available land resources in the command area of the irrigation canal but a deficit of funds and manpower for developing this land. The government of the Georgian Republic received one of the first concessions for a plot of about 10,000 *dessiatinas* located along the new canal of the Hunger Steppe Irrigation System. The term of lease was fifteen years. During the first year, the Gruzarenda (the representative of the Georgian government) was exempted from any fee according to the lease agreement, but then a progressive fee (up to the tenth year of lease) had to be paid in kind or in cash. Cotton had to be sown on one third of the irrigated area. The Georgian Republic was obliged to build all necessary irrigation and drainage networks over this eight-year period while transferring the water infrastructure (after termination of the lease agreement) to the People's Commissariat of Land Resources. Apart from the Gruzarenda, Mr Penkov made an application for a concession of 3,000 *dessiatinas* at the 40 km point of the canal for the inhabitants of Yangi Village (1,000 *dessiatinas*) and the co-operative of peasants (700 *dessiatinas* in the command area of the Canal L-1). In 1923, the amount of land with irrigation networks increased to 12,496 *dessiatinas*. As a result, the number of settlers considerably increased and there were 5,490 farms by the end of that year.

The Irrigation System Management granted these concessions on behalf of the Turkistan Water Administration. In 1923, apart from land leased to the Gruzarenda, about a further 22,500 *dessiatinas* were leased and, in spite of rather late terms of signing the leasing agreements, about half this area was irrigated in that year. These concessions were subsequently restructured into collective farms or state farms. In particular, the famous Pakhta-Aral state farm in the Hunger Steppe was later created on the plot leased to the Gruzarenda (this farm is now located in the territory of Makhtaaral district in South Kazakhstan).

Establishing Soviet power and finding a suitable form for integrating existing rules and regulations of water use based on centuries-old practice and traditions with the collectivization of farms gave birth to new water and land reclamation associations (partnerships). They started to be established in 1919 and 1920, but the Soviet government only issued the special law "On Land Reclamation Associations" in 1921.

The government decided to share responsibility with water users and actually implemented the strategy set out in the last regulations of the tsarist government. The government was responsible for field surveys, designing the radical reconstruction of irrigation systems, and new construction of large-scale irrigation projects. In the short term, the rehabilitation of damaged irrigation infrastructure also was the responsibility of government as this was an impossible task for the local population. However, water users (organized into land reclamation associations) were responsible for the operation and

maintenance of operable irrigation systems. The government provided them with technical assistance and, if needed, long-term low-interest credit.

Nevertheless, when comparing land reclamation associations with present-day Water User Associations (WUA), it should be noted that land reclamation associations had a much wider scope. Their remit included land management, agricultural methods and farming techniques, land reclamation works, and, of course, water use.

From the outset, land reclamation associations were supported by managers of irrigation systems in the lower reaches of the rivers where the conditions of water supply and water use were quite complicated. Here, water users expressed their willingness to be responsible for the operation and maintenance of irrigation systems, given reimbursement of all expenses for improving their water supply (see *Bulletin of Integration*, No. 3, 1923, and No. 9, 1924). In practice, initially the target of the new law was to establish a clear legislative basis for existing irrigation communities that were functioning based on old traditions. The role of irrigation communities was described in a report of the Economic Bureau of IRTUR (Shastal 1923) as follows: "*For the purpose of legitimate water distribution among farms which divert water from a common canal* (ariq) *for irrigation of their plots, the population was organized into irrigation communities. These are associations of separate water users. These associations exist only on those* ariqs *where water use is based on an established sequence of water supply. They have spontaneously arisen and become quite widespread in the Turkmen Province. The number of these associations on one* ariq *varies depending on the carrying capacity of an* ariq *and the quantity of farms supplied with water from this ariq within the territorial boundaries of a settlement or community, which is the administrative unit.*"

The number of farms or separate water users that formed an irrigation community also depended on the irrigated area, cropping pattern, soil quality, size of taxes and other factors. In other words, it depended on the size of the water use unit and the water limits allocated to one water user, taking into account all the above-mentioned factors. Thus, separate groups of farms or irrigation communities used water from an *ariq* for irrigation of their plots according to the established priority sequence and allocated water limits. Irrigation communities, in turn, distributed water among their members also according to the established priority sequence and allocated water limits. However, sometimes water was distributed within an irrigation community without any sequence but in amounts that did not exceed the water volumes allocated to this irrigation community according to the water use plan. An allocated amount of water for irrigation of a unit area is usually a local unit rate, the name of which differed between regions. Its size could also vary between districts and regions, because it depended on water availability, the carrying capacity of an *ariq,* and the irrigated area within its command area, as well as the technology used to distribute water among water users.

The *mirab* was elected by all water users and, at the same time, appointed as an official representative of the local water management organization for water supervision. He was responsible for water distribution among and within irrigation communities. He was also responsible for adjusting the basic principles of water use to actual current conditions, and for the resolution of disputes and conflicts between water users, as well as technical supervision over the irrigation system under his jurisdiction. This provided the framework for a self-governing irrigation community. However, it is necessary to

note that in some cases clans were quite powerful and the irrigation communities were built not only according to territorial and hydrological principles but also according to clan principles. The most striking examples of such clan relations could be found in the Turkmen (Trans-Caspian) Province.

The Resolution of the Council of the People's Commissars of the Turkistan Republic of May 2, 1923 stated that the existing irrigation communities could be a cornerstone for establishing land reclamation associations. Based on this resolution, the organization charter of land reclamation associations was developed and approved by the People's Commissariat of Land Resources. According to the resolution, land reclamation associations needed to be established for joint operation and maintenance of existing irrigation systems and, if necessary, the modernization of these system based on co-operative principles. Land reclamation associations should also be responsible for implementing new irrigation works, the construction of drainage networks, river training work, taking measures to prevent soil erosion (gullies), sand fixation, use of hydropower, developing plots unsuitable for farming, and the construction and maintenance of water supply facilities. So the purpose for which the land reclamation associations were organized completely coincided with the land reclamation activities based on irrigation.

Consequently, although the law speaks about land reclamation associations, lawmakers quite definitely had in mind the combination of activities related to both water management and land reclamation that could be undertaken through the implementation of irrigation projects. Article 2 of this resolution, which specifies those features that should be taken into consideration when establishing the associations, confirms this to an even greater degree. For example, it states that: "*The population creating an association should integrate the whole water system (or the relevant part of it) without any gaps, and when establishing the association itself it is necessary that all that use the water resources of this water system (or the relevant part) should participate in the association framework. It is necessary to avoid a situation in which one part of the population, even the majority, that uses water from this water system for irrigating their plots wishes to implement land reclamation work and establish an association for this purpose, but another part of the population that also uses water from this water system does not wish to participate. Under all conditions they will benefit from the land reclamation work implemented by the majority, even that undertaken outside their will, because they all are located within the same system and therefore they should co-operate.*"

On the basis of these considerations, and for the benefit of land reclamation itself, Article 2 stated that associations could only be formed when four fifths of the water users and land users within a certain water system expressed a wish to participate in the land reclamation association, so that "enrichment" of the minority at expense of the majority does not take place. Once the necessary fourth fifths of users agree to form an association, the other water users and land users who did not wish to participate were also considered as members of the association with all the attendant rights and duties because it was deemed that they would also share the overall benefits (Shastal 1923). This interference of the state into the relationship of neighbors (water and land users) for the purpose of establishing an optimal regime of water use over the whole (or part) of an irrigation system was not only acceptable from a legislative point of view but was also rational in essence.

The measures undertaken for developing the water and land reclamation associations had a positive effect. Table 3.3 taken from the work of Prof. Kilchevsky (1927) shows the rather fast growth in the

Table 3.3 Distribution of land reclamation associations by number of household members.

Republic	Number of Members		Up to 50	50–150	150–500	500–1500	1500–5000	5000 plus	Average
	date 1.10.1924	date 1.10.1925							
Uzbek	14,893	113,383	3	3	11	15	13	6	2,223
Kazakh	5,134	31,692	4	9	15	15	5	–	660
Turkmen	2,042	14,919	25	9	2	3	1	1	364
Kyrgyz	5,919	10,067	2	1	3	–	1	1	1,333
Tajik	–	7,441	–	2	7	4	1	–	531
Total	23,988	177,502	34	24	38	37	21	8	1,075

Table 3.4 Growth in the number of associations.

Level of activity	Number of associations			
	As of 1.1.1924		As of 1.1.1925	
	Quantity	%	Quantity	%
Province	3	1.0	4	1.2
Province (area of former governor)	43	14.3	20	5.9
Larger than district (uyezd)	42	13.9	103	30.4
Uyezd (district)	155	51.5	154	45.4
Smaller than a district (uyezd)	58	19.3	58	17.1
Total	301	100.0	339	100.0

members (households) of these associations, especially in Uzbekistan. It is interesting to note the administrative level of these associations. While initially associations were formed at a provincial level, by 1925 associations at a district level had begun to be established because there was a need (similar to today) for co-ordination of the hydrological and administrative borders and interests.

Prior to the national demarcation across Turkistan as a whole, there was a certain shift in priority in establishing a proper system of water use, with top priority for repair of the head waterworks and organization of the management system both at the top and bottom levels.

By the end of 1924, the total area under irrigation, taking into account Bukhara and Khiva regions, amounted to 2,823,000 hectares (Schutter, Dukhovny and Tuchin 2000) – less than 80% of the potential irrigated area according to assessments cited above. It should be noted that the irrigation network over this area was serviced by up to 8,590 people, including 950 people obtaining their salary from the state budget, 537 people in receipt of operational credits and 7,103 people with funds collected from the local population. Thus, water users themselves incurred most of the operational costs. Nevertheless, at some sites in the Zarafshan Valley, Tashkent Province and Fergana Valley, the head waterworks had become ever more complex engineering constructions.

3.4 After Demarcation of the National Republics

A resolution creating new Soviet Socialistic Republics in Central Asia was adopted at a session of the Central Executive Committee of Soviets of the USSR on October 27, 1924. In the beginning the Uzbek SSR and Turkmen SSR were created – the Tajik autonomous republic became an administrative unit of Uzbekistan, the Kyrgyz autonomous province an administrative unit of the Russian Soviet Federative Socialist Republic and with the Karakalpak autonomous province as an administrative unit of Kazakhstan.

Two years later, the Kyrgyz autonomous province was reformed into an autonomous republic and then in 1935 into a Soviet Socialist Republic. In 1929, the Tajik Autonomous Soviet Socialistic Republic (ASSR) was reformed into a Soviet Socialist Republic, and in 1932, the Karakalpak autonomous province was reformed into an autonomous republic within the administrative boundaries of Uzbekistan.

Substantial reform of the higher echelons of the water management system took place in the region. The Turkistan Water Administration was restructured into the Central Asian Water Administration, and was responsible for the implementation of large-scale survey and design works, and constructing projects of inter-republic significance. In each republic, water authorities were established for the purpose of organizing the operation and maintenance of the irrigation systems and preparing the land and water reform that was declared on December 2, 1925. The aim of this reform was to restructure the land-use system. It limited irrigated plots to an area of 3 *dessiatinas* in order to allocate irrigated land to the poor and landless peasants. Table 3.5 is based on data from the People's Commissariat of Land Resources, and shows the distribution of land resources in 1923.

Undoubtedly, there was an aspiration to not only eradicate rich landowners in Central Asia (*bais* and former khan's officers) but also to distribute plots as much as possible to landless peasants (members of the Koshchi union) who actively supported the Soviet power. With this in mind, and given the restricted repartition of rich landowners' land, the focus was on reconstructing the so-called old

Table 3.5 Distribution of irrigated land in 1923.

Province	Percentage of each category of land users (%)		
	Landless peasants	Land-poor peasants (less than 3 *dessiatinas*)	Peasants having 3 or more *dessiatinas*
Fergana	3.4	83.1	13.5
Tashkent	12.5	48.4	39.1
Samarkand	2.1	51.5	46.4

irrigation systems, developing fallow lands and also, in some places, using state land reserves. About 186,400 hectares were confiscated, but by 1928 the overall area redistributed amounted to only 517,174 hectares, shared among almost 90,000 farms.

By that time, land reclamation associations had become responsible not only for on-farm but also partly for inter-farm irrigation and drainage networks, and they had gained in influence and significance. Under the leadership of water authorities, they were engaged in reconstructing the irrigation systems, constructing small head intake structures and check structures, and even developing virgin lands. The rapid growth in the number of land reclamation associations in Uzbekistan after the national demarcation in Central Asia attracts attention. While only 51 land reclamation associations existed in Uzbekistan in 1925 (see Table 3.3), the number of associations had increased to 223 in 1928.

There were some radical changes in the structure of capital investments into the water sector in the region. A typical example was given by Fayzulla Khodjaev (1970), the former chairman of the Council of the People's Commissars in Uzbekistan, who was an outstanding politician but was arrested in an act of political repression in 1937 and killed the following year.

In the late 1920s, as Table 3.6 shows, the federal budget for irrigation increased three times over three years and become the dominating financial contribution. By comparison, the level of contributions by water users was practically constant. These investments financed large-scale works such as rehabilitation of the Djun and Khan canals and the left bank of the Karasu Canals Tashkent Province, the reconstruction of the Dargom, Palvan, Kurtyk, and Payariq canals in Samarkand Province, the construction of the Jilvan Irrigation system in Bukhara Province and many other works. The largest irrigation project was the construction from 1927 to 1929 of the Ravatkhodja Reinforced Concrete Dam on the Zarafshan River (Figure 3.6b), which had a transit flow rate of 1,350 m^3/sec and two water outlets on both riverbanks. Around the same time, the water authorities revisited the ideas in the decree signed by Lenin on the development of the virgin lands (of about 40,000 hectares) in the Dalverzin Steppe.

Table 3.6 Investment on irrigation works in Uzbekistan [in rubles].			
Source	**1927–28**	**1928–29**	**1929–30**
All-union budget	3,908,730	7,994,208	12,522,882
Republic budget	1,713,438	726,289	9,878,858
Local fund	199,653	689,943	2,228,400
Local budget	–	–	1,169,000
Water fund	3,457,566	5,455,640	4,990,000

Figure 3.6b View of the newly constructed Ravatkhodja Reinforced Concrete Dam on the Zarafshan River (Sadykov 1975).

One of achievements of this time was the construction of a few large water projects in the Fergana Valley, including the Kampyr-Ravat Hydro scheme in the Karadarya River. These waterworks were built in 1934 under the leadership, and according to the designs, of the outstanding Russian hydraulic engineer and later Academician, V. Poslavsky, who developed and implemented the so-called Fergana type of diversion structure.

The design of this diversion structure is based on the principle of using the transverse circulation of the river flow to prevent the bed load sediment being fed into irrigation canals (see Figure 3.7).

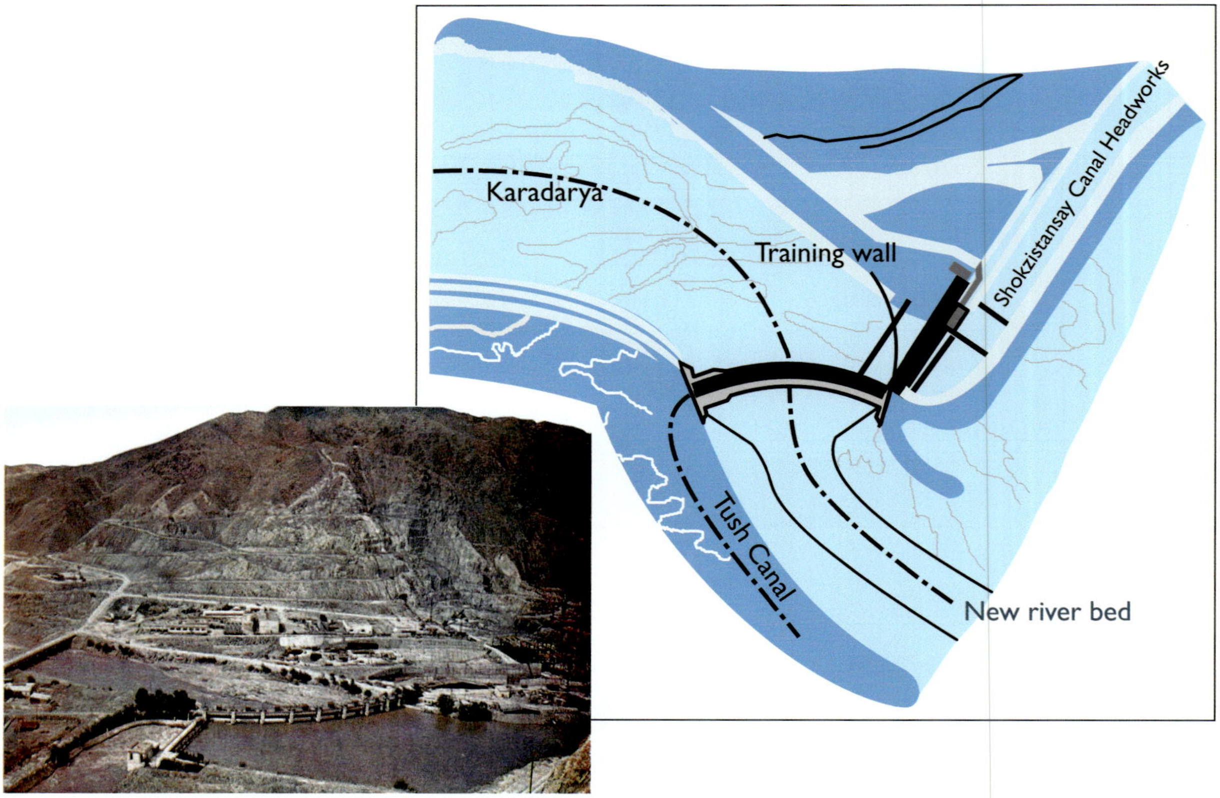

Figure 3.7 Scheme of the Kampyr-Ravat hydrostructure [47, p115 and 48, p 88].

The head regulator is placed on the left concave riverbank opposite to the direction of near-surface currents, and the bottom spillway is opposite to the direction of near-bottom currents that are released through the side galleries of the dam. Monitoring has shown that even under a maximum sediment load amounting to 13 kg/m^3 and a maximum size of sediment particles of up to 400 mm, the bed load does not come into the irrigation canal under conditions of diverting 68% of the maximum river flow. Essential components of this hydraulic structure are a sill in the shape of a curvilinear wall and a sediment-deflecting wall, which increases the factor of safe water diversion up to 0.97. In subsequent years, this type of water diversion structure was widely applied on many rivers in Central Asia, including the Kuiganiar Dam on the Karadarya River, the Akkadarya hydro scheme and the Pervomay Dam on the Zarafshan River, the Gazalkent hydro scheme on the Chirchik River, the Takhiatash hydro scheme on the Amudarya River, and many others.

The creation of the Central Asian Experimental Research Institute for Water Management in 1924 (later known as SANIIIRI – Central Asian Scientific and Research Institute of Irrigation), which for many years was the "factory" of scientific personnel and "bearer" of advanced ideas for the whole region, was one of the great events in the history of the water sector in Central Asia. Prof. Zhurin, the driving force behind the project to construct the laboratory buildings, become the first director of SANIIRI (Figure 3.8).

Figure 3.8 SANIIRI – old building.

At the beginning of SANIIRI the institute had a laboratory of hydraulic research, a hydro-engineering department, department of construction technologies, department of development of irrigation methods, operations and maintenance department, and economics department. Later, as the institute developed, more departments and laboratory facilities were added.

The Central Asian Applied Research Institute for Water Management implemented the first experiments on models of the Ankhor and Burjar straight drop structures (the Chirchik Irrigation System). They also engaged in a study of models of the regulators of the Dalverzin Canal (on the Syrdarya River), of the Yangiaryk Canal (on the Naryn River) and of the Ravatkhodja Dam on the Zarafshan River. Through regular field studies of irrigation systems, researchers developed the theory behind water use planning as well as several practical applications.

One can firmly say that no large hydraulic structure nor large-scale project in Central Asia was subsequently constructed without modeling tests done in SANIIRI, where thanks to the intellect and "golden hands" of scientists and technicians a solid basis was created to deliver scientific and technical progress in the water engineering sector. In particular, the scientific development and practical application of scheduled water distribution within the irrigation systems in 1929–1930 (with a plan based on modeling experiments) can be considered as one of the great achievements of SANIIRI's scientists and specialists in the water sector in Uzbekistan. Between 1931 and 1935, scheduled water distribution was introduced everywhere in Uzbekistan and, later, in the other republics of Central Asia.

The considerable increase in financial investment along with the application of new theoretical approaches to irrigated farming facilitated a rapid growth in the area of irrigated land in Uzbekistan.

According to Khodjaev (1970), this increase in irrigated land reached a rate of 100,000 hectares a year in the period 1927–1930. However, this increase was not based completely on a new engineering base. The first work implemented was the construction of water intake structures and straightening of main and inter-farm irrigation canals. If possible, water diversion structures from irrigation canals with many heads were eliminated by constructing unifying structures. The drainage network consisted of open drainage canals less than 2 m deep, with quite a large intermediate spacing, but land leveling had still not been implemented on most of the irrigated areas. As a result, all these irrigation systems had to be reconstructed by the 1950s.

The final secession of Tajikistan from the Uzbek Republic coincided with the beginning of the construction of the largest hydro-engineering complex in Central Asia, the Vakhsh Irrigation System. The decision to proceed with this project was made in 1927 after successful experiments in cultivating fine fiber varieties of cotton (Kholdjuraev 2003). The project provided for the construction of a diversion structure with a carrying capacity of 154 m^3/sec for irrigating 94,000 hectares, including 25,500 hectares of previously developed land. Construction work was initiated at the end of 1929, and the work included, for the first time in the irrigation practice, construction of a railway and a network of vehicle roads. The construction work used excavators, scrapers and graders (of both Russian and foreign manufacture), as well as the first locally produced ditch excavators. This enormous hydro-engineering complex came into operation in 1933.

These large-scale construction works resulted in the development of virgin lands and many people needed to be resettled into the new regions. Henceforth, any irrigation development that was accompanied by development of virgin lands claimed much attention from the Soviet government and the republican authorities, because it required the creation of new collective or state farms. Collective farms as voluntary co-operative organizations were usually created on land that had been developed for some time. During 1928–1935, collective farms were mainly established on the basis of land reclamation associations. The government assisted a new collective farm not only with credit but also with direct investment to improve the irrigation networks formed by joining together the existing farms that would be combined into the new farm. The land reclamation associations were often reformed into production units of collective farms and the inter-farm irrigation networks required more attention and investment.

State farms, as governmental organizations, received considerable funding for construction of production facilities, irrigation systems and road networks. They organized water use systems and other works related to agricultural production themselves. A typical example of irrigation management at farm level is the Pakhta-Aral state farm in the Hunger Steppe. This farm, with a gross irrigated area of 6,200 hectares in the period 1924–1930, had become a highly productive farm that increased cotton productivity 1.7 times and achieved a total output of about 10,500 tons of raw cotton a year. Its water-saving practice (water consumption was less than 5,300 m^3/ha), scientifically substantiated the crop rotation patterns and application of mineral fertilizers that gave this farm a leading position in the cotton sector of the USSR (Dukhovny et al. 1976). This state farm became a field-based learning centre for many agriculture managers in the arid zones of Central Asia.

The Bayram-Ali state farm in Mary Province (Turkmenistan), the Bayaut state farm in Syrdarya Province (Uzbekistan) and many others showed similar high efficiency indicators for agricultural

production. During this period, special water management bodies established under a Machinery and Tractor Stations[14] contract were responsible for water-sharing between collective farms and individual farmers (water users).

It proved more successful to establish collective farms, based on available land reclamation associations and rural settlements, than state farms. The creation of state farms lagged considerably behind the progress made in setting up collective farms, and work in this respect was mainly aimed at establishing advanced pilot farms rather than establishing large-scale agricultural practice. The government could not create the required preferential conditions for state farms, and so progress was slow. During the post-war period, collective farms were the major suppliers of raw cotton and other agricultural products, and provided most of the agricultural well-being for the rural population.

Weinthal (2002) has argued that in contrast to the reforms of the irrigation sector in the West after the French revolution, the centralized planning framework that was introduced following the Socialist revolution in Russia resulted in an inefficient system or governing water resources and agricultural production which was subject to too much administrative pressure. However, the production indicators of the Pakhta-Aral state farm are evidence to the contrary. When comparing the leading farms in Central Asia with similar collective farms in Israel, it is obvious that, in the 1930s, the productivity and profitability indicators of state farms and collective farms were much higher, and the same can be said about crop yields and irrigation water consumption. According to these indicators, the region only began to lag behind Israel in the 1970s, when Israel succeeded at a national level in reforming the water resources management system based on efficient irrigation methods (with drip and sprinkler irrigation).

Developing the Central Asian Hydropower Master Plan was an essential achievement of the first five-year plans. The Master Plan provided for several projects, including several the large hydropower stations (HPS) – Charvak on the Chirchik River, Farkhad on the Syrdarya River, Toktogul on the Naryn River, Nurek on the Vakhsh River and Tuyamyun on the Amudarya River. These hydropower stations would allow the development of pumped irrigation projects in the Uzbek SSR and in the Tajik SSR. At the same time, the HPS reservoirs would increase the areas under irrigation and improve the sustainability of the water supply in Uzbekistan, Tajikistan and Turkmenistan. Actual implementation of these projects only became possible a few decades later.

The so-called method of people's construction, which in fact rehabilitated the tradition of large-scale participation of local water users in the construction and operation of water infrastructure works (*ḳhoshar* and *ḳazu*) discussed earlier, opened a new page in the history of hydro-technical activity in Central Asia. The construction of the Big Fergana Canal is undoubtedly the most impressive example of this approach. The large main canal intercepted all small streams on the left bank of the Syrdarya River within the Fergana Valley (Figure 3.9) This allowed irrigators to even out the water availability, by adapting to the diversification of high water periods and replenishing in periods of water shortage, at the expense of the water resources of the large tributaries of the Syrdarya, the Naryn

[14] A special organization that implemented mechanized works and services in collective farms on the basis of a contract in the USSR in that period.

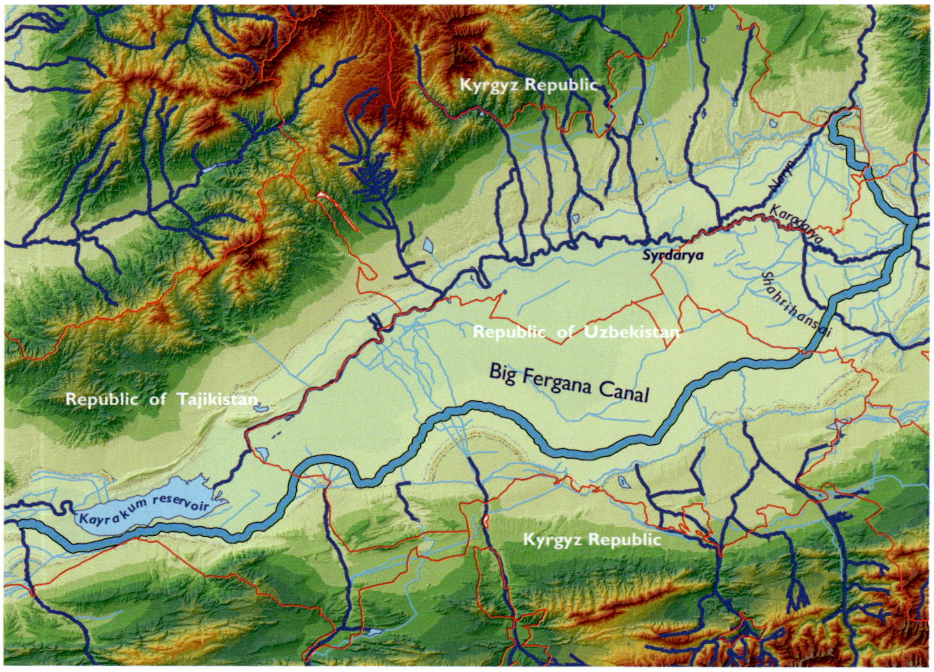

Figure 3.9 The Big Fergana canal scheme.

and the Karadarya. The project was developed under the leadership of three outstanding hydraulic engineers, A. Lebedev, A. Askochensky and V. Poslavsky, and has supplied water to irrigated land in Kyrgyzstan, Uzbekistan and Tajikistan. Construction of this canal started on August 1, 1939. Some 18.2 million m^3 of earth was moved in 45 days by an army of 160,000 peasants assembled from the three countries of the region. All work on the 48 large structures and 275 small structures (42,200 m^3 of concrete work and 1,106 tons of metal work) was also implemented during a three months period (Figure 3.10).

The daily rate of concrete work approached 5,000 m^3 per day in spite of the low level of mechanization. The work used batch mixers with a working capacity of only 400 liters, simple conveyer belts and simple vibrator compaction needles. The leader of Uzbekistan, Usman Usupov, directly lead this construction working from the headquarters of the project. Targets were reached, and the project provided a sustainable water supply for more than 300,000 hectares of irrigated land.

A report about this grandiose achievement in *Pravda* (published December 30, 1939) stated: "*The Big Fergana Canal will be remembered in the history books as one of its most fine pages and as a gem of people's creative work. The initiative of the Fergana collective farms has found broad response among the working people over the whole country and provided an incentive for initiating new and similar people's construction works. The significance of the Big Fergana Canal consists in the fact that it is the expression of the titanic forces of the working people who are ready for new achievements in the form of voluntary socialistic works . . .*"

Figure 3.10 Construction site at the Fergana channel [48].

This initiative became widespread over all of Central Asia. In the same year, the Big Zarafshan Canal was built in the same way in Samarkand Province. In Khorezm Province the Shavat and Palvan canals were rehabilitated, and many new canals were built – the canals of the Shakhrud Irrigation System in Bukhara Province, the Big Gissar Canal in Tajikistan, the Tashkeprik Dam in Turkmenistan and the Ural-Kushum Canal and Tugay Lateral Canal of the Kirov Canal in Kazakhstan.

The construction works, with their atmosphere of people's initiative and selfless labor and with a high collective responsibility and enormous creativity, will rest in the Soviet people's memory forever. Workers, engineers and designers all felt themselves a single part of the organism operating to the maximum intensity of the forces mobilized by the aspiration to achieve progress. It was not necessary to hurry anybody, because each knew his or her place on a construction site and each felt his or her own responsibility for the timely execution of an assignment. These could be geodesic surveys in the burning sun, design work that was often prepared during sleepless nights, earth excavation works, concrete works, or delivery of construction material. All these activities were equally important, and all work had to be completed on time. However, such enthusiasm and initiative, based on the centuries-old traditions of joint activity, were (and to some extend still are) incomprehensible to foreign analysts with a background in western efficiency.

This movement has been described in some books as the "unpaid labor" of 160,000 people (generally Uzbeks) under the supervision of engineers and technicians (generally Russians) (Mamedov 1965). This is how some critics attempt to misrepresent Soviet and Russian history, showing an evidently nationalistic bias. It is necessary to understand the particular atmosphere of the mass impulse in order to realize the spirit and joy of this work. Writers as far back as Rudyard Kipling have noted that "the East is the delicate concern incomprehensible for each …"

Resettling considerable numbers of settlers accompanied the extensive development of virgin lands and hydro-technical constructions. In the beginning, mainly during the pre-Second World War period, this process had a positive effect because of the small plot size awarded to local peasants. For example, 30 new rural settlements and two industrial centers (Bayaut and Mirzachul) were developed in the Hunger Steppe on newly developed virgin lands. In the period 1936–1941, about 36,000 people were resettled in a region reaching from the Fergana Valley mountain districts of the Turkistan Ridge to the neighboring districts of Tajikistan. Some 274 collective farms were also organized on lands prepared for irrigation in the Vakhsh Valley in Tajikistan. The total number of settlers exceeded 28,000 people.

The Great Patriotic War (the Eastern Front of the Second World War) did great damage to further development of the water sector, but at the same time the conflict initiated development in new directions. For example, due to a decline in the supply of flour and grain from Ukraine and the Chernozem zone of Russia, it was necessary to increase the area of irrigated land devoted to cereal production.

The proven method of people's construction works was successfully used again. In 1942 and 1943, as a result of *ḳhoshars,* some 242,000 hectares of intra-scheme fallow land were developed into small plots for sowing wheat and barley. At the same time, a few strategic projects were initiated to produce foodstuffs for refugees from regions occupied by fascists or otherwise affected by the war. Uzbekistan opened the door to more than 1,000,000 people and provided them with jobs in the many industrial enterprises that had been evacuated to this region. For example, there was one aircraft factory, two tractor plants and other sites where military equipment, ammunition and airplanes were manufactured, sometimes under the open sky. It was necessary to supply not only food for workers but also power for these industrial enterprises. The six hydropower stations of the Bozsu Hydropower Stations Cascade were put into operation and the Kattakurgan Reservoir on the Zarafshan River was built for seasonal regulation of the river flow in order to provide a sustainable water supply for Bukhara Province.

Constructing the large Farkhad hydropower complex in only three years (1942–1945) under the leadership of the talented organizer Akop Sarkisov was a great heroic achievement. This multipurpose complex had to supply electric power to the Begovat Industrial Unit, including the large metallurgical works and a cement mill, as well as provide a sustainable water supply for irrigating 500,000 hectares in the Hunger Steppe and 50,000 hectares in the Dalverzin Steppe by gravity. This hydro scheme was the first built on the Syrdarya River (Figure 3.11).

This structure consists of a non-overflow earth dam, 28 m high, and a spillway dam, 25 m high, with eight gates each 10 m wide. The estimated carrying capacity amounts to 4,430 m^3/sec with an emergency discharge of about 5,800 m^3/sec. The head diversion structure of the Dalverzin

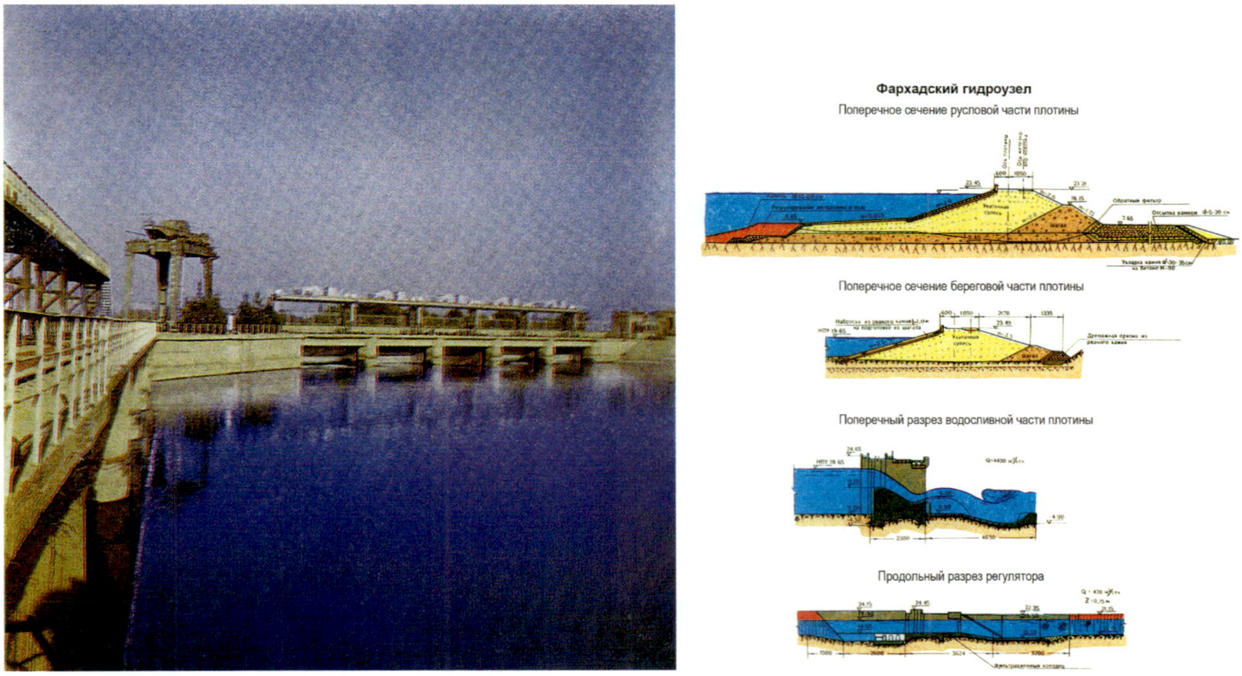

Figure 3.11 Large hydropower complex of the Farkhad HPS [48, 1975].

Canal, with a flow rate of 67 m^3/sec, is located on the right riverbank and the head structure of the HPS diversion canal, with a flow rate of 470 m^3/sec, was built on the left bank. Subsequently this diversion structure has provided water for the pumped irrigation in Tajikistan and diverted water into the South Golodnayasteppe Canal in Uzbekistan and the Dustlik Inter-State Canal (in the past this canal was named after Kirov) that distributes water between Kazakhstan and Uzbekistan.

Improving the irrigation systems characterized the post-war period and the reconstruction of irrigation systems and infrastructure was a continuous activity. The first (weak) actions involved replacing traditional irrigation systems by engineering (or semi-engineering) based irrigation systems. This was undertaken immediately after establishing Soviet power. In 1913 traditional irrigation systems served 95% of the irrigated areas, but by 1932 their share only amounted to 42% (Chelguzov 1934). These activities mainly consisted of reconstruction of the diversion facilities (by replacing brushwood and stone barrages and spur dykes with concrete structures with regulators), construction of drainage networks, and the redesign of irrigation fields with a rectangle configuration suitable for developing fallow and abandoned land.

The transition towards collective and state farms in the agricultural sector required a new stage in the reconstruction and rehabilitation of irrigation systems. Under these new conditions, it was not practical to limit water management activities to reconstructing the main water infrastructure and to leave further aspects of the use of irrigation water to self-reliant farms. Instead, it became necessary to

rehabilitate the whole irrigation network, including the systems down to the field level. Reconstruction of the irrigation systems under these new conditions required:

- construction of a collector-drainage network that would not only remove excessive surface water but also drain waterlogged or salt-affected irrigated land
- construction of settling basins in order to prevent siltation of irrigation systems and to reduce the huge requirement to clean sediment from the irrigation canals (some of which had to be cleaned two times a year)
- the decommission of numerous water intakes within the state and collective farms, and the creation of a configuration of irrigation networks well-adapted to scheduled water supply and water use
- increasing the land use factor and overall irrigation efficiency.

At this time, inspection roads and communication lines also became standard components in the regular design of irrigation systems.

In August 1950, the government of the USSR adopted a resolution on the transition towards new irrigation systems that provided for improved use of irrigated lands, larger irrigated units (fields) up to a size suitable for mechanized land treatment (8–10 hectares), the replacement of permanent field irrigated networks with temporary irrigated ditches, and (most importantly) the leveling of irrigated lands.

To implement this resolution required a few million of cubic meters of earthworks over an area of 1,500,000 hectares over a period of 3 to 5 years. The government allocated more than 12,000 ditchers of various types to Uzbekistan alone. Water management organizations in Central Asia needed to receive about 20,000 earth-moving machines in total. Some 38 earth-moving machinery stations were created in the Soviet republics to service agricultural machinery and to implement earthworks on collective farms and state farms on a contract basis.

Developing these activities in the post-war period under implementation of the fourth and fifth five-year plans (1946–1950 and 1951–1956) was accompanied by enhanced mechanization of construction and repair works, as well as the introduction of machinery for cleaning of the irrigation canals.

Before summarizing developments in the irrigation systems in Central Asia during the initial period of Soviet rule (as described in detail in Sections 3.3 and 3.4 of this book), it is necessary to note that in this period there was a moderate increase in the area under irrigation combined with a rather considerable enlargement of areas under cotton and improved yields of foodstuff supplies (see Table 3.7).

1. The irrigation systems were improved from an engineering point of view over the entire territory of former Turkistan, especially at the level of inter-farm canals, hydro structures and the collector-drainage network.
2. In the Soviet period, the water use system overcame a long reorganization during the transition from community owned and private land use to land reclamation associations. Public participation, although rather a burden for low-income farmers, was maintained to a sufficient degree. The subsequent development of collective farms and state farms formally preserved a democratic

Table 3.7 Trends in the development of the water and agriculture sectors in Central Asia, 1918 to 1960.

Indicator/Index	Year		
	1918	1935	1960
Population (million)	6.21	9.3	14.2
Able-bodied population (million)	1.8	1.86	2.3
Irrigated area (million ha)	3.2	3.5	4.5
Including area under cotton (million ha)	0.54	1.8	2.2
Raw cotton production (million tons)	0.59	2.2	3.82
Water consumption, (km^3/year)	43.2	50.3	61.0
Water consumption (m^3/ha)			
Food consumption per person (kg)	13.5	14.4	13.6
Meat	8	13	18
Milk	56	43	110
Corn	190	290	280
Fruit	32	34	36
Cotton processing in the region (million tons)	0.16	0.21	0.35

form of governance at the farm level. However collective and state farms were actually transformed from community and co-operative organizations into agricultural enterprises under strict government control. This caused a loss of opportunities for public participation in water resource management and the financial planning of agricultural and water management activities, and it excluded use of market principles in irrigated farming.

3. In understanding the significance of sustainable water resource management as the basis for further development and long-term prosperity, the Soviet government took responsibility for the organization and financing of the water sector. In a difficult time for the country, the "people's construction work" involving enormous numbers of people allowed the water sector to develop and helped Central Asia compensate for the vast lost production areas in the Ukraine and in

the Chernozem zone of Russia. During the post-war period, the government established a well-organized framework of water governance and management, which had to ensure sustainable maintenance of the irrigation systems for years to come.

4. In spite of the need to rehabilitate the damaged national economy after the end of the war, during the next two five-year periods the government prepared and started to implement a large-scale program for development of the hydropower sector and for irrigated farming in Central Asia. In this respect, they took into consideration the huge manpower available and the natural resources and aspiration of the local population of the region, who definitely wanted to keep up their agricultural activities and their specific lifestyle.

3.5 The Water Sector of Central Asia – a New Uplift for Agricultural Development

In contrast to other regions of the Soviet Union, the population of Central Asia, especially the people of South Kazakhstan, Kyrgyzstan, Tajikistan, and Uzbekistan, were deeply tied to agricultural production, irrigated farming and a settled rural style of life. In some regions of the non "black soil belt" or even in the Volga region, the Soviet authorities herded peasants into collective farms because former landless peasants (so-called kolkhoz peasants) did not adjust to the way of rural life that was inherent to rich peasants (after dispossession of so-called kulaks). In the republics of former Turkistan, the situation was completely different. The communal joint land use roots were very strong. Peasants were connected with the system of irrigation water use, and the rural population's love and the affection for their clans was strong. Today the funeral of a person of rural origin, even after complete urbanization and even when he has become a high-level official, still takes place on the cemetery where the ashes of his forefather's rest.

After the failure to raise the level of agricultural production all over Russia, the Soviet government had to use the huge potential for developing irrigated farming in Central Asia. Some foreign experts have pointed only to the aspiration of the Soviet government to create a base for producing raw cotton in the USSR, but this interpretation can be easily challenged. Already by the 1960s, the production of raw cotton in the country completely met the Soviet textile industry's needs. The target consisted of developing the agricultural potential for exporting raw cotton to foreign markets and producing other crops for the domestic market. It made sense to use the highly professional agricultural skills of the peasants in the region, where their love and affection for their land makes their labor most productive. In spite of the fact that the industrial sector was already quite developed in Uzbekistan, including aircraft construction, chemical industry, instrument making and mechanical engineering, it was difficult to involve the indigenous population in industrial production. The majority of industrial workers came from the Russian-speaking part of the population. Along with the need to execute the Soviet Union's plans, there was the challenge of providing employment to a rapidly growing rural population. Given that the cost of creating one workplace in the industrial sector was between 16,000 and 35,000 rubles compared with 3,000 to 5,000 rubles for a job in the agricultural sector even under

conditions of highly mechanized agricultural production, the expediency of increasing employment in the agriculture sector is evident.

Focusing only on the developing cotton sector in Central Asia, Western economists (for example Weinthal 2002) have stressed that the tsarist government could not implement these agricultural plans in Turkistan by means of private investment because it could not offer real rights of ownership. The Soviet authorities however "*undertook the economic integration of Turkistan by centralized means rather than with private capital…The restoration and subsequent enlargement of the irrigation systems served two purposes for the Bolsheviks. First, it helped them to meet the economic challenges associated with increasing cotton productivity. Second, it provided the means to appease the peasants in order to bring them under more central control and accelerate the process of collectivization*" (Weinthal 2002).

This gives the impression that collectivization was only necessary for cotton production, but what then can be said about collective farms with other specializations such as horticulture and grain or rice production? Didn't these collective farms form one economic framework along with the cotton-growing farms? In her research Weinthal (2002) did not hide the enormous social and economic significance of irrigated farming including for growing cotton. However, the figures she gives were evidently exaggerated. For example, her statement that 40% of the labor force was engaged in the cotton sector in 1980 is absolutely incorrect. At that time the area under cotton cultivation amounted 1,600,000 hectares (table 3.7), which required some 1,000,000 people. At the same time, the total working population in Uzbekistan was approximately 11,000,000 people. Thus only around 9% of able bodied population of that time was engaged in the cotton sector.

The statement that "cotton policies directly led to the exploitation of the indigenous population and the natural environment" distorts the real impact of mechanized agricultural production including cotton. For example, the average annual salary of agricultural workers in the state farms in the Hunger Steppe amounted to 1,700 rubles over the period 1968–1975 while machine operators made around 2,420 rubles. These figures are comparable to the average salary of urban industrial workers and were higher than the salaries of officials in the cities. The current salaries in the agricultural sector under the present conditions of "a new unsettled capitalist market" shows that, when converted to US dollars, the salary of highly qualified agricultural workers today is about four times less than during the time of the "socialist system", which many Western writers call "the Soviet colonial system".

All over the world, the construction of water infrastructure has created opportunities for new large town planning and this has been the experience existed in Central Asia as well. New large settlements were established and a new industrial complex developed through constructing the Kampir-Ravat Hydro scheme (Topolino Settlement). A new city, Bekabad, came up opposite the old village when constructing the Farkhad Hydropower Station on the Syrdarya River, and other projects had similar impacts. The construction of water infrastructure resulted in the creation of urban components and involved the local population in construction work, transport and other services. These wider socioeconomic targets rather than simply cotton production alone were the basis for implementing the

large-scale program for the development the water sector and irrigated farming in the second half of the twentieth century.

In the USSR, the middle of the twentieth century was marked by the adoption of the Great Communist Construction Program, which was mainly aimed at developing the hydropower sector and at land reclamation work. The great wave of projects, such as the Volga Hydropower Station Cascade (Kuibyshev, Stalingrad and Cheboksar HPS), the Votkyn HPS on the Kama River, Kakhov and the Kanev HPS on the Dnepr River, the large scale irrigation in South Ukraine and creation of the giant forest shelter belt across the Volga steppe region, had to reach Central Asia. In 1950, construction of the Kayrakkum Hydropower Station was started on the Syrdarya River. This had a reservoir with a storage capacity of 4 km^3 to provide seasonal regulation of the river run-off in its middle reach. At the same time, a master plan for the integrated use of the Syrdarya hydropower resources, as envisaged and designed earlier by Alexsandrov, was developed.

Later work on the large and still untouched Amudarya River started. The Karakum Canal that diverts water from this river was the largest waterwork in Central Asia of that time. The initial resolution of the Council of Ministers of the USSR about the construction of the Main Turkmen Canal (from the Amudarya to Krasnovodsk) was adopted on September 12, 1950, when Stalin was still alive. It was planned to build this 1,100 km canal according to the options suggested by Morgunenkov and Tsinzerling, with a water intake structure at the Takhiatash site on the Amudarya. It was proposed that it follow a route bypassing the Sarykamysh Depression and onwards along the ancient Uzboy channel across the Kara Kum Desert. The plan was to construct three hydropower stations, with power systems installed with a capacity of 100,000 kW, as well as to support irrigation of 1,300,000 hectares in the southern districts of the Trans-Caspian plain in Western Turkmenistan. This meant a planned water withdrawal of 350 to 400 m^3/sec with a future increase of up to 600 m^3/sec (Gerasimov and Gindin 1976).

Towards the end of 1950, large-scale works were organized near Takhiatash by the Sredazgidrostroy[15] state trust (established in Urgench) under the Ministry of Internal Affairs. There was a sufficient labor force because three special purpose prisoner camps (2,000 prisoners in each camp) were established near the future construction site. First, the Charjou to Khodjeyli railway was built, with a branch line to Takhiatash. By 1952, there were 10,000 workers at this construction site. However, after the death of the Father of Peoples[16], construction work was suspended and then completely ceased because the State Engineering Commission (SEC) had proposed a second option for the Main Turkmen Canal, with a water diversion near Basarga Village located in the middle stretch of the Amudarya.

Arduous discussions took place even at government level. Representatives of construction organizations argued for the advantages of the initiated option proposed by Morgunenkov. The Academy of Science of the USSR supported these representations, and contended that under implementation

[15]"Sredazgidrostroy" is a typical Soviet abbreviation that means specialized organization for constructing waterworks in the Central Asian region. In Soviet practice, a trust means an organization created for executing specialized construction work.

[16]During one period of Soviet power, Stalin was called the Father of Peoples.

of this first option for the water diversion structures, the river keeps its hydrological parameters unchanged practically down to its discharge into the Aral Sea. In case of the second option for the Main Turkmen Canal (as proposed by M. Ermolaev), the Amudarya could lose its water supply role in its lower reaches, damaging the downstream economy and the natural environment because of permanent water deficits. In addition, Academy members V. Kovda and I. Gerasimov attempted to prove that supplying water into the ancient Amudarya delta, especially into the area with ancient irrigation systems including a water diversion from the Atrek River, would be more efficient from an economic and ecological point of view. Here more valuable fruits and vegetables could be cultivated due to the typical subtropical climate. Many scientists, including the leading hydro-geologist V. Kunin, argued that the new route along the Kelif Uzboy would cause considerable water losses due to the deep groundwater table (30 m below the land surface). However, the route from the Takhiatash site would run across the territory of the ancient delta with a shallow groundwater table (Figure 3.12) (Gerasimov and Kovda 1952a, 1952b).

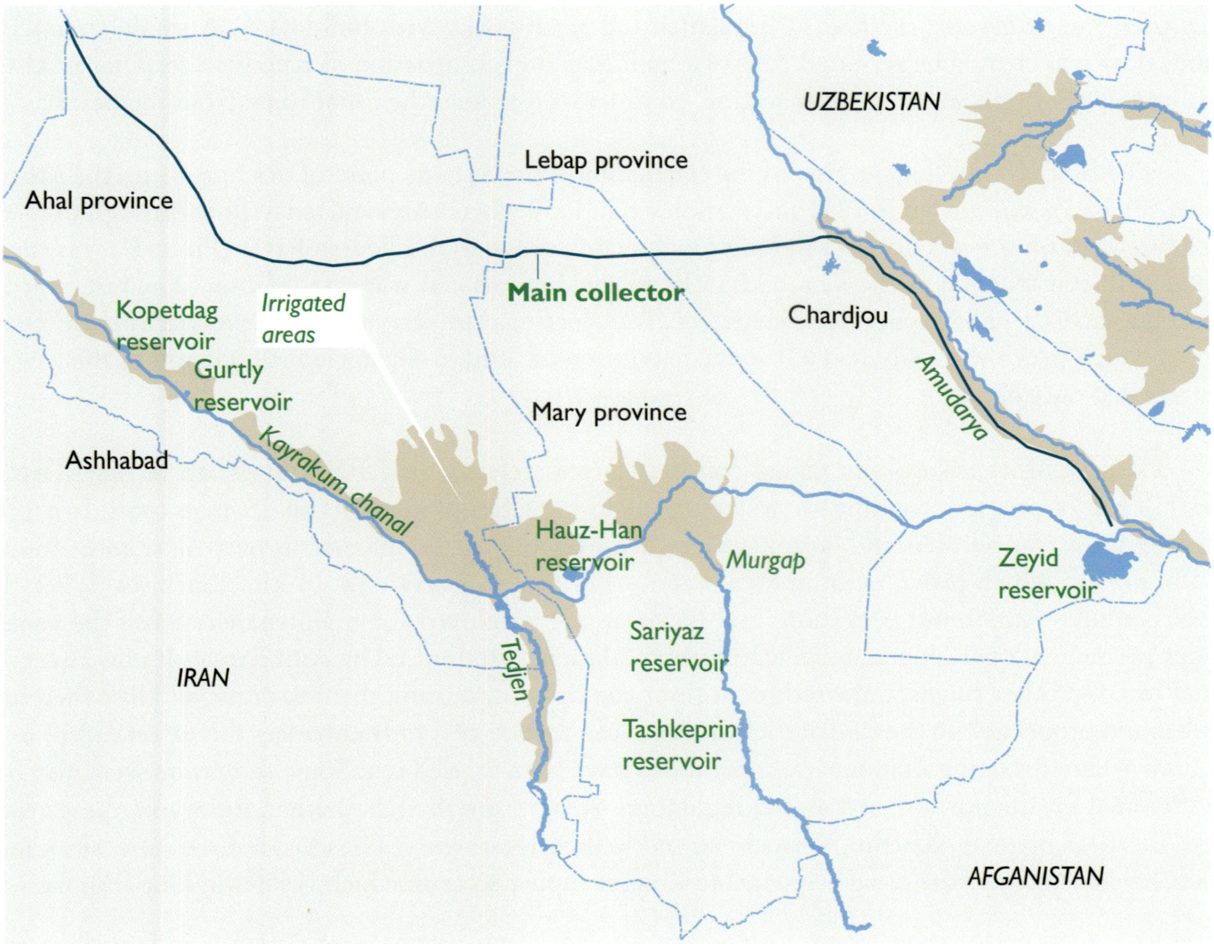

Figure 3.12 Original map of the Karakum canal (Karakum Darya) after Rojenko E, SIC-ICWC.

Two factors eventually played a key role in rejecting the original route. These were the persistence of the Turkmenistan government, which wanted the southern option for the route of the canal, and the antagonism in the Soviet government under the leadership of Khrushchev and Malenkov, who tended to be against all former decisions made by the government under the leadership of Stalin.

By the end of 1953, the Sredazgidrostroy trust had moved into Mary City and established a few construction camps along the route of the future canal. From the very outset, contractors faced great hardship in constructing the canal under conditions of drift sands. The first 52-km section of the canal run through the channel of the ancient Basarga-Kerky Canal and then an additional 48 km run through the Kelif Uzboy. However, at the section downstream from the Kelif lakes neither excavators nor scrapers could make quick progress with the earthworks because those sections of the canal excavated during the day were more than half filled the next morning with sand blown in by the wind. Specialists applied two specific technologies to tackle this problem. First, they organized a water flow through the already completed pioneer channel. Second, they did not use strictly rectilinear sections for the canal but tried to accumulate water within depressions between barchans (sand dunes) and then made cuts linking these depressions, thereby releasing water to create conditions for scouring. At sites of ridged barchan sands, they used the transverse method of implementing earthworks with bulldozers. A stable channel for water flow was created by repeated runs of a bulldozer and compaction of sand by caterpillar tracks. In this way, step by step and section by section, contractors regained the canal route from the desert.

The first 400 km long section of the canal, almost down to its point of discharge into the Murgab River (Phase I), was started in 1954 and completed in 1959 when water started to flow through the canal. In 1960, work on the second 144 km long canal section down to the Tedjen River (Phase II) was started along with construction of the Khauzkhan Reservoir for seasonal water regulation. The reservoir had a storage capacity of 833 million cubic meters. Transferring Amudarya water resources into the Tedjen River, which previously only had a low capacity, allowed land in the ancient floodplain of this river to be irrigated once again.

At present, the Karakum Canal consists of two sections; an 1,100 km stretch of open channel and 110 km of pressure conduit for water supply to the cities located on the Caspian coast. It is probably the most complicated and largest waterwork ever built. The maximum head flow rate amounts to 550 m^3/sec, which provides an annual water withdrawal of about 12–14 km^3 (Sarkisov 1992). The canal is under year-round operation, and dredgers are employed for maintenance. Since the canal is under permanent operation, a stable waterproof sediment film formed by Amudarya silt now covers the canal bed. On other canals, these sediment films can dry out, causing them to crack and lose their anti-infiltration properties. In the case of the Karakum canal, each year only enhances this effect except in the sections where dredging maintenance operations have been carried out. Some reservoirs were also built on the canal for the purpose of seasonal regulation. Apart from the Khauzkhan Reservoir, these are the Zeyd (head), Kopetdag, Kurtlin, Vostochnoe and Balkhan reservoirs. The irrigated area now amounts to 1,000,000 hectares and the canal is navigable along its upper section, which is 450 km long (Figure 3.13).

In spite of the death of Stalin, special emphasis was put on developing the water sector in the sixth five-year economic plan (1956–1960). On August 6, 1956, the Council of Ministers of the USSR issued a

Figure 3.13 Karakum canal (photograph by G. Poltarev).

resolution "On Irrigation of Virgin Lands in Golodnaya Steppe[17] in the Uzbek SSR[18] and in the Kazakh SSR". This resolution proposed the extension of irrigated areas by 200,000 hectares in the Uzbek part and by 100,000 hectares in the Kazakh part of the steppe. The resolution was adopted for implementation under one of the directives of the sixth five-year economic plan, which stated: "For the purpose of further considerable increase in cotton production, it is necessary to create the largest national cotton-growing district by means of implementing the required works for irrigation and development of the fertile virgin lands in the Golodnaya Steppe."

At the start of implementation of the sixth five-year plan, the country made a considerable leap forward in grain production following the development of virgin and fallow land in Kazakhstan and Siberia, which was used for growing especially valuable wheat varieties. At the same time, a rise of productivity was not observed in the cotton sector for a long time – it had remained at around 1.8–1.96 ton/ha since 1948 – and the increase in cotton production had mainly been achieved through the extension of small-scale irrigation projects.

A considerable increase in the area under irrigation in the cotton-growing zone of the country would not only allow the area under cotton cultivation to be extended but would also extend the area under alfalfa and other grasses, and, by introducing crop rotation systems, it should also avoid the threat of monoculture effects. It would also facilitate an increase in the productivity of rice cultivation on long-existing irrigated lands.

Virgin land in the Golodnaya Steppe was selected for development on the basis of a variety of promising factors. The proximity to industrial centers, such as Tashkent, Samarkand and Leninabad,

[17]The Hunger Steppe.
[18]Uzbek Soviet Socialistic Republic.

and the available power resources provided by the Farkhad Hydropower Station located on this steppe were important. Also significant were the existence of the highway and the Tashkent to Ashkhabad railway as well as the possibility for water diversion from the Syrdarya by gravity and availability of sufficient water resources thanks to completion of the Karakum waterworks. In addition, surplus labor resources were available in the nearest densely-populated oases in Uzbekistan, Kazakhstan and Tajikistan, and these could be used for developing this project, which would located on the territory of these three republics. Finally, the Golodnaya Steppe triangle (Figure 3.14) was a region that had been extensively studied, surveyed and covered by scientific investigations at the time.

With the wounds of the war healed, the state was able to allocate considerably more funds and technical means for the quite expensive development of this new virgin land for cotton production. The country could already manufacture the hundreds of excavators, thousands of scrapers, bulldozers, trucks and the other machinery necessary for these large-scale works in the water sector at that time.

This resolution contained two absolutely new concepts. The first was the much wider scope of the project. Prior to 1956, irrigation of virgin lands in general and in the Golodnaya Steppe in particular was planned and implemented by the Ministry of Agriculture of the USSR only within its remit for the construction of water infrastructure. However, the development of these lands together with agricultural development in general was the responsibility of organizations with the individual republics, as

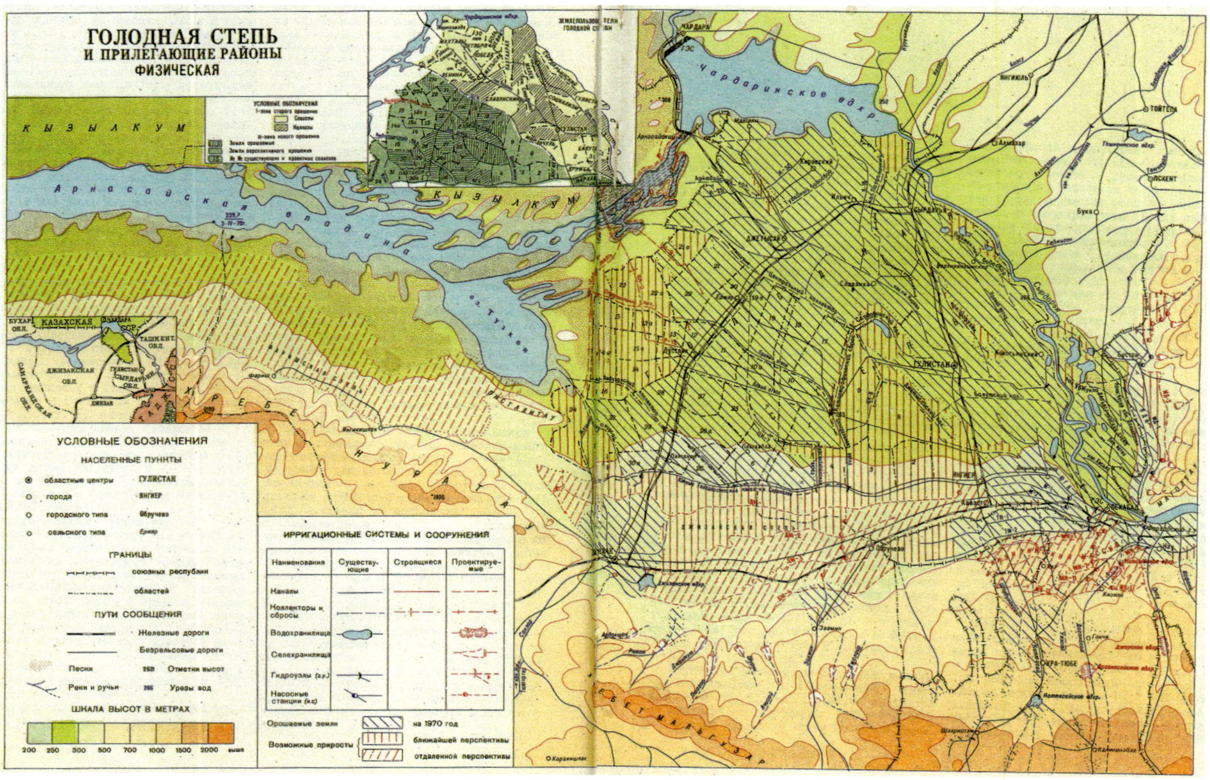

Figure 3.14 The Golodnaya Steppe scheme (after SIC-ICWC).

well as of the farms themselves in accordance with their abilities and resources. It should be note that there existed no real proper co-ordination between these different organizations and agencies, which resulted in a gap between water and agricultural construction, and especially between water infrastructure development and agricultural development. The level of sophistication of agricultural construction was quite low, and was mainly limited to the construction of primitive housing for settlers without roads, power supply lines and social infrastructure. Development and agricultural construction plans were only presented schematically, and this was even the case in the irrigation and land reclamation project reports. The necessary capital investment for the works was not confirmed, and it was often greater than the general cost estimate, meaning that the work was not financed in the framework of the integrated overall project. It is evident that such an organizational approach to development could not produce the required results.

The dynamics of irrigated area development in the Golodnaya Steppe in previous years had shown that, even in the most favorable periods of land development, the average increase in newly developed lands did not exceed 7,000–7,500 hectares a year. More than 40 years would be required to develop 300,000 hectares at such rates. To accelerate the development of virgin land in the Golodnaya Steppe and provide conditions of sustainable fertility required a radical reform of the fundamental principles of organization and technology of construction work based on enhanced industrialization.

It was necessary to avoid underperforming irrigation and land development and to not repeat the mistakes of previous schemes, which mainly related to land reclamation in general and the control of water losses from the irrigation networks in particular. For this purpose, new technical solutions had to be elaborated for design and construction under industrial conditions of work. To meet these challenges, it was proposed to make a shift towards an integrated system of implementation, which co-ordinated the different kinds of construction work (mechanization, irrigation systems) and social infrastructure within the task of developing the new lands.

It was decided to also include in the project the construction of temporary engineering utilities and production centers, and the establishment of companies to produce industrial and construction materials. Other elements related to the development of the virgin land of the Golodnaya Steppe that came with the scope of work included settlements and the production facilities, irrigation and drainage networks of state farms.

It became therefore necessary to build housing and a social and cultural urban infrastructure. Special attention in the resolution of the Central Committee of the Communist Party of the Soviet Union was given to construction of the main engineering infrastructure (vehicle roads, power supply lines, communication lines, etc.), stressing the need for a comprehensive and advanced development of the new irrigation schemes.

The second principle provision in this important and even historical document (which set the scene for all works related to the development of new land) was the policy towards developing this region by means of construction of new state farms. It was only planned to establish collective farms on 20,000 hectares out of the 300,000 hectares, mainly on long-time fallow land within the territories of existing farms. It was planned to construct state farms on the other 280,000 hectares. Some 34 new state farms had to be constructed, including 11 state farms in Kazakhstan and 23 state farms in Uzbekistan. This was an absolutely new approach that marked a significant shift in organization of the agricultural production to the state.

To create incentives to attract construction and agricultural workers to the Golodnaya Steppe, on September 24, 1956 the Central Committee of the Communist Party of the Soviet Union issued a decree that provided for some fringe benefits. All construction organizations were referred to the highest rank of construction works for the remuneration of labor, and an additional 15% wage rise was offered (for living under desert and low water supply conditions). These benefits were introduced for all workers and employees in the newly developed area. High three-month resettlement benefits were allocated to settlers who were sent to this region for a permanent job. Settlers who arrived into collective farms and state farms were exempted from agricultural tax and income tax and from the obligation of supplying their agricultural output for four years.

Initially, during 1956 and until the first half of 1958, work in the Uzbek and Kazakh part of the Golodnaya Steppe was carried out separately by two organizations established in each republic. These were the Glavgolodnostepstroy construction management office in Uzbekistan and the Kazgolodnostep-stroy office in Kazakhstan. A major achievement of this period was the formation of well-qualified work collectives, especially in the Glavgolodnostepstroy construction management office, as well as the construction of Yangier City as the future administrative center for the development of the whole Golodnaya Steppe along with railway stations with large storehouses for construction materials and equipment in Obruchevo City and Jizak City. In addition, new irrigation and drainage methods and technologies were introduced. For example, a separate irrigation system of 4,500 hectares was built on the land of the Bayaut Irrigation Scheme. This consisted of two main lined canals and a subsurface distribution network constructed of asbestos-cement pipes. For the first time, a system of subsurface drains was built over an area of 2,000 hectares to prevent soil salinization. The efficiency factor of this irrigation system amounted to 0.82, which compares well with a typical efficiency factor of 0.55 for the irrigation systems built earlier.

In spite of the support given to all union and republic organizations involved in the project, inter-republican and cross-sectoral barriers severely hindered co-ordinated planning, development, procurement and technical progress in the Golodnaya Steppe. In 1958, managers and specialists in the Glavgolodnostepstroy construction management office, together with the designers who were directly involved in the work related to irrigation and development of the virgin land, prepared proposals for the integrated construction, irrigation and development of the Golodnaya Steppe. They argued for a more rational approach under the co-ordination of the Ministry of Agriculture of the USSR.

Based on these proposals, the Central Committee of Communist Party of the Soviet Union and the Council of Ministers of the USSR issued a joint resolution on July 18, 1958. In this resolution they noted the need to establish enterprises for the construction industry and organize integrated cotton-growing state farms as advanced and highly mechanized enterprises with state-of-the-art irrigation systems as well as with comfortable housing and social infrastructure in the Golodnaya Steppe. This joint resolution directly pointed out that the design and construction of irrigation infrastructure in the Golodnaya Steppe required new irrigation methods, construction of drainage tubewells and subsurface field drains, as well as wide application of lined canals and other anti-infiltration measures. It was equally important to introduce automated facilities for distribution and monitoring of irrigation water and use the most advanced and economic designs and methods of construction. The functions of the Glavgolodnostepstroy construction management office were considerably extended, and from 1958 this organization begun to implement all work according to well co-ordinated and approved plans. These plans included the complete scope of works related to irrigation and development of virgin lands in the Golodnaya Steppe together with the temporary operation and maintenance of all infrastructure. Prior to completing all the key work related to irrigation and development of the land, including one cycle of cotton-alfalfa crop rotation, the newly constructed state farms stayed within the frame of a single management unit together with the construction organizations.

This new set-up for the integrated implementation of all construction work related to developing the virgin land in the Golodnaya Steppe marked a transition from the earlier segmented method of construction towards a diversified complex of interventions implemented according to a rigorous time schedule.

As a result, special consideration was given to matters related to the development of the building industry, construction of roads and railways, power supply systems, water supply, gas supply and heat supply, construction of repair workshops and other facilities necessary for the state farms as well as issues related to irrigation and land reclamation. Other work that was to be implemented at the same time related to land-leveling operations, planting of forest shelter belts, and construction of the social infrastructure (markets, community centers, hospitals, cafeterias, etc.).

The shift from a scattered design of separate small-scale irrigation systems and state farms towards integrated projects that covered all necessary components for transformation of desert areas into irrigated land with highly mechanized agriculture was indeed urgently required. All these components were not sufficiently considered in earlier projects.

For implementing this work, the Glavgolodnostepstroy construction management office was restructured into an inter-republican organization active in Kazakhstan, Tajikistan and Uzbekistan. This office united all directorates of the project under common and specialized construction organizations (trusts) and numerous existing subdivisions for construction and installation works. The storehouses for construction materials and equipment attached to railway stations, power utilities, road-transport carriers, and many other production units also came under its responsibility. The Glavgolodnostepstroy construction management office was also responsible for establishing and developing the organization that would operate and maintain the irrigation and drainage infrastructure, as well as for providing guidance to farms and taking responsibility for agricultural activities.

Asphalt roads, power transmission lines, communication lines and trunk main lines crossed the former desert. Modern comfortable settlements on new state farms emerged in the steppe (Figure 3.15). In 1960, the state farms developed in the Golodnaya Steppe harvested their first yield of

Figure 3.15 Development of the Golodnaya steppe [Sadykow 48].

cotton, vegetables, melons, and corn. Each following year 15,000–20,000 hectares of newly irrigated land came into operation.

However, the water sector of Central Asia did not simply focus its efforts on the Golodnaya Steppe. Considerable work related to irrigation and the development of new land was initiated in the Fergana Valley (Central Fergana), where thanks to the reconstruction of the Big Fergana Canal, which increased its carrying capacity from 100 m^3/sec to 175 m^3/sec, about 36,000 hectares of additional irrigated land came into operation in a region that already suffered from overpopulation.

A number of important waterworks were constructed on the Chirchik River, including the Gazalkent Hydroelectric Scheme with the head structure of the diversion canal linked to the Bozsu Hydropower Cascade. The Kuyluk reinforced concrete dam for the Upper Chirchik Hydro Scheme was also developed at this time. Uzbekgidrostroy, the organization that specialized in construction of waterworks, started construction of future key hydro schemes in the Chirchik River basin with

the Charvak Dam (a rockfill dam 168 m high and a reservoir with a storage capacity of 2.6 km^3, see Figure 3.16).

In 1963, the Chimkurgan Reservoir on the Kashkadarya River (with a storage capacity of 500 million m^3) was put into operation, allowing irrigation of an additional 60,000 hectares in Kashkadarya Province that suffered from water shortage.

Large-scale construction projects were also initiated in the other republics. Kazakh hydro builders started to construct the Chardara Dam on the Syrdarya, downstream from the Golodnaya Steppe. This had a storage capacity of 5.7 billion m^3 and had an emergency spillway that allowed water to discharge into the Arnasai Depression. Construction of two further projects were started almost simultaneously: the Toktogul Hydropower Scheme (a concrete dam 215 m high and a multi-year regulation reservoir with a storage capacity of 19 km^3) on the Naryn River, and the Andijan Reservoir (storage capacity of 1.9 km^3) with a unique arch dam 211 m high on the Karadarya River.

Figure 3.16 Charvak control structure and reservoir (photograph by G. Poltarev).

The 1960s were marked by the high emphasis placed on the development of the water sector in Tajikistan. The excellent conditions for developing the hydropower sector, horticulture, viticulture and cultivation of unique varieties of perfume geraniums attracted the attention of the national government.

In 1961, construction of two hydro schemes was started on the Vakhsh River. These were the Nurek Hydropower Station, a unique hydro scheme that included a reservoir and diversion structure that supplied water to the Dangara Irrigation System, and the Baypaz Hydropower Station. The Baypaz Hydropower Station (capacity of 600,000 kW) was constructed in a record time of only eight years. The dam is 62 m high and it was actually built in one day (March 29, 1968) when, by means of a well-directed explosion (2,000 tons of explosives were used), the Vakhsh River's channel was dammed by a man-made obstruction some 2 million m^3 in volume. This was the first time in the world that this type of explosion was applied in the practice of hydro-technical construction work. This fact was mentioned by the US Minister of Agriculture Robert Long in his welcome book, in which he noted: "*During my trip to the USSR, I understood how some difficult problems are solved here. This considerably outruns the technical solutions developed in the USA. Therefore, we hope to apply your experience in the field of irrigation and hydro-power engineering.*"

The unique Yavan tunnel, a 7.3 km long structure, 5.1 m in diameter with a maximum carrying capacity of 70 m^3/sec, was constructed at the same site. This structure was extremely significant. Although the Yavan and Obikyik valleys were located close to the Vakhsh River which has abundant water, they were separated from the river by the Karatau Ridge. After construction of tunnels, it was possible to irrigate 41,000 hectares in these fertile valleys.

Construction of the Nurek Dam, the world's highest, created the basis for another important irrigation project in Tajikistan. The Dangara irrigation project covers about 100,000 hectares in area. This large 300 m high rockfill dam has created the conditions for generating 14 million kWh of electricity each year and has the head necessary for diversion of water into the Dangara tunnel. The first electric power was generated by this hydropower station in 1974.

In the same period, large-scale work was in progress in the north of this republic. In 1958, two floating pumping stations (Samgar and Khodja-Bakirgan) with a total command area of 19,000 hectares were put into operation using the water resources of the Karakum Reservoir.

A great contribution to developing the virgin lands in the Golodnaya Steppe was made by Tajik builders of hydropower stations and land reclamation specialists. Two pumping stations constructed by the Tajiktselinstroy company lifted water from the diversion canal of the Farkhad Hydropower Station and supplied water into two parallel canals TM-1 and TM-2, allowing the irrigation of 39,000 hectares of lands that were located at a higher elevation than the South Golodnaya Steppe Canal. This land was transferred from Uzbekistan to Tajikistan by a decree issued by the All Union government. The principle of integrated development of new land was employed here as well, with the construction of state farms with all public utilities and irrigation systems. These systems had a high efficiency factor due to the use of subsurface self-pressurized pipelines with a certain degree of hydro-automation. Developing land for cotton production in the Tajik part of the Golodnaya Steppe was challenging given the very complex soil and hydro-geological conditions, but already by 1963 a high yield was harvested. The Tajik

construction organization was also in the process of constructing the Kayrakkum Hydropower station and was also engaged in the development of intra-farm fallow land within the Samgar Irrigation System. More than 10,000 hectares were developed in Kirov, Nau and Asht administrative districts within existing farms and irrigation systems.

The first successes in developing new land in the Golodnaya Steppe encouraged the national government to maximize use of the integrated approach to irrigation and land development in other regions. The integrated method of water infrastructure construction had proved its viability and efficiency. For the integrated irrigation and development of large new areas in Central Asia and Kazakhstan, the General Central Asian Board for Irrigation and Construction of State Farms (Glavsredazirsovkhozstroy) was therefore established in 1963. Branch offices were created for implementing irrigation projects in each region, including Golodnostepstroy and Karshistroy in the Uzbek SSR, Karakumstroy in Turkmenistan, Tadjiktselinstroy in Tajikistan and Glavrissovkhozstroy in Kazakhstan. In 1966, Glavsredazirsovkhozstroy was also charged with implementing works related to irrigation and land development in the lower reaches of the Amudarya and the Karakalpakirsovkhozstroy regional branch was established for this purpose.

The Central Committee of the Communist Party of the Soviet Union and the Council of Ministers of the USSR commissioned a range of additional tasks from Glavsredazirsovkhozstroy. These included the temporary operation and maintenance (during the period of agricultural land development) of irrigation systems, the provision of pumping stations, motor roads, power transmission lines and other facilities as well as overall management of newly established state farms during the initial stage. Its duties also included developing and putting into practice advanced irrigation and drainage methods to ensure rational use of irrigation water and to prevent soil salinization. Finally, this organization was tasked with the organization of research and development activities related to development of the building industry, construction of irrigation systems and establishment of cotton growing state farms .

For the purpose of implementing these tasks, Glavsredazirsovkhozstroy and the national branches had to plan and organize construction works, develop new land and manage the production activities of new state farms. This approach allowed them to carry out the entire scope of work related to irrigation and land development on the basis of integrated plans and common technical policies in a much more co-ordinated way than before.

The Sredazgiprovodkhlopok design and survey institute, which specialized as a design engineering bureau for irrigation projects and worked with associated firms to introduce new irrigation techniques, the directorates of construction projects and the Promstroymaterials trust, which was responsible for the development of new industrial technologies and construction materials, later became part of Glavsredazirsovkhozstroy.

As a result, and for the first time in the region, there was a co-ordinated system of organizations uniting all related works, starting from preparation of a feasibility study and continuing through planning, design and construction work, and eventually leading towards an overall integrated agricultural development effort.

3.6 Integrated Development of Deserts in the USSR – the Golodnaya Steppe and Other Lands

At the time of establishing Glavsredazirsovkhozstroy (the Central Asian Head Directorate for Irrigation of Virgin Lands and Construction of State Farms) in 1963, there was a considerable amount of engineering work being implemented along with construction of the main irrigation and drainage canals, the South Golodnaya Steppe Canal, Central Lateral and Central Golodnaya Steppe Collector Drain. This work encompassed vehicle roads and railways, power transmission lines, communication lines, water mains, etc. In the same period, the highly productive industrial construction materials sector not only provided the necessary construction materials and prefabricated structures but also helped transform construction sites into assembly sites. The construction sector in the USSR came to be based on industrial, mass production principles.

The industrial infrastructure included four integrated plants and four factories with an established capacity of more than 400,000 m^3 of precast concrete and claydite-concrete, 86,000 m^3 of silicate wall blocks, and 400,000 m^2 of gypsum rolling products, 70,000 m^3 of expanded clay, 1,760,000 m^3 of graded inert aggregates, 1,300 km of tile pipes, 42 million bricks, and about 200,000 m^3 of joinery products (Figure 3.17).

Figure 3.17 Industrial enterprises in the Golodnaya Steppe [Sadykov 48].

Five new settlements developed in the Golodnaya Steppe in that period, with enterprises for the building industry, key construction organizations, technical facilities, housing estates, public utilities and social infrastructure.

This approach to the organization of construction work was very new and daring. In contrast to existing practice, work related to the irrigation of virgin land was not started immediately. Initially, part of the allocated funding was spent on establishing a powerful basis for the building industry in the form of facilities and settlements. The aim was to speed up construction and development by ensuring that construction materials and prefabricated components were produced in sufficient amounts, and that labor resources were available in the newly created settlements.

Creating a base for the building industry within a framework of an integrated construction and development organization structure allowed to some work to be undertaken at the plant rather than on the construction site. In addition, the integrated management of the building industry and the construction work created a clear-cut system for the procurement of construction supplies in accordance with a well-planned construction schedule.

After this restructuring in 1961, it became possible to initiate large-scale irrigation and land reclamation work and the construction of state farms, and to organize agricultural production on the reclaimed new land. From its inception to the beginning of the 1980s, this powerful organization implemented work worth 2,442 million rubles. This sum includes 1,812 million rubles spent on civil works and infrastructure. The total capital investment was 2,134 million rubles. Over this period, 310,000 hectares were reclaimed and some 818,000 m^2 of housing facilities built. Other facilities that came into operation included school buildings (for 34,000 students), children's pre-school institutions (for 14,400 students), 97 Production and Commercial Centers (PCC), 1,260 field camps, 6,331 field machine repair workshops, livestock sheds for 15,358 heads of cattle, 5 ginneries, and 29 cotton purchasing centers.

Of the 46 state farms initially planned in this project, four started their agricultural activities in 1961 and ten started in 1965. There were 57 state farms in existence in 1979[19]. Later 19 state farms were transferred to the Ministry of State Farms of the Uzbek SSR and one state farm to the Ministry of Agriculture of the USSR. An additional ten state farms were transferred to the same ministries in 1980.

When analyzing the activities in the different sectors in this integrated system, it is possible to distinguish some specific trends and changes in the scope of work as well as cumulative economic effects over time. Table 3.8 presents some the indicators.

The period between 1956 and 1960 saw the development of different construction sectors – building industry, transport and procurement facilities. During this period, the investment had a very low cost effectiveness. In the same period, construction support facilities, service lines, settlements

[19]The number of state farms in the Golodnaya Steppe increased against the initial project plan since the size of the state farms was reduced from 8,000–10,000 hectares down to 5,000–6,000 hectares in order to improve management.

Table 3.8 Changes in the production of different sectors involved in the integrated development of the virgin land of the Golodnaya Steppe, million rubles.											
Year	**Gross production per sectors**								**O&M at the expense of the budget**		**Total**
	Construction works	**Building industry**	**Transport**	**Procurement**	**Agriculture**	**Services**	**Output processing**	**Trade**	**Water sector**	**Roads**	
1956	1.8	0.3	0.2	0.9	–	–	–	0.6	–	–	3.8
1957	19.4	1.5	2.8	10.6	–	–	–	7.1	–	0.2	41.6
1958	37.3	3.3	5.7	19.1	–	–	–	13.2	–	0.5	79.1
1959	33.0	5.1	5.8	20.1	–	–	–	14.7	–	0.5	79.2
1960	30.8	5.3	5.9	20.4	–	–	–	15.3	–	0.6	78.3
1961	42.0	9.5	7.0	27.1	4.8	0.3	5.1	19.4	0.3	0.6	116.1
1962	46.5	11.8	8.3	32.8	10.0	1.1	11.2	21.1	0.1	0.6	143.5
1963	59.1	14.7	10.1	39.6	22.5	2.7	24.2	24.2	0.1	0.6	197.8
1964	62.2	18.4	11.2	45.8	28.9	2.9	31.6	28.7	0.3	0.6	230.6
1965	72.6	25.6	12.8	50.1	30.0	3.1	34.2	31.6	0.7	0.6	261.3
1966	79.8	30.9	14.3	24.2	38.3	3.7	48.7	37.4	0.7	0.6	278.6
1967	82.5	33.8	15.0	42.4	46.3	4.2	51.3	41.2	1.1	0.6	318.4
1968	102.7	42.3	17.1	72.8	43.8	3.8	50.1	48.3	5.7	0.1	387.3
1969	93.6	46.5	17.7	62.1	59.0	6.1	68.1	55.8	5.1	0.7	414.7
1970	114.0	55.0	20.0	72.1	81.4	7.8	93.3	56.4	5.1	0.9	509.0
1971	123.2	62.4	20.4	77.2	103.3	9.7	112.1	67.8	5.4	1.8	583.3
1972	136.3	69.0	21.8	82.4	125.6	11.3	136.3	77.5	6.3	2.0	668.8

(*Continued*)

Table 3.8 (Continued)

Year	Gross production per sectors								O&M at the expense of the budget		Total
	Construction works	Building industry	Transport	Procurement	Agriculture	Services	Output processing	Trade	Water sector	Roads	
1973	161.4	75.5	25.1	93.4	150.7	21.1	164.2	75.5	6.8	2.5	762.2
1974	169.4	81.8	25.6	85.5	153.1	13.6	170.1	83.8	7.4	2.5	792.8
1975	177.1	85.6	26.4	93.5	144.2	14.5	156.2	85.5	8.1	2.5	793.6
1976	169.0	88.3	29.0	93.1	223.0	18.2	247.1	93.9	8.2	3.3	978.0
1977	169.8	88.3	27.7	99.6	219.6	19.6	241.1	105.0	8.3	3.4	982.4
1978	174.5	88.6	29.4	93.4	253.0	21.2	286.5	116.6	8.1	3.5	1074.8
1979	172.1	88.4	30.1	94.1	279.5	21.1	301.4	129.0	8.4	3.4	1128.3

for construction workers, and main lines were under development. This stage can be considered as a preparation period.

During the period of large-scale integrated construction work, when construction of state farms was progressing at high rates followed by development of agricultural production, construction work was still the prevailing activity. To further develop the virgin land, technical solutions were selected for the specific areas and their performance in practice was monitored. Although the income from economic activities was increasing, this still did not cover the investment allocated for construction work in these years. During this period, similar objects were being constructed at common assumed planning rates because all related sectors had to match the rate of development of the newly irrigated land.

During a period of such intense development of newly reclaimed land, agriculture had to become the dominating economic activity and had to generate the output to recover the investment according to macroeconomic principles. Therefore a major task in the region was to increase the efficiency of key agricultural production and that of all auxiliary sectors. Initially only the leading agricultural sectors which provided the main economic effect (cotton growing and rice growing) were developed, but gradually the need to develop a multi-product agricultural sector increased. The general aim was to

better satisfy the food supply of the population from local production, by, for example, growing fruit trees in newly planted orchards, increasing the percentage of grass lands for forage within a crop rotation system and encouraging other products to be cultivated.

The next period in the integrated system approach, as envisaged by the master plan, required the completion of the construction of all auxiliary facilities and processing enterprises. This meant inspecting and revealing construction defects and project shortcomings with a proper follow-up, and putting a structure in place to maintain sustainable operation of all production units in the integrated system.

Separate time frames that correspond to the initial and finalization period can be distinguished. Large-scale integrated construction work took place between 1962 and 1970, and there was a period of intense development between 1970 and 1978. The period between 1978 and 1986 marks the completion of the integrated systems in the newly developed areas.

Progress, and the efficiency of the integrated approach, mainly depended on the exact and well-organized implementation of various construction work in co-ordination with the development of new land. This required the optimal programming of the construction work and further development of new land.

The construction of each state farm including all necessary works (irrigation and drainage network, land leveling, etc.) was implemented over a four-to-five year period. On average 17,000 to 20,000 hectares of new irrigated land came into operation each year. This meant that between three and five new state farms with the necessary irrigation and drainage systems and production facilities had to be built annually. Each state farm required 200–350 km of on-farm lined canals with laterals (precast elevated flumes or buried pressure pipelines), 200–500 km of subsurface field drains and 50–100 km of collector drains and 40–50 km of on-farm roads. These needed to be constructed, and there was a considerable scope of work relating to land leveling and leaching of salt-affected lands which also had to be implemented.

Enthusiasm, purposefulness and a clear understanding of the need to considerably increase irrigated agriculture efficiency was used to transform the Golodnaya Steppe into a testing area for advanced irrigation and drainage technologies. This work also drew on a combination of the older generation's experience and the creative search of the youth. Here, for the first time in the USSR, several advanced technologies were introduced on a large scale.

- On-farm irrigation networks were made from precast flumes, 0.6 to 1.3 m high, with joints sealed by squeezing foam rubber packings in the assembling process.
- The lining of irrigation canals was assembled by laying precast slabs over a 200 μm thick polyethylene film (Figure 3.18).
- The canal bed was prepared for the placing of polyethylene film by a multi-bucket cross-digging excavator.
- Two new types of machines were used for putting in field drains (Figure 3.19). A trenchless machine with a tricuspid vertical plough (for a drain depth of 2.5 m and with a capacity up to 2 km a day including placing of a sand and gravel filter) and a trencher with digging attachments that allow a narrow trench to be dug (with a digging depth down to 2.5 m and a trench width of 0.20 m).

Figure 3.18 Lining canals with prefabricated concrete on a polyethylene film (Dukhovny 1981, p. 421).

Figure 3.19 Construction of horizontal drainage with a trenchless machine (Dukhovny 1981, p. 338).

- Tubewell drainage systems were used (tubewell diameter of 1000 mm; and well yields up to 35 l/sec) (Figure 3.20).
- Advanced irrigation methods were introduced, including the use of flexible plastic hoses for distribution of irrigation water over furrows and polyethylene pipes.

State farms were established immediately after the completion of construction work and reclamation of a sufficient area of land that was prepared for irrigation and development. The management of state farms had to implement the next immediate requirements:

- to use the land prepared for development as much as possible
- to cultivate crops provided by the project in order to achieve maximum efficiency of agricultural production
- to combine a high level of mechanization in all agricultural work with advanced farming practice
- to introduce and demonstrate state-of-the-art methods of crop cultivation.

The experience obtained in the Golodnaya Steppe has shown that the integrated approach to development of new land allowed state farms to meet these requirements.

Figure 3.20 Vertical drainage system in operation (Dukhovny 1981, p. 112).

When in 1970 the state farms came under the Glavgolodnostepstroy Territorial Construction Management Office, they produced about 370,000 tons of various agricultural goods with a value of some 180 million rubles; by 1980 this had increased to more than 1,800,000 tons of agricultural goods with an estimated value of 488 million rubles.

Vegetable growing, horticulture, melon and gourd cultivation, the livestock sector and poultry were successfully developed along with cotton. Some dairy production was based on the forage resources provided due to the crop rotation schemes in the cotton-growing farms, but large specialized farms for cattle, poultry and fruit production were also developed in the same period. As a result, a former desert provided not only all these agricultural products to the local 120,000 population but also annually exported 220,000 tons of melons and watermelons, 15,000 tons of fruits and grapes, as well as poultry, milk, vegetables and other agricultural products, bringing prosperity to the region.

As Table 3.8 shows, by 1980 the newly-developed irrigated areas in the Golodnaya Steppe had been transformed into a vast economically developed district with a total gross production amounting to 1,128 million rubles. Complying with the integrated approach, the Glavgolodnostepstroy had not only constructed irrigation and drainage systems, numerous engineering services, settlements and cities but also organized the operation and maintenance of the complicated irrigation and drainage systems, community facilities, as well as the water mains, gas mains and power transmissions. In 1961, the Irrigation Systems Administration (ISA) was established for the Operation and Maintenance (O&M) of Irrigation and Drainage (I&D) systems, as well as for planning water distribution through canals under operation and repair. The Land Reclamation Inspectorate was formed under the framework of the ISA. It responsible for monitoring soil conditions, groundwater tables and salinity, as well as for planning and supervising the cleaning of the drainage network. In 1968, the Golsovhozremont specialized trust was established from existing energy and communication departments. There were specialized divisions for the O&M of the sanitary engineering infrastructure (water supply system, heating mains and sewage systems), and two offices were responsible for repair and maintenance of electric networks and substations. The road-repair service was also under the umbrella of this trust. All these organizations were serving the state farms on a contract basis. After three-to-five years of actual operation, this O&M service had turned itself into a well-organized system, which not only provided good maintenance but also improved the existing infrastructure.

The project in the Golodnaya Steppe is a good example of purposeful improvement of natural conditions through the impact of irrigation. The experience obtained in the Golodnaya Steppe has shown that a well-conceived system of management and monitoring can considerably improve the natural and economic conditions of a former desert.

Comparing climatic indicators is evidence of the fact that under a process of transition from "virgin" land towards a modern "oasis" the air temperature of the surface layer may go down as much as 2–5°C. The absolute humidity of the air may increase by 5–11% and average monthly values of relative air humidity may rise 2–2.5 times (from 18–24% up to 49–56%). Soil temperature may decrease by 1.6–3.5°C. Eventually the transition process from virgin land to irrigated fields may reduce wind speeds by 0.6–3.3 m/sec (as measured over a monitored period of three years) when vegetation develops.

The radiation balance and evaporating capacity also became subject to sudden changes. In developing new land, the evaporation capacity is reduced by 10–12% during the first year, by 18–21% during the second year, and by 20–25% during the third year according to the monitoring data from the Golodnaya project (Data of Dukhovny, 1973, (6)).

Proper design of the drainage systems and minimizing water consumption have allowed the sustainable control of groundwater tables and gradual groundwater desalinization, preventing soil salinization, reclaiming salt-affected soils and, above all, achieving a progressive rise of irrigated land productivity. Since 1961, an area of heavily saline soils in the newly-developed zone was reduced from 36,400 ha to 6,800 ha, and another area of medium saline soils was reduced from 21,200 ha to 11,600 ha. The annual volume of soluble salts leached out from the root zone varied between 12.6 tons/ha to 29.6 tons/ha on average, under application of a leaching fraction of 1200–1500 m^3/ha. The efficiency of the irrigation system amounted to 0.81, which was 1.5 times higher than the average in Central Asia at that time.

The investment in irrigation and land development in the Golodnaya Steppe was already repaid by the project's eleventh year before extensive new construction work took place. Later, it generated annual revenues of about 60–100 million rubles for the state, and net revenues were permanently increasing in spite of the ongoing increase of capital investments. The average annual income for a worker engaged in development of new land amounted to 2,947 rubles (averaged over the period of development) against 1,987 rubles in the republic as a whole (equivalent to US$4,911 and US$3,311 respectively). These indicators show the considerable socioeconomic effect that this integrated approach to the development of new land produced.

Comments by high-ranking visitors at that time concerning the integrated method for the reclamation of new land are quite notable. In 1967, Suleiman an acquaintance of the integrated development policy for these irrigated lands and an irrigation specialist himself by education and experience, wrote in the visitor's book: "*This affluence of the steppe is evidence of the success of your struggle against natural forces. The engineers, workers, managers and designers who planned, designed and constructed these structures deserve a high assessment and congratulations. This steppe has been turned into a prosperous area. I wish the Uzbek people, using the grandiose forces of modern science and engineering, to develop the remaining part of this steppe, where about four millions of donums*[20] *are already under irrigation. Let the whole Golodnaya Steppe be turned into a prosperous country.*"

The experience of integrated irrigation and development of virgin land in the Golodnaya Steppe was applied again and improved, first in the Karshi Steppe and then in the Jizak Steppe and the lower reaches of the Amudarya in Uzbekistan. Later, a similar approach was adopted in the Kyzylkum Steppe in Kazakhstan and in some irrigated schemes in Turkmenistan and Tajikistan.

The irrigated area in the Karshi Steppe occupies about 1,000,000 hectares. However no more than 200,000 hectares could be irrigated using local water resources from the Kashkadarya and Guzar rivers. The Esky Angar Canal was built in 1955 and diverted water from the Zerafshan River, which allowed

[20]Donum is Turkish unit of area (1 donum = 1000 m^2)

an additional 40,000 hectares to be irrigated. Therefore, the irrigation of other land in the Karshi Steppe could only be facilitated using the water resources of the Amudarya River. The Sredazgiprovodkhlopok design institute developed a scheme for the supply of irrigation water with a diversion at a site near Kyzyl-Ayak Village through the Karshi Main Canal. This had a carrying capacity of 200 m^3/sec and lifted water up to a height of 132 m by use of six pumping stations. These stations were equipped with unique pump units at an installed capacity of 450 MW. The Talimarjan Reservoir (with a storage capacity of 1.4 km^3) was constructed for seasonal regulation of the water supply diverted from the Amudarya into the Karshi Main Canal (design flow rate of 350 m^3/sec) according to the established schedule of irrigation water supply. Two branches of the Karshi Main Canal supply water to the irrigation scheme, which has a total area of 500,000 hectares (Figure 3.21).

On August 8, 1969, the Council of Ministers of the USSR approved the terms of reference for the Karshi Irrigation Phase-I Project. For developing this territory the Karshistroy Territorial Construction

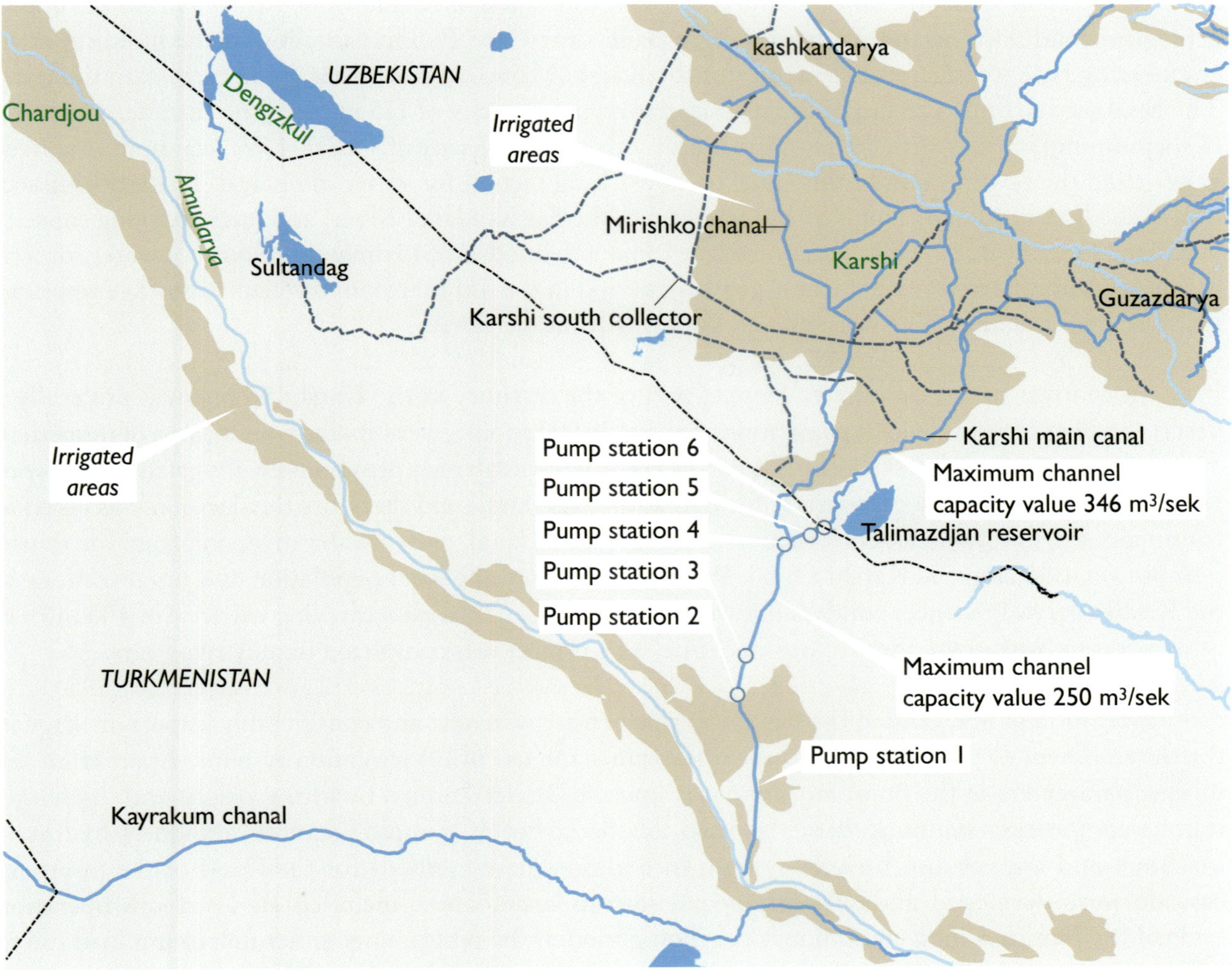

Figure 3.21 Irrigation scheme of the Karshi Steppe (after E. Roshenko).

Management Office was established in the organizational framework of the Glavsredazirsovkhozstroy company. In this region, in contrast to the Golodnaya Steppe, a powerful construction production support base was developed in Karshi City. This unit included construction machinery, a trucking company with 1200 trucks, a workshop for repairing construction equipment, a factory for manufacturing precast concrete (the largest in Uzbekistan with a production capacity of 220,000 m^3/year) and many other facilities.

This was the first time in the history of the national land reclamation works sector that such a concentration of diverse industrial enterprises was gathered on one site and it facilitated a planned increase in capacity. It successfully increased the rate of construction work for a minimal investment. It also realized a reduction in the costs of the O&M of the boiler houses, warehouses attached to the railways, and railway transport. Well-planned settlements for the construction workers were built in the Karshi Steppe prior to the land reclamation work.

The most complicated task was organizing the water supply into the Karshi Steppe. This required a water diversion structure on the Amudarya, and this water then had to be raised 132 m. Favorable geological conditions on the Amudarya right bank near Cape Pulizindan allowed the building of an engineering type water intake structure for the Karshi Main Canal without the need to dam the river. The head section of this 78 km long canal has a carrying capacity of 175 m^3/sec. The cascade, consisting of six pumping stations with 36 pumping units (with a total capacity of 450,000 kW), was built to convey water from the head section of this canal to its working section for water discharge. The working section of the Karshi Main Canal is 87 km long (up to the Kashkadarya River) and has a carrying capacity of 350 m^3/sec. The initial 48-km section of the canal was lined. Approximately 1.4 km^3 of water for irrigation is accumulated during the non-growing period in the Talimardjan off-channel storage reservoir for seasonal regulation. This was built 86 km away from the canal.

In contrast to the Golodnaya Steppe, part of the territory in the Karshi Steppe was not really a desert area. Settlements, with a population of about 20,000 people, were spread over an area of more than 50,000 hectares. About 26,000 hectares could be considered as already provisionally irrigated. There was, however, a need to supply more water for irrigating these lands and therefore the decision was taken to commence the simultaneous construction of the Ulyanov Canal, with a water diversion from the fourth pumping station, and the Karshi Canal. With support from the local population and organizations of the Kashkadarya Province, builders put this 70 km long canal (with a carrying capacity of 100 m^3/sec) into operation within the shortest possible time. The project was completed in only two years.

Irrigation of new land in the Karshi Steppe hence saw a new and considerably important stage in the development of the water sector in Uzbekistan – the use of lift irrigation systems. Apart from the unique parameters of the pumping stations (Figure 3.22), determined by lifting height and discharge, unique engineering solutions were also used for the layout of pumping stations, design of hydraulic machines and the vacuum breaker system. In addition, new methods for the O&M of the pumping cascade were developed and successfully put into practice. These included an 11-month operation cycle of the pumps, a long continuous operation period of the pump units under maximum load (up to 2.5 months), an operation mode under conditions of the close inter-location of the pumping stations, and methods for maintaining a sustainable water intake.

Figure 3.22 Construction of a typical Karshi canal pumping station (Dukhovny 1981, p 85).

Of course, the best practices tested and adopted in the construction of irrigation networks in the Golodnaya Steppe were widely applied in the Karshi Steppe. These included the lining of inter-farm canals with in-situ cast concrete or precast slabs laid over a polyethylene film and installing an on-farm irrigation network made from precast elevated irrigation flumes or buried asbestos-cement pipes.

Along with the irrigation network, a drainage system consisting of field drains (50 meters per hectare on average), on-farm collector drains and main drainage canals was constructed. The South Main Drainage Canal, some 180 km long and with a carrying capacity of 60 m^3/sec, discharged drainage water into the Amudarya. Construction of settlements for new state farms and other production facilities took place at the same time as the development of the irrigation systems and land development. More than 700,000 m^2 of housing, as well as schools, pre-school institutions, clubhouses and movie theatres were built. The settlements were equipped with modern service lines, public utilities and amenities, and green spaces for recreation.

3.7 The Plenary Session of the Central Committee of the Communist Party of the Soviet Union in May 1966 and the Great Leap Forward

During the Soviet period, in a procedure that seem rather strange to western observers, all key decisions were jointly made by the Central Committee of Communist Party and the Council of Ministers of the

USSR. The development program was presented to the congress or plenary session of the Central Committee of the Communist Party of the USSR, and a heated discussion (often programmed beforehand) was organized to demonstrate the democratic principles of decision-making. Sometimes the programs of coming congresses or plenary sessions were sent in advance to the regions where a more realistic (less regulated) discussion took place on the basis of local opinions and wishes. Discussions at the provincial and district level were sometimes rather critical, with opinions reflecting reality, and they were rather democratic and free. After the decision was taken by the Central Committee of the Communist Party of the USSR, a joint resolution of the Central Committee of the Communist Party and the Council of Ministers of the USSR was officially published. The 1970s were especially rich in these important political events.

The USSR Agriculture Intensification Program, with its focus on raising crop yields, was adopted at the twenty-third congress of the Central Committee of Communist Party of the USSR. The integrated land reclamation program was acknowledged as one of the key strands of this intensification of agriculture. Following the resolutions of this congress in May 1966, the program was transformed into a plan of action at the plenary session of the Central Committee of the Communist Party of the Soviet Union. This plan allocated considerable funds for irrigation, drainage, water supply, chemical amelioration works, forestry, construction of storage reservoirs, etc. The plan focused on the tasks needed to create large-scale, irrigation-based grain production districts in the North Caucasus, South Ukraine, the Volga Region and Kazakhstan. Special emphasis was put on further development of the land reclamation work in Central Asia for the purpose of increasing production of cotton, fruit and grapes. To completely meet the national needs for rice, it was proposed to construct new hydraulic engineering systems in different regions including the lower reaches of Amudarya and Syrdarya rivers. As a result of these decisions, the rate of bringing irrigated land into operation in the Soviet Union considerably increased to more than one million hectares per year. The area of irrigated land in Central Asia increased accordingly, from 6,846,000 hectares in 1970 to 7,861,000 hectares in 1980. In Uzbekistan, up to 130,000 hectares of newly developed irrigated land came into operation in these years.

What actually distinguished the new program of agriculture development from the previous initiatives? On the one hand, there was a clear focus on land reclamation as the basis for solving the country's food production problems, but, at the same time, there was also a focus on increasing the production of cash crops.

Figure 3.23P shows productivity growth indicators for agriculture on irrigated lands. These were planned in the framework of this program and were quite typical of the growth achieved. A main focus was to increase crop yields. For example, it was planned to raise the production of rice threefold (though the area of land under rice cultivation would only double from 248,000 hectares to 495,000 hectares). The cotton production area was raised by 40% and 20% in stages and vegetables by 90% and 50% respectively. Forage production had to be tripled.

The need for comprehensive attention to problems in the water and land reclamation sector was another typical feature of this program. The whole national water sector was restructured in order to enhance scientific and technical design principles and logistics. Organizations engaged in land

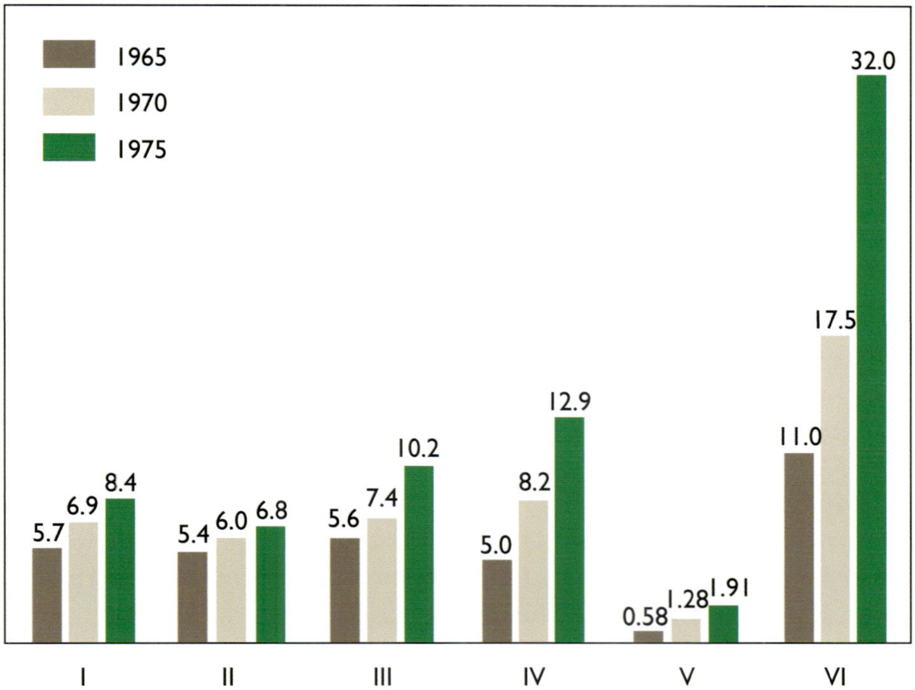

Legend

I	Cotton	IV	Grain
II	Sugar beet	V	Grain (including rice)
III	Vegetables	VI	Fodder (without concentrated food)

Figure 3.23 Growth of key agricultural output on reclaimed lands over the period 1965 to 1975 (in tons, or million tons for fodder crop units, according to an analysis by the authors).

reclamation work and water resource management were united into an independent economic sector under overall supervision of a union-wide ministry. Internally this sector was equipped with the appropriate structure of both republic and union level institutions and organizations. These were some of the key features of the organizational structure of the newly established ministry.

- There were ministries of water resources and land reclamation in all republics, which were responsible for the operation and improvement of the water sector and farming on existing irrigated lands as well as new development of irrigation and water infrastructure. These were funded from the individual republic budgets.
- Large water organizations like Glavsredazirsovkhozstroy were assigned to implement specific large-scale works in the various regions and separate. Glavrissovkhozstroy, in particular, was responsible for irrigation of virgin land and construction of state farms in the rice-growing zones in Central Asia and Kazakhstan. Glavkarakumstroy had the same responsibilities for the command area of the Karakum Canal.

- Specialized organizations for development of rural water supply such as Souzselkhozvodosnabzhenye concentrated their efforts on construction of rural water mains. The main works were in Kazakhstan, in the area of the former rain-fed unused lands that were now being developed for grain production. Water for public and domestic water supply was not available before in these vast areas. To provide a water supply here, two water mains 1,700 km long and each with a carrying capacity of 50,000 m^3/day (the Ishim and Bulaev water mains were the largest in the world at that time) were built. It was planned to construct an additional 30 water mains for rural water supply with a total length of 20,000 km and a total carrying capacity of 360 million m^3 of water per year. In addition, Souzselkhozvodosnabzhenye was engaged in supplying water to pasture areas (about 100,000,000 hectares in Kazakhstan). More than 6,000 tubewells and about 30,000 dug wells equipped with mechanical water-lifting devices, watering places, and storage tanks for emergency water supply were built.

The transformation of the water sector of the USSR into an independent branch of the economic system, under the leadership of the Ministry of Land Reclamation and Water Resources with appropriate organizations in the republics, contributed considerably to the increased investment in land reclamation work that took place in the late 1960s and early 1970s.

A single system was established to standardize the design of works, which was used by the more than 70 specialist design institutions active in this sector. This approach efficiently solved the problems that might have resulted from the intensification of design work for large-scale projects, and put systems in place to ensure close co-operation between ministries and directorates in the sector. Rather than many ad hoc ministries, this approach created not only a water infrastructure but also a socioeconomic infrastructure. Table 3.9 presents the investments made in the water sector prior to and after the planning decisions made by the plenary session of the Central Committee of Communist Party of the Soviet Union in May 1966.

As Table 3.9 suggests, the sector was developing in a rather integrated manner – investments in the water infrastructure were accompanied by investment in apartment houses, public utilities, social infrastructure and production and processing facilities on state farms. While prior to 1966, some 86.7% of the total investment in the sector was for construction of water infrastructure, this fell to 75.9% after the May 1966 plenary session. Total investment more than doubled during the next five-year period. The Table 3.9 shows the increases for 1971–1975 and 1961–1965.

The increased investments in the development of the manufacturing capacity for construction work and O&M services was especially remarkable (by 1971–1975 it was more than 15 times higher than in 1961–1965). The Ministry of Land Reclamation and Water Resources of the USSR had actually become the largest industrial ministry, one that manufactured one million cubic meters of construction materials and thousands of tons of metal ware and pipelines for the water sector. The ministry had created its own specialized engineering industry. Specialized factories manufactured screw-type machines for continuous digging of irrigation canals, drainage machinery (trenchers and trenchless machines) and sets of machines for the lining of irrigation canals.

Table 3.9 Investment in land reclamation and water resources management, million rubles.

Construction work	1961–1965	1966–1970	1971–1975		
			Plan	Increase 1961–1965	Increase 1966–1970
Water infrastructure	3,558	6,702	13,518	3.8 times	2 times
Apartment houses, public utilities, social infrastructure	352	1,129	2,056	5.8 times	1.8 times
Production facilities on state farms	136	436	1,082	8 times	2.5 times
Industrial manufacturing capacity for construction work	60	552	883	15 times	2.5 times
Total	4,106	8,819	17,539		

As a result, by 1970, 98% of earthworks were mechanized. The corresponding figures were 80% for concrete works and 77% for field assembly operations. The increase in the volume of earthwork, concrete works and field assembly operations over the 1966–1970 five-year period was some 215%, 176% and 224% respectively.

The ministry had managed to establish a uniform O&M system for irrigation and drainage systems based on administrative principles of governance. Provincial Irrigation System Administrations (PISA) were established in each province, and District Irrigation System Administrations (DISA) in administrative districts accordingly. Organizations responsible for cleaning of inter-farm and on-farm drainage networks and repair of waterworks were brought under the operational management of the PISA. They undertook work related to the reconstruction of irrigation systems, improving water supply, land leveling and construction of collector drains. However, they were mainly engaged in repairing on-farm irrigation and drainage systems in the collective farms and state farms.

In 1971, following a ministerial proposal, the Council of Ministers of the USSR approved "The Regulations of Operational Service in the Water and Land Reclamation Sector". In accordance with these regulations, the key tasks of these operational services became: i) organizing O&M of inter-farm irrigation and drainage networks to provide conditions necessary for harvesting valuable and sustainable yields on irrigated or drained lands; ii) monitoring irrigated and drained land conditions; iii) maintaining optimal water regimes and implementing measures necessary for preventing land salinization

or waterlogging; iv) organizing rational water use within the irrigation systems; v) providing timely and trouble-free supply of water to collective farms, state farms and other water users in accordance with the established schedules; vi) improving the irrigation and drainage systems and waterworks by raising the technical level of their O&M and widespread adoption of the latest achievements and best practices in science and technology; vii) providing technical assistance to collective farms, state farms and other water users in maintenance of the on-farm irrigation and drainage networks and waterworks; viii) reducing operating costs and achieving high labor productivity; ix) implementing the measures aimed at raising the efficiency of irrigation systems; and x) putting advanced techniques of water application into operation.

Special attention was paid to monitoring the condition of reclaimed lands. Specialized hydrogeological and land reclamation field surveys were established in all provincial irrigation system administrations. They engaged in regular monitoring of groundwater tables and groundwater salinity, the volumes and salinity of drainage water, operating conditions of subsurface field drains, collector drains and tubewell drainage systems, and trends in soil salinization. They were also responsible for preparation of reports containing proposals for developing, repairing and cleaning drainage networks, leaching operations on salt-affected lands, and reuse of drainage water. These reports were submitted to the provincial irrigation system administrations for review and to inform the planning of O&M work.

The operational service, tasked to preserve sustainability of the water system, improve operational services and introduce state-of-the-art technologies depended highly on the level of O&M work in each republic. This O&M work was financed by each republic and was strictly regulated according to approved work plans. Therefore, huge and creative initiatives had to be taken by the ministries and provincial and lower level water organizations in order to implement all measures and works necessary in accordance with the regulations. Finance was always planned according to the principle of "levels achieved". However, this approach limited creative initiatives and did not take the specific character of each water management district into consideration. From this point of view, Weinthal (2002) has correctly stated that "in Central Asia the water resources management system – water sharing, water protection and delivery to water users – was extremely centralized, vertically integrated and exceptionally bureaucratic". One cannot but agree with this assessment, but at the same time it is necessary to understand the mechanism behind this system.

As Weinthal wrote, the water consumption limits for each republic were actually approved by the State Planning Committee of the USSR against the applications made by union-wide ministries based on their production programs. At the same time, the Ministry of Water Resources prepared the plans for water sharing within the main river basins. It did this on the basis of parameters in the integrated water resource use schemes developed for these river basins, as well as by using the annual forecast of the river runoff prepared by the Hydrometeorological Service. It would however be a great mistake to think that republic ministries and local authorities were completely ignored during this process (Weinthal 2002 and others). In fact, water limits and the integrated water resource use schemes themselves were elaborated and approved only after co-ordination and consultation with the republic planning bodies and water management authorities. Based on annual water limits, the individual republican water

authorities shared their water limits among provinces, large irrigation schemes, etc. The criticism that this system of planning lacks market approaches usually comes from people who have never engaged in water resource management, because even in countries with a market economy, market forces are only taken into account when a water auction is proposed. This takes place very infrequently in the western world (the USA, Canada), because even in the west certain obligations have to be meet, such as preserving first user rights, protecting ecological flows, and following the priority sequence of water diversion under conditions of scarcity (as in Europe).

Of course, Central Asia also had to plan for and accommodate the rapid developments in water resources management and land reclamation that followed the establishment of the Ministry of Water Resources. In the 20-year period from 1945 to 1965, some 5.6 billion rubles were invested in water resource management and land reclamation in Central Asia and Kazakhstan. During the following 15 years (1965–1980) 12.3 billion rubles were spent on developing water infrastructure and on land reclamation in Uzbekistan alone. As a result, very large scale water management projects were implemented in the Amudarya river basin. This included the construction of the Tuyamuyun Reservoir, with a storage capacity of 7.3 km^3, and the remodeling of the headwork systems of the Kyzketken and Soviet-Yab canals in Turkmenistan. Also in this period, work started on the development of irrigation in the Khorezm Oasis, where the carrying capacity of the Tashsaka Main Canal was increased to 295 m^3/sec. Most of the irrigated lands in Bukhara Province were shifted to irrigation systems using water supplied through the Amu-Bukhara Main Canal with a pumping cascade. In 1967, Amudarya water was supplied to the newly-irrigated lands in the Surkhan-Sherabad Valley. A combined system of irrigation water supply was created in this valley, based on the Amu Zang Canal sourced by a water diversion from the Amudarya and on the drawdown of water reserves from the South-Surkhan Reservoir constructed on the Surkhandarya River. In Turkmenistan, ongoing work included the construction of the Karakum Canal (Phase III) as well as development of new land in the Khauzkhan and Geoktepe areas located in the south of the republic. New land in the Dashkhovuz Oasis, located in the north, was also developed in this period.

It is significant that during the Soviet period 51 dams over 10 m high (with a total storage capacity of 76 km^3) were built in Central Asia (Figure 3.24). These dams had a storage capacity of more than half the total storage capacity of all present-day reservoirs, and most came into operation between 1976 and 1980. Two of the largest hydro schemes constructed were the Toktogul and Andijan projects.

- The Toktogul Dam on the Naryn River is 215 m high with a storage capacity of 19.5 km^3. The hydropower station's capacity is 1,200,000 kW and it has a designed seismic stability to Richter magnitude 9. The concrete dam has one central section and three abutment sections on each side with a back balancing device for the water. The dam axis is curved towards the tailwater. The water is released through two bottom spillways, each with a carrying capacity of 1,200 m^3/sec, and a surface spillway.
- The Andijan hydro scheme on the Karadarya River consists of a 115.5 m high concrete buttress dam, comprising 33 sections each 25 m wide. The dam has a total length of 965 m and a discharge outlet with a capacity of 250 m^3/sec. The storage capacity of the reservoir is 1.75 km^3. The hydropower station has a capacity of 140,000 kW and has a shaft spillway with a tailrace tunnel.

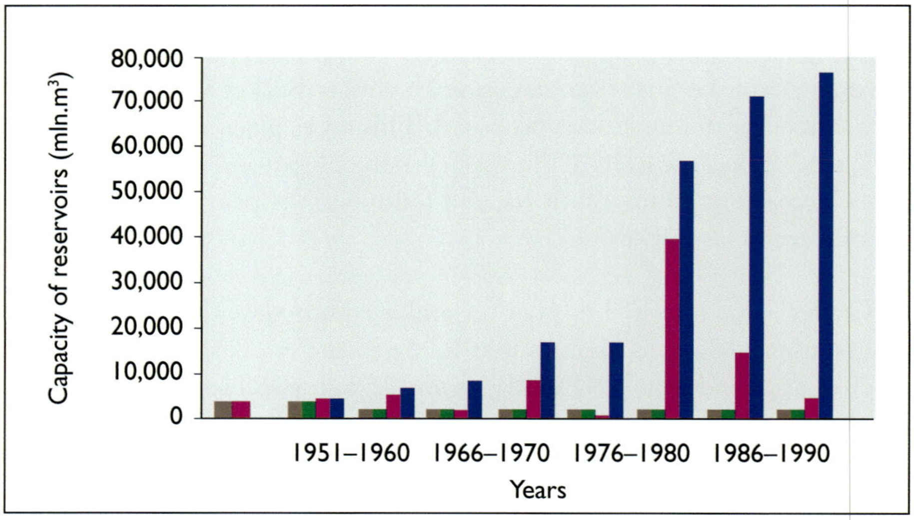

Legend

- Number of dams constructed over 5-year periods
- Cumulative number of dams
- Total capacity of storage reservoirs put into operation per 5-year period
- Cumulative capacity of storage reservoirs

Figure 3.24 Trends in construction of large dam and storage reservoirs in Central Asia according to an analysis by the authors.

The construction of large reservoirs allowed complete seasonal regulation of the Amudarya river runoff. This was achieved through a rather peculiar combination of regulating the river itself and the intra-system flows between the reservoirs that were constructed on the large main canals. The Khauz-khan, Kurtlin and Kopetdag reservoirs, with a total storage capacity of 1.335 km^3, were built on the Karakum Canal. The Talimardjan Reservoir, with a storage capacity of 1.525 km^3, was built on the Karshi Main Canal and three reservoirs with a total storage capacity of 1.879 km^3 were constructed within the system of the Amu Bukhara Canal. The factor of regulation of the Syrdarya river runoff reached 0.92 due to the construction of a few large reservoirs, which guaranteed all-year regulation while maintaining the design schedule of water releases (Borowets et al. 1969).

In this postwar period, the total power potential of Central Asian countries was increased by 7.8 billion kW through hydropower generation and an additional 2.5 million hectares of land came under irrigation. The social implications can hardly be exaggerated. A few million people were resettled to newly-developed irrigated land from densely populated regions (Zeravshan, Fergana, Surkhandarya, Gissar and Chimkent). The newly irrigated land in the Golodnaya Steppe is a typical example. In 1956 the local population amounted to 10,000 people, but today more than a million people live there.

The effectiveness of this work in meeting the original objectives is clearly demonstrated by the considerably increase in the output of all kinds of agricultural products. As Table 3.10 shows, by 1980

Table 3.10 Trends in agricultural output in Central Asia, '000 tons.

Country	Year						
	1940	1950	1960	1965	1970	1975	1980
Cotton							
Kazakhstan	93	62	49	86	305	317	302
Kyrgyz Republic	95	120	126	167	205	208	87
Tajikistan	172	289	399	609	810	906	917
Turkmenistan	211	276	363	553	1,011	1,130	1142
Uzbekistan	1,416	2,282	2,949	3,903	4,895	5,359	5159
Total in Central Asia	1,987	3,029	3,886	5,318	7,226	7,920	7,607
Grain							
Kazakhstan	2,502	4747	18,693	7,595	21,662	27,497	21,321
Kyrgyz Republic	588	434	649	560	1,087	1,371	1,344
Tajikistan	324	209	256	226	203	294	321
Turkmenistan	124	84	40	83	157	257	309
Uzbekistan	615	443	720	637	1,053	2,297	2,450
Total in Central Asia	4,153	5,917	20,358	9,101	24,162	31,716	25,745
Vegetables							
Kazakhstan	172	182	390	583	894	1,040	1,159
Kyrgyz Republic	45	45	84	134	271	342	439
Tajikistan	44	26	49	72	257	333	426
Turkmenistan	32	25	68	121	185	260	291
Uzbekistan	315	164	384	490	1,127	2,076	2,506
Total in Central Asia	608	442	975	1,400	2,734	4,051	4,821

(Continued)

Table 3.10 (Continued)

Country	Year						
	1940	**1950**	**1960**	**1965**	**1970**	**1975**	**1980**
Potato							
Kazakhstan	394	1,158	1265	1128	1,819	1,947	1,952
Kyrgyz Republic	105	135	113	248	300	259	309
Tajikistan	38	36	31	51	101	133	165
Turkmenistan	6	5	5	9	12	14	18
Uzbekistan	114	113	165	170	195	216	305
Total in Central Asia	657	1,447	1579	1,606	2,427	2,569	2,749
Melons							
Kazakhstan	–	–	–	–	–	–	308.5
Kyrgyz Republic	36.1	–	–	–	–	83	85.9
Tajikistan	37	–	–	36	53	86	132
Turkmenistan	–	–	–	–	–	–	201
Uzbekistan	332	153	270	363	549	787	1,046
Total in Central Asia	–	–	–	–	–	–	1,773
Fruit							
Kazakhstan	–	60	70	95	279	396	475
Kyrgyz Republic	42	32	34	76	106	244	240
Tajikistan	170	75	84	212	240	423	370
Turkmenistan	–	17	28	40	60	66	77
Uzbekistan	268	295	287	423	696	1,014	1,198
Total in Central Asia	–	479	503	846	1,381	2,143	2,360

the production of raw cotton had increased by 2.6 million tons or by almost 50% on 1965 levels, cereal output more than tripled, and vegetable, potato and fruit output was up by 2.7, 1.5 and 3 times respectively.

Many international publications have been devoted to the Soviet development of the Central Asian water sector. These works mainly repeat the thesis first presented for judgment by a world audience by Micklin (1991) and then by Michael Glantz (1998). They say that "cotton monoculture" was to blame for the later water management problems and that the Ministry of Water Resources was set up specifically to meet the need for raw cotton. Even Erika Weinthal (2002) who objectively attempted to assess the development of the water sector in Central Asia during the Soviet and post-Soviet period could not resist writing that "the political economy of Soviet Central Asia becomes synonymous with cotton monoculture; yet in order to gain the compliance of the indigenous population, the authorities concurrently needed to invent a mythology of 'king cotton' in which cotton was equated with the 'blood of life for Central Asia'". One cannot but agree with this assessment of the role of cotton in the life of the Central Asian republics. However, any assessment of this period needs to made not only on the basis of a perceived monoculture but on an understanding of the times and the relations between economics, production, social welfare and the environment.

The need to develop irrigation and to develop the water sector as a whole required a Ministry of Water Resources – this was a demand made at the time – not only for producing raw cotton but also for producing rice, cereals, vegetables and fruit (as Table 3.10 confirms). By 1933, the cotton sector already met the internal demand for raw cotton from the textile industry of the USSR *(Cotton growing in the USSR,* 1972). However, central government had to solve the employment problem of a rapidly-growing rural population with maximum economic efficiency. It found that cotton was the most profitable crop. If the export price for cotton fiber was kept at an assumed level of US$1,700 per ton, it became clear that the cotton producing sector provided the highest direct profit for the federal government.

Another misconception that needs challenging is that cotton consumes a lot of water compared to other crops. Table 3.11 shows the average water consumption per hectare and per ton for different crops in Uzbekistan. It is clear from this table that cotton consumes (through evapotranspiration) the same amount of water as corn and somewhat more than wheat, orchards and vineyards, but much less than many vegetables and alfalfa. However, cotton requires quite high conjugate expenses, both for cultivation (manufacturing specialized machinery and implements, and production of fertilizers and agricultural chemicals, etc.) and processing (fiber to yarn to textile fabrics, and seeds to raw oil to refined oil). The labor input ratios for production steps I, II, and III (preparation; production; processing) equal 0.12, 1.0, and 2.56 according to net income. "This economic chain" was very efficient, both from the point of view of generating state revenues and from the point of view of creating jobs. The intensive development of cotton growing in Central Asia help to create a powerful manufacturing sector in specialized agricultural machinery, which produced 80,000 tractors, 21,600 mechanical cotton pickers, more than 650,000 cotton sowing machines, as well as cultivators, levelers, rippers and other processing machinery. A considerable contingent of highly skilled producers of agricultural machinery was established in Central Asia, especially in Tashkent, Uzbekistan.

The cultivation of cotton led to a high degree of mechanization in the agricultural production process. Between 1970 and 1975, all sown areas in the Syrdarya, Tashkent, Jizak and Kashkadarya

Table 3.11 Rates of water consumption per unit area and per ton of different crops produced (estimated by G. Stulina and G. Solodky using the CROPWAT program).

Crop	Net irrigation rate [m^3/ha]	kg/m^3
Cotton	5000	0.6–0.8
Wheat	4600	0.8–1.0
Corn	4900	0.8–1.5
Alfalfa	7600	0.6–1.5
Cabbage	5700	1.4–7.0
Onions	5700	3.2–7.0
Rice	18000	0.33
Orchards	5100	–
Vineyards	3500	24.0

*All data for Fergana Province (District V zone of crop water requirement).

provinces of Uzbekistan were cultivated using specialized machinery and some 70–75% of the cotton yield was harvested using mechanical cotton pickers. This was considered to be very important social progress in the regional agricultural sector.

Although writers have pointed to the "adverse effects of monoculture" and "environmental problems" in Central Asia, there are many similar examples all over the world. One should compare the situation in California and other western states in the USA. The Colorado River was turned into an ecologically dead stream, very similar to the Syrdarya. The USA is lucky that the Colorado flows into the Ocien rather than into a land-locked water body like the Aral Sea. There was also a catastrophic drop in water levels of Mono Lake in California. In fact, Mono Lake is to be considered a "younger brother" of the Aral Sea in view of their similar environmental disasters. Today's assessments are the result of new ecological insights that date from the 1980s and which changed views all over the world, rather than the attitudes towards the development of a cotton monoculture that were commonly held at the time in the postwar world. Even "the Soviet System", in its twilight years, established a State Committee for Nature Protection on January 7, 1987.

This section of the book has stressed the purposeful and creative activities of the Ministry of Water Resources and the Soviet government as a whole in the ten-year period that followed the May 1966 plenary session of the Central Committee of Communist Party of the Soviet Union, however it is also necessary to mention the key shortcomings that eventually resulted in growing problems for the water sector and for the life of the entire country and region.

1. Management of the water sector paid considerable attention to investment in new construction, focusing on physical indicators and the reconstruction of irrigation systems, rather than analyzing problems in respect of water delivery to consumers, the effectiveness of water use, and the need to reduce water losses. In spite of the integrated structure in the water sector, managers first attended to irrigation and drainage and ignored (or paid less attention to) the need to organize sustainable and reliable water supply, river basin management (especially of trans-boundary rivers) and, especially, the environmental aspects of water resource management (water resource protection, environmental flows). The impacts on the environment (not only the Aral Sea but also the river deltas) were grossly underestimated. Co-ordination with other economic sectors (hydropower, navigation, etc.) was conducted on a union-wide level rather than on a river basin level.
2. Early on the water sector did not develop the financial and economic instruments to support its own existence and governance. The problem was not only that no payment was required for water or services related to water delivery. The water management sector did not set up appropriate financial and contractual agreements between water organizations and water users, which could have regulated intra-sectoral and cross-sectoral relationships in order to ensure the sustainable operation of the water sector. The water sector relied on large budget allocations, and it immediately collapsed after the transition towards market relations in the newly independent states. This does not mean that the newly independent states completely refused to finance the water sector, but both water organizations and water users were not ready for a new economic environment.
3. The sector invested into water infrastructure without monitoring that planned indicators for water saving, crop yields and final outputs were achieved.
4. Stakeholder participation was limited to the leaders of the republics, provinces and ministries, and the opinion of water users was only taken into consideration incidentally (for example, in the Golodnaya Steppe) (Dukhovny 1976).
5. The national governments of the newly independent countries could not maintain and manage the irrigation systems after the collapse of the USSR. As a result, the amount of irrigated land in the complete territory of the CIS countries has reduced by 6.5 million hectares in recent years.

3.8 Climbing a Road that Goes Down

Seen from the outside, and with reference to planned gross indicators, the national economy was developing. However, some rather severe counter indicators accompanied the increase in the land area under irrigation and the growth in agriculture production. The key problem was the growing water consumption and the consequent exhausting of the water resources of the region.

The first wake-up calls were heard in 1974 and 1975, when two successive dry years caused a critical reduction of river flows. These were natural events that were assumed to occur only three times over a 100-year period! In spite of the additional use of 2 km^3 of return waters and some 1 km^3 of groundwater (some additional 1300 irrigation tubewells with a discharge of up to 40 l/sec were drilled) and enormous efforts directed at controlling the water shortage, only the production plans for raw cotton and wheat were implemented. Production of all other crops, especially secondary crops, were lower than planned. In addition, preventive leaching was not done over an area of more than 1,500,000 hectares, which caused a shift from seasonal soil salinization to a permanent situation for an area of more than 1,000,000 hectares. Even in the newly developed area in the Golodnaya Steppe, in spite of the intensive drainage of irrigated land and an optimal water and salt regime applied in the root zone, the amount of salt in the aeration zone increased by up to 7 ton/hectare each year during the two years of drought. By comparison, the salt content in the soils decreased by some 5–15 ton/hectare annually in years with normal water availability. In some farms, where intensive salt exchange with artesian waters occured because of existing hydro-geological conditions, the annual accumulation of salts in the aeration zone exceeded 15 ton/hectare resulting in the complete disappearance of non-saline soils (on soil maps they were reclassified into the category of slightly saline soils.)

It should be noted that the foundation for intensifying the water deficit was laid in the planning documents for future regional development (schemes for integrated water use in the Syrdarya river basin for 1975–1978 and in the Amudarya river basin for 1978–1980 respectively), which were drawn up by the Sredazgiprovodkhlopok Institute and approved by the State Planning Committee of the USSR (Sredazgiprovodkhlopok Institute, 1974). Both schemes ignored the need for water not only in the Aral Sea but also in the Pre-Aral (coastal) regions – the deltas of both rivers on their way to the Aral Sea. The schemes provided for sanitary and environmental flows of 3.1 km^3 and 3.2 km^3 respectively, while the water requirements of both deltas were 8.6 km^3 for the Amudarya delta (Schutter et al. 2000) and 6.82 km^3 for the Syrdarya delta together with the Small Sea (Weidel et al. 2004) according to present-day estimates.

The planned area of irrigated land in both river basins was based on the assumption that comprehensive reconstruction, which would raise the efficiency of irrigation systems and reduce water consumption per unit irrigated area, would be implemented in parallel to developing new irrigated land. In fact, this enabled the republic governments, the provincial water organizations and also water users to ignore the environmental needs for water and divert almost all water resources from the rivers to the fields, releasing a considerable amount of that water into collector drains due to inadequate water management. It is quite instructive to compare the unit water consumption for irrigation in different provinces in Uzbekistan. Table 3.12 shows that excluding Jizak, Syrdarya, Kashkadarya, and Samarkand provinces – areas that were developed using integrated approaches and where the unit water consumption for irrigation did not exceed 14,000 m^3/ha, in other provinces the unit water consumption for irrigation varied from 18,000 to 36,600 m^3/ha.

Table 3.12 Unit water consumption for irrigation in different provinces of the Uzbek SSR, 000' m^3/ha (data of the Ministry of Water Resources) (Dukhovny et al. 1985).

Province	1970	1975	1976	1977	1978	1979	1980
Karakalpakstan	28.8	32.5	30.6	28.6	36.9	33.1	36.6
Andijan	12.8	13.7	14.7	14.5	15.8	19.2	19.1
Bukhara	14.7	15.0	15.1	15.4	18.2	18.2	18.1
Jizak	–	9.0	12.2	11.7	10.4	9.5	9.3
Kashkadarya	10.5	10.7	15.2	12.7	14.2	16.1	13.5
Namangan	–	16.3	11.8	15.9	17.8	20.6	18.2
Samarkand	10.6	8.5	9.1	9.6	12.1	12.2	10.3
Surkhandarya	20.8	14.6	16.4	14.9	17.2	20.4	16.5
Syrdarya	11.9	10.0	11.6	12.6	12.6	12.2	13.7
Tashkent	13.3	13.6	12.9	15.4	15.1	15.9	14.7
Fergana	15.9	15.6	15.9	17.8	17.6	20.4	19.2
Khorezm	30.5	28.8	29.3	25.2	36.9	32.0	33.0
Uzbek SSR (average)	15.4	15.1	17.1	16.0	18.2	19.9	18.2

Let us therefore compare the planned and actual rates of reconstruction of the irrigation systems in the period of 1965 to 1980.

Although according to the data presented in Table 3.13 the plans for reconstruction work were exceeded, in fact the reconstruction of irrigation systems was unco-ordinated and non comprehensive. Instead only indicators related to three different activities were monitored which were water availability, the condition of reclaimed land and land leveling. According to the planned comprehensive reconstruction included in "the schemes of integrated water resource use", the efficiency of irrigation systems had to be raised up to the level achieved in the Golodnaya Steppe (efficiency coefficient 0.75–0.78). The amount of water used for leaching operations and desalinization of irrigated land had to be reduced and advanced irrigation methods had to be introduced. In fact, reporting was based on the physical work implemented rather than actual indicators achieved. The statistics showed that the planned physical work was fully implemented or even exceeded, but the actual efficiency of this work was considerably lower.

Table 3.13 Planned* and Actual** Rates of Reconstruction of the Irrigation Systems.

States	1976–80	1981–90	1985–90
Kazakhstan	73.4/26.0	152/117.5	163.4/118.9
Kyrgyzstan	–	42.3/17.75	44.5/21.2
Tajikistan	29.6/16.2	92.6/20.55	234/62
Turkmenistan	18/11.0	200/83.6	200/85.0
Uzbekistan	92/46.0	312/130	783/246
Total	213/63.2	1012/239.15	1414.9/533.1

*Planned data are taken for the level of reconstruction of hydraulic complexes with a cost of $2,500/ha from the *Complex scheme of water use in Syrdarya and Amudarya*.
**Actual data are from the *Statistic view on development of reclamation of land in the USRR, 1966-1985*, Volume 1 and 2, Minvodhos USSR, Moscow 1986, and some from the *Review for 1985–90*. Actual data are based on the statistical report on transfer of quantities by the WUAs to the meta database and on the basis of a transfer coefficient for drainage of K–0,35; network improvement K–0,2, and land levelling K–0,12. This coefficient is based on the cost of each type of modernization relative to the cost of rehabilitation of the whole complex.

Table 3.14 reviews one of three of the monitored indicators – notably the one that relates to the condition of reclaimed land. The statistics show that over 973,300 hectares of irrigated land were reclaimed in Uzbekistan between 1971 and 1980, but the question is what really happened to the condition of this land in this period? In 1970, there were some 1,374,700 hectares of saline and waterlogged lands in the republic, and by 1980 this area had reduced to 1,101,100 hectares. This means that only 273,000 hectares were reclaimed in the decade (less than one third of the area presented in the statistical report). In 1966–1980 some 2.7 billion rubles were spent on improving the irrigation systems in Uzbekistan, but the unit water consumption on irrigated land increased from 15,400 m^3/ha in 1970 to 15,100 m^3/ha in 1975 and to 18,200 m^3/ha in 1980. In the same period, surface main field drains and collector drains were designed and constructed to control soil salinization and water logging, resulting in increased volumes of drainage water discharge and reducing the effect of leaching operations. Note that about 15 km^3 of water in Uzbekistan and about 24 km^3 of water in Central Asia as a whole are consumed annually for leaching operations on salt-affected irrigated lands.

Next to the increase in unit water consumption in this period there was an excessive removal of soluble salts from the soils (up to 50 ton/ha) and further deterioration of the water quality in the rivers. Water salinity in the Syrdarya increased from 0.5 g/l up to 1.3 g/l, and in the Amudarya River from 0.7 g/l to 1.2 g/l on average (see Figure 3.25).

There were several main reasons why the unit water consumption increased and why the planned decrease was not achieved.

Table 3.14 Areas of irrigated lands affected by salinization or waterlogging that needed to be reclaimed (data of the Ministry of Water Resources as of November 1 of each year) in '000 ha.

Province	1970	1975	1976	1977	1978	1979	1980
Karakalpakstan	188.9	151.5	164.1	168.1	151.9	160.9	166.3
Andijan	185.0	128.2	131.9	124.6	123.4	120.9	117.7
Bukhara	102.7	119.8	120.0	121.2	121.0	113.6	122.4
Jizak	–	12.1	20.0	29.3	27.2	37.1	57.5
Kashkadarya	6.2	71.6	97.0	107.7	114.2	126.9	132.2
Namangan	184.0	88.5	78.0	177.5	77.6	72.1	70.0
Samarkand	186.3	106.2	103.7	101.1	97.8	95.0	81.0
Surkhandarya	19.8	65.0	88.1	85.1	85.4	80.6	71.7
Syrdarya	184.6	33.4	44.9	34.9	38.0	42.6	36.7
Tashkent	113.3	87.5	97.9	94.6	91.9	81.9	89.8
Fergana	103.5	91.4	89.6	87.6	84.0	79.0	74.9
Khorezm	100.4	79.4	81.2	79.5	79.7	77.7	80.3
Uzbek SSR, on average	1,374.7	1,034.6	1,116.4	1,111.2	1,092.1	1,103.3	1,101.1

- Up to 60% of all irrigated land that came into operation with government financing was not of an advanced engineering level with regard to the efficiency of the irrigation and drainage systems and water application techniques.
- The work was unco-ordinated. It was purely aimed at raising the level of water availability. The reclamation of irrigated land often only dealt with certain on-farm operations and failed to produce any real savings in unit water consumption due to comprehensive and constant reconstruction requirements.
- Nobody in the regional republic governments was responsible for introducing the water application techniques required to improve water resources management at the field level, nullifying all efforts aimed at developing the irrigation and drainage systems. Work was implemented at

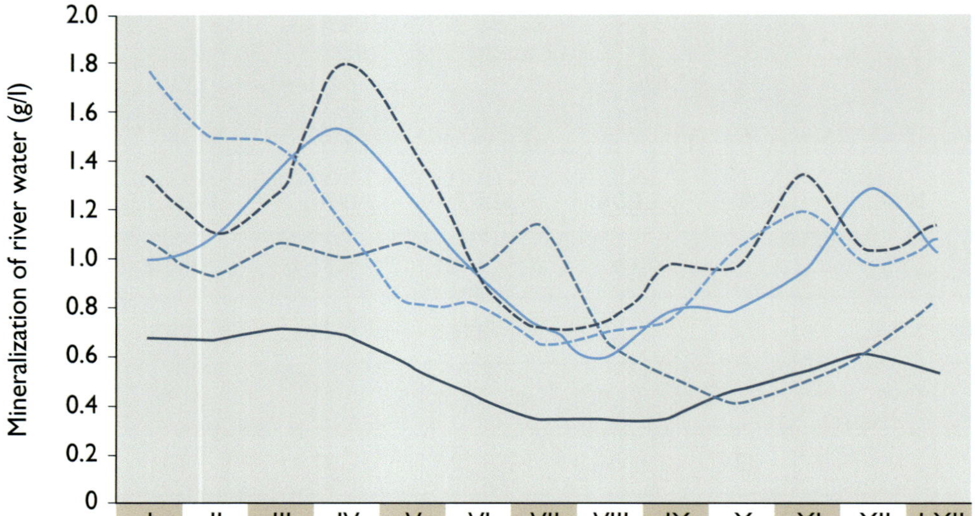

I	II	III	IV	V	VI	VII	VIII	IX	X	XI	XII	I-XII	
0.67	0.67	0.71	0.68	0.56	0.43	0.34	0.33	0.36	0.46	0.54	0.60	0.53	1960–1963
1.07	0.92	1.07	1.01	1.07	0.95	1.13	0.64	0.50	0.42	0.50	0v62	0.83	1985
0.99	1.12	1.37	1.53	1.26	0.90	0.72	0.58	0.78	0.79	0.97	1.29	1.02	1995
1.35	1.11	1.28	1.80	1.45	0.93	0.70	0.75	0.98	0.97	1.35	1.02	1.14	2000
1.76	1.51	1.47	1.10	0.82	0.81	0.65	0.71	0.76	1.05	1.19	0.98	1.08	2002

Figure 3.25 Mineralization changes in the Amudarya (after Iskander Ruziev, SIC-ICWC).

the junction of the inter-sectoral interests of the Ministry of Water Resources and the Ministry of Agriculture and, as a result, there was no appropriate ministerial responsibility.

- There was a dramatic lagging behind of the development of on-farm irrigation and drainage systems compared to the inter-farm systems, and this trend of on-farm underperformance intensified over time.

As a result, it could be argued that water availability on irrigated land was actually artificially reduced!

Many scientists and advanced practitioners focused attention on these negative trends in water use practice in the region and were alert to the need for radical restructuring of the water resources management system. Rabochev (1973), Dukhovny (1974), Dukhovny and Litvak (1976), and other authors made recommendations for changing the methods of land reclamation in Central Asia and shifting from a practice of using excessive water towards the adoption of more optimized and water-saving technologies for irrigation and drainage. For example, a study of the irrigation efficiency on the lands drained by shallow drainage systems in the Khorezm Region and the irrigation efficiencies on the lands drained by deeper drainage systems in the Golodnaya Steppe in Uzbekistan and Kazakhstan and in the Khodji-Bakirgan and Samgar irrigation systems in Tajikistan, clearly showed the advantages of the lat-

ter and the need to shift towards deeper drainage systems in combination with an appropriate schedule of water applications. Table 3.15 presents some data adopted from Dukhovny et al. (1985) that were used for proving the need of a shift towards advanced methods of land reclamation.

The significance of shifting towards advanced drainage systems (especially subsurface horizontal drainage) was clearly demonstrated in some large-scale projects. This was desirable not only from the point of view of sustainability – maintaining an optimal land reclamation regime with lower operating costs and lower total water consumption – but also because it reduced the amount of salts deposited into the rivers with the return water. Introduction of the proposed innovations all over the region would save 5–7 km^3 of water that was being used for leaching operations while still maintaining the existing irrigation schedule.

The enhancement of irrigation systems efficiency was promoted at this time. Studies of water losses from irrigation canals had been conducted early on in the twentieth century. Kostayakov led field surveys of crop water requirements that resulted in a suggested formula for calculating water losses from an irrigation canal, based on design parameters and soil properties. Later, from 1925 to 1927, Yanishevsky carried out research work that allowed him to specify values for the efficiency of piedmont irrigation systems (up to 0.36) and plain irrigation systems in the Golodnaya Steppe and Bukhara (up to 0.49) (Dukhovny 1981). Data produced by A. Rachinsky on the operational services for the Zeravshan Valley (see Table 3.16) are quite illustrative in this respect.

In 1965, the average efficiency of irrigation systems in the Syrdarya river basin was 0.53 and in the Amudarya river basin 0.5. After 1965, there was the widespread introduction of technical measures for preventing seepage losses from irrigation canals. This was initiated for most irrigation systems in the Syrdarya river basin as well as, to a lesser degree, those in the Amudarya river basin.

Table 3.15 Comparing the different methods of land reclamation in Central Asia.

Indicator	Units	Khorezm		Golodnaya Steppe
		Pilot project: Pravda collective farm	Average over the province	Average over the region
Gross irrigation rate	'000 m^3/ha	16–18	26–30	8–9.5
Per unit of raw cotton	m^3/ton	4100	8000	3200
Drainage rate	l/sec/ha	0.33–0.21	0.49–0.36	0.26–0.12
Removal of salts	ton/ha	32–60	26–50	10–35
Drainage operation costs	ruble/ha	6	20.6	9

Table 3.16 Efficiency factors typical for irrigation systems in the Zeravshan Valley from 1928 to 1936 (Archives SANIIRI).

Irrigation system	**Year**		
	1928	**1931**	**1936**
Upper part of the valley	0.35	0.56	0.56
Lower part of the valley	0.47	0.61	0.54
Districts			
Right-bank districts	0.48	0.61	0.58
Left-bank districts	0.60	0.65	0.55
Karakul	0.44	0.56	0.60

High efficiency was achieved for the first time in Central Asian irrigation systems in the newly developed areas in the Golodnaya Steppe. Here some 715 main and inter-farm canals were built between 1960 and 1972. Only 115 km of these canals had an earthen bed, and all the other canals were lined (3720 km of canals were constructed using precast irrigation flumes). The sites had calculated land slopes, and buried self-pressure irrigation distributors (which comprised 12.7% of the total irrigation network) were constructed using asbestos-cement and reinforced concrete pipes. As a result, a high efficiency was achieved, with low water losses (including operational spills) and a drastic decrease in unit water consumption. The efficiency of the inter-farm irrigation network was 0.84 to 0.92 and the on-farm network efficiency reached 0.86 to 0.92. The overall efficiency of the irrigation systems varied between 0.727 and 0.850, or 0.788 on average (Dukhovny et al. 1985).

Although, many irrigation projects and reconstruction works were subsequently built following the Golodnaya Steppe project, by 1975 the percentage of lined irrigation canals in the network was only 26% in Uzbekistan, 16% in Kazakhstan, 25% in Kyrgyzstan, 42% in Tajikistan and 5% in Turkmenistan.

The other approach that could facilitate a considerable reduction in total water withdrawal was the improvement of water application techniques. Raising the efficiency of water application techniques offers great potential for water savings in the agricultural sector. Unfortunately, initiatives to improve water application techniques were mainly designed to enhance labor productivity – a good example here is the use of sprinkler machines with far-reaching field cover. However, enhancing irrigation efficiency at field level should be a main objective of any reconstruction work. Sprinkler irrigation is quite productive only under the specific conditions of rather high air humidity that is typical for climatic zones with

an unsteady weather regime or for semi-arid steppes where humidity is approximately 55–65%. Under such conditions the water application efficiency can reach 0.75–0.81. However, in the arid climates of deserts and semi-deserts, where the air humidity is below 30–45%, the efficiency of sprinkler irrigation does not exceed 0.7.

Pilot projects with sprinkler irrigation in the Karshi Steppe and in the low foothills in Tashkent Province have shown that the water application efficiency was rather low (0.62). Therefore, furrow irrigation remained the main irrigation method in these climatic zones. The efficiency of this method can be increased when using appropriate techniques to around 0.62 to 0.82. To improve irrigation efficiencies, it is necessary to select optimal parameters for the siting of furrows and the rate of water inflow into the furrows, and to ensure a high quality of land leveling, automatic water distribution into furrows, reuse of return water, and daily regulation of water distribution within the farms.

For distributing water into furrows, SANIIRI[21] and the State Specialized Design Bureau for Irrigation (the SSDB for Irrigation) developed different specific engineering means including buried perforated pipes, flumes, portable rigid and semi-rigid perforated pipes, flexible perforated hoses, movable irrigation machines, single border irrigation ditches and horizontally leveled irrigated plots, undischarged syphons, and special irrigation tubes (see Figure 3.26). Some of these advanced irrigation techniques, such as rigid portable perforated pipes, flexible perforated hoses and buried perforated pipes, were introduced extremely slowly. Drip irrigation systems and subsurface irrigation systems were only used on some 10% of the irrigated area.

The efficiency of water application methods is determined not only by mechanization and improved standardized methods of soil moisturing but also, and equally importantly, by reducing non-productive water losses. However, as a result of adopting more advanced irrigation methods in the Golodnaya Steppe (flexible perforated hoses and portable rigid perforated pipes made of aluminum for delivering water from the precast elevated canals into the heads of the furrows), water application efficiency here reached around 0.82. Due to the introduction of different types of canal lining, new irrigation methods, and advanced systems of subsurface drainage and tubewell drainage, the water and salt balances of this irrigated land differed considerably from all the other irrigated lands in Central Asia (Dukhovny et al. 1985).

Table 3.19 presents data that allows the performance in the Golodnaya Steppe to be compared with the indicators of overall water use in the Aral Sea basin over the period 1965 to 1975 (Dukhovny and Litvak 1976). The average unit water diversion into the irrigation system of the Golodnaya Steppe was some 40–45% less than the average unit water diversion over the whole region. The efficiency was some 20% higher, and salt removal was minimal. These indicators are evidence of the high economic efficiency of this type of irrigation system. Unfortunately, the introduction of advanced irrigation systems on irrigated land in Central Asia was never fully implemented, reaching about 50% in Uzbekistan, 100% in Turkmenistan and some 80% in Kazakhstan.

[21]The Central-Asian Research Institute of Irrigation – SANIIRI is the transliteration of the Russian abbreviation.

Figure 3.26 Demonstration of new irrigation techniques (pictures, archive Dukhovny V.A.).

Table 3.17 Estimated water balance of the newly irrigated areas in the Golodnaya Steppe, '000 m^3 net per hectare (Dukhovny 1973).

Components of the water balance	1966			1967			1968			1969			1970		
	Zone		Total	Zone		Total	Zone		Total	Zone		Total	Zone		Total
	I	II		I	II		I	II		I	II		I	II	
Inflow															
Rainfall	2.22	2.22	2.22	2.57	2.57	2.57	3.24	3.24	3.24	5.71	5.71	5.71	2.86	2.86	2.86
Irrigation	10.6	7.34	8.64	11.1	8.26	9.54	8.21	6.74	7.5	7.52	5.6	6.28	10.1	7.22	8.2
Outside inflow	0.12	0.12	0.12	0.12	0.12	0.12	0.12	0.12	0.12	0.12	0.12	0.12	0.12	0.12	0.12
Total	12.94	9.68	10.98	13.79	10.95	12.23	11.57	10.1	10.86	13.35	11.43	12.11	13.08	10.2	11.48
Outflow															
Drainage	–	0.9	0.5	–	1.44	0.82	–	2.01	0.94	–	2.14	1.32	–	1.81	1.18
Surface tailwater discharge	–	0.75	0.41	–	0.58	0.33	–	1.03	0.48	–	0.66	0.40	–	0.58	0.38
Evapotranspiration	7.48	7.74	7.62	7.95	8.27	8.10	8.11	8.29	8.20	8.45	8.67	8.55	8.8	8.95	8.9
Increased groundwater storage	2.23	–	1.0	3.30	–	1.40	1.56	–	0.83	2.57	–	0.98	1.77	–	0.61
Percolation to groundwater	2.07	0.28	1.08	2.5	0.12	1.14	1.37	-0.81	0.35	2.33	–	0.89	2.81	0.93	0.36
Total	11.78	9.67	10.61	13.75	10.41	11.79	11.04	10.52	10.80	13.35	11.47	12.14	13.38	10.41	11.43
Difference, percentage	+1.16	+0.01	+0.37	+0.04	+0.54	+0.44	+0.53	-0.42	+0.06	–	–0.04	+0.03	–0.3	–0.21	+0.05
	+9.0	+0.1	+3.35	+0.2	+4.9	+3.6	+4.56	-4.16	+0.5	–	–0.3	+0.26	–2.7	–2.0	+0.46

Table 3.18 Salt balance of the newly irrigated areas in the Golodnaya Steppe, '000 tons (Dukhovny and Litvak 1976).

Components of the salt balance	**Year**				
	1966	**1967**	**1968**	**1969**	**1970**
Inflow					
Irrigation	757	1546	1354	700	1171
Rainfall	31	42	59	113	60
Capillary rise	93	117	165	207	123
Underground inflow	216	252	275	306	324
Total	1097	1957	1853	1326	1688
Outflow					
Drainage	720	1201	1542	2818	2021
Surface tailwater disposal	287	231	496	576	377
Total	1007	1432	2038	3394	2398
Salt accumulation	+90	+525	–	–	–
Reduction in soil salt storage	–	–	185	2068	710
Salt accumulation/reduction per hectare of irrigated land	+0.6	+3.6	–1.0	–11.5	–4.0
Salt accumulation/reduction per hectare of irrigated land with GWT higher than the drainage lines	+2.8	+10.4	–3.0	–29.6	–12.6

It should be noted that the comprehensive reconstruction of irrigation systems that was planned in the schemes of integrated water resource use and protection was not implemented, and the plan was even ignored to some extent. The plan set out individual and separate indicators for land reclamation, but these did not cover the efficiency of water and land use. The schemes for the Amudarya and Syrdarya river basins involved huge works, which were not implemented in full. For example, *The Scheme of Integrated Water Resource Use and Protection in the Syrdarya River Basin* (published in 1979) involved the

Table 3.19 Changes in water use in the Aral Sea Basin (without the rivers not discharging into the main streams of the basin).

Indicator	Year		
	1965	1970	1975
Water withdrawal, km^3	72.7	87.9	103.5
Syrdarya river basin	36.4	40.2	45.8
Amudarya river basin	35.3	47.7	57.7
Including: for irrigation, km^3	66.2	79.6	91.1
Syrdarya river basin	29.4	35.4	38.5
Amudarya river basin	33.8	44.2	52.6
Return water, km^3	10.5	24.0	26.8
Syrdarya river basin	3.3	16.2	17.0
Amudarya river basin	5.4	7.8	8.8
Non renewable water consumption, km^3	63.2	63.9	76.7
Syrdarya river basin	33.1	24.0	28.8
Amudarya river basin	30.1	39.9	47.9
Including: for irrigation, km^3	46.3	59.9	69.7
Syrdarya river basin	17.6	22.3	25.0
Amudarya river basin	28.7	37.6	44.7
Irrigated area, million hectares	3.82	4.41	5.13
Syrdarya river basin	1.90	2.10	2.40
Amudarya river basin	1.92	2.31	2.73
Unit water withdrawal, '000 m^3/ha	16.54	18.05	17.76
Syrdarya river basin	15.47	15.86	15.04

(*Continued*)

Table 3.19 (Continued)

Indicator	Year		
	1965	1970	1975
Amudarya river basin	17.00	19.30	19.26
Unit water consumption, '000 m^3/ha	12.12	14.49	15.54
Syrdarya river basin	9.23	10.62	10.42
Amudarya river basin	14.00	16.98	16.37
Irrigation system efficiency	0.52	0.55	0.58
Syrdarya river basin	0.53	0.56	0.63
Amudarya river basin	0.50	0.53	0.55
Gross agricultural output for irrigated lands, billion rubles	2.83	3.97	5.43
Syrdarya river basin	1.32	1.76	2.15
Amudarya river basin	1.51	2.21	3.28
Water withdrawal, km^3 per billion rubles of output	22.3	20.05	16.78
Syrdarya river basin	22.3	20.10	17.90
Amudarya river basin	22.4	20.00	16.00
Water consumption, km^3 per billion rubles of output	16.36	15.09	12.64
Syrdarya river basin	13.33	12.67	11.62
Amudarya river basin	19.00	17.00	13.60

reconstruction of irrigation and drainage systems covering 1,067,900 hectares of the 2,664,700 hectares under irrigation, including 25% of the irrigated areas in Kyrgyzstan, one third of the irrigated areas in Uzbekistan, 25% of the irrigated areas in Tajikistan and 62% of the irrigated areas in Kazakhstan.

The integrated development of new land in the Golodnaya Steppe, the Karshi Steppe and in some regions of Tajikistan, Kyrgyzstan and Kazakhstan had set a precedent for the construction of advanced

irrigation and drainage systems. However, the state lacked the funds for the considerable scope of work necessary for a comprehensive reconstruction of the irrigation systems. Table 3.20 shows that only 23.8% of the irrigated land had advanced irrigation systems in Kazakhstan, and the respective figures were 17% in Kyrgyzstan, 29.6% in Tajikistan, 10.5% in Turkmenistan and 12.9% in Uzbekistan. Advanced drainage systems were constructed on 8.9% of the irrigated land in Kazakhstan, 5.4% in Kyrgyzstan, 5.4% in Tajikistan, 1.4% in Turkmenistan and 13.4% in Uzbekistan respectively.

The growth of water diversion from rivers unfortunately continued unrestricted (with the exception of dry years such as 1975). This was for two reasons. The first cause was the continued intensive development of new irrigated land. The second cause was the increasing rate of installing drainage

Table 3.20 Key Indicators of the technical level of irrigation systems in Central Asia as of 1980.

Indicator	**Units**	**Republic**				
		Kazakhstan	**Kyrgyzstan**	**Tajikistan**	**Turkmenistan**	**Uzbekistan**
Irrigated area, including:	'000 ha	1,949	944	618	942	3,454
Open canals	'000 ha	1,735	928	493	919	3,397
Precast flumes	'000 ha	250	146	58	76	392
Buried pipelines	'000 ha	214	16	125	23	57
Efficiency of irrigation systems		0.65	0.59	0.65	0.56	0.64
Efficiency of on-farm irrigation networks		0.80	0.73	0.77	0.73	0.73
Efficiency of land use		0.85	0.82	0.84	0.81	0.81
Area with sprinkler irrigation	'000 ha	495	88	1.0	–	9.0
Area with drip irrigation	'000 ha	–	–	–	–	2.0
Drainage systems, including:	'000 ha	470	151.0	51.0	57	2,200
Subsurface drains	'000 ha	22	47.0	10.0	10.0	138
Tubewell drainage systems	'000 ha	152	4.0	2.30	3.0	326

Table 3.21 Growth rates of irrigated land in the republics of Central Asia and Kazakhstan, '000 ha and as a percentage of the irrigated land in 1965 (Dukhovny 1976).

Republic	Year				
	1965	**1970**	**1975**	**1980**	**1983**
Uzbekistan	2,575 100%	2,751 106.8%	3,006 116.7%	3,407 132%	3,652 141.8%
Tajikistan	463 100%	518 110%	567 121%	605 129%	631 135%
Turkmenistan	514 100%	643 125%	855 166%	942 183%	1,061 206%
Kyrgyzstan	861 100%	883 103%	911 106%	975 113%	1,008 117%
Kazakhstan	1,368 100%	1,451 106%	1,630 119%	1,930 141%	2,086 152%
Total in Central Asia	5,786 100%	6,246 108%	6,959 120%	7,861 136%	8,438 146%

systems which resulted in more intensive use of groundwater to achieve the water and salt balance in irrigated land and which increased the need for leaching operations. Both factors resulted in additional water diversion from rivers.

The huge increase in water diversion from the rivers becomes understandable if one takes into consideration the large-scale construction of surface drainage networks over an area of 2.4 million hectares in the region after 1960. Drainage water was disposed into the rivers through newly constructed drainage canals. Renewable water use did not increase, but increased water salinity in the Amudarya and Syrdarya rivers was definitely observed. From this point of view, the trend for increasing volumes of return water to enter the upper and lower reaches of the Syrdarya within the Fergana Valley is quite indicative (Dukhovny 1986).

It is significant that although there was only a 44% increase in land under irrigation, the volume of return water had increased 12 times. This delivered an irreparable blow to the environmental equilibrium in the Syrdarya river basin! The scientific sector, and SANIIIRI in particular, tried to counteract this development, and concentrated on improving the operation and maintenance (O&M) of the irrigation and drainage systems.

Table 3.22 Correlation between return water and other factors in the Fergana Valley.

Years	Inflow into the valley (km^3)	Irrigation (km^3)	Area ('000 ha)	Return water (km^3)	Recorded water releases (km^3)	Difference between return water and water releases (km^3)	Percentage of return water	Unit length of drainage systems (m/ha)	Efficiency of drainage system	Percentage of recorded water releases	Percentage of non-recorded water releases
1936–1940	22.29	10.8	822	0.72	0.10	0.62	7.0	5.6	0.45	1.0	6.0
1941–1945	26.7	11.3	860	1.89	1.40	0.49	16.7	7.8	0.45	12.3	4.4
1945–1950	31.1	12.0	877	3.96	2.89	0.93	33.0	14.8	0.47	24.0	9.0
1951–1955	29.4	15.1	952	6.08	4.44	1.64	40.02	16.2	0.49	29.4	10.8
1956–1960	23.3	16.4	983	7.02	5.31	1.71	42.8	18.8	0.52	32.3	10.5
1961–1965	23.3	16.4	983	7.02	5.31	1.71	42.8	18.8	0.52	32.3	10.5
1966–1970	30.9	17.8	1022	8.40	7.80	0.6	47.1	22.8	0.56	43.8	3.3
1971–1973	25.5	18.5	1187	9.0	7.36	1.64	48.6	24.9	0.58	39.7	8.8

SANIIIRI elaborated an O&M improvement program that envisaged a transition towards a functional structure of operational services, including specialized organizations for water resources management at the river basin level (BWOs). It also suggested a transition towards preventative maintenance of the irrigation and drainage systems instead of simply undertaking emergency repairs, as well as charging for appropriate water services. A two-tier tariff was proposed, with a charge for each cubic meter of water delivered and for each hectare with properly maintained irrigation and drainage networks (Dukhovny 1976). This system aimed to improve the incentives for water users to save water and for water organizations to provide quality services, because payment for each hectare would depend on the quality and scale of servicing the irrigation and drainage systems.

The existing system of planning strongly worked against innovations, based as it was on "achieved level" and with a budget determined by notorious indicators such as gross production, gross operation of production facilities, and gross volume of repair and maintenance works rather than actual results. Nevertheless, some elements of the proposed new system were introduced on a step-by-step basis.

However, the data and research findings convincingly show that the vision of the Soviet government, and especially of the planning agencies, was deficient, especially with respect to ecological effects, and too focused on chasing momentary indicators.

Let us compare the actual values and trends in water use with an alternative scenario based on the government taking a decision to stop the use of gross production indicators and to proceed with constructing advanced irrigation systems. This would have required a reduction in the rate of bringing new irrigated land into operation and the construction of highly-productive irrigation systems on an integrated basis like the systems in the Golodnaya Steppe.

Figure 3.27 shows the level of irrigation water supply in the Uzbek SSR (from 1965 to 1975) in comparison with the irrigation water supply in the Golodnaya Steppe. Of the total 17,200 m³/ha water

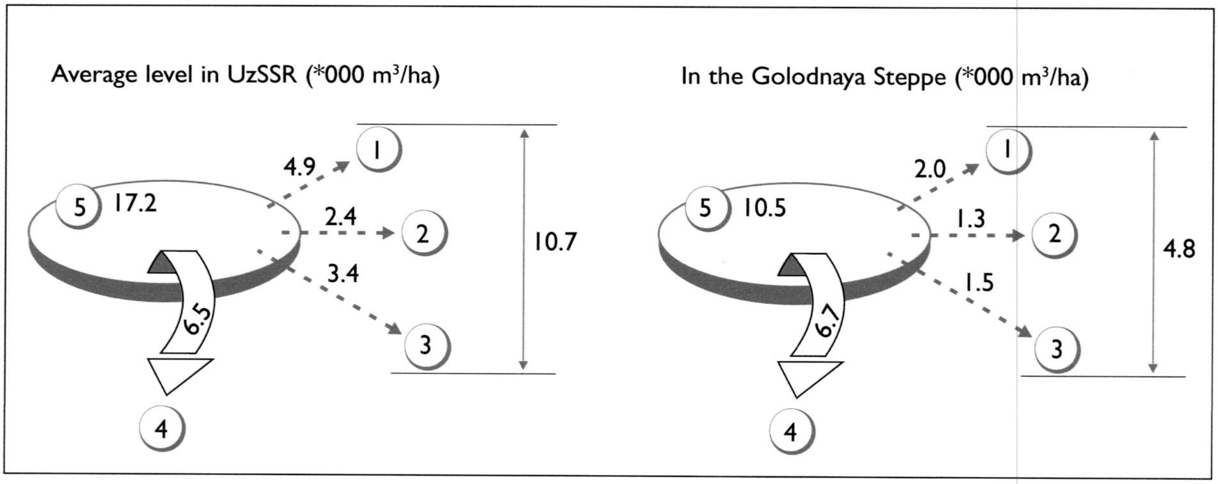

Legend

1 Water losses in the conveying system
2 Water losses on field level
3 Leaching share
4 Productivity share
5 Total water supply per hectare

Figure 3.27 Comparison of irrigation water use efficiency in the Golodnaya Steppe (advanced irrigation systems) and average use throughout the Uzbek SSR (1965–1975).

supply in the republic, irrigation water losses accounted for 10,700 m^3/ha, including 4,900 m^3/ha lost in the water conveying system, 2,400 m^3/ha on the fields and 3,400 m^3/ha for leaching. Thus, less than 40% of the water diverted from the rivers was actually used. Irrigation water losses in the Golodnaya Steppe were much lower at 5,400 m^3/ha, including 3,600 m^3/ha in the water conveying system, 1,300 m^3/ha on the fields and 1,500 m^3/ha for leaching. Almost 60% of diverted water was actually used. This means that at 1980 levels instead of the needed 103 billion m^3 of water for the irrigation of 7 million hectares, only 73.5 billion m^3 would be used and the total water withdrawal would not exceed 85 billion m^3. We'll revert to this issue in the next section, when we review an alternative to "the Soviet way of development".

Quite inconsistent processes were deployed in the Central Asian water and land reclamation sector in the final decade (1980–1990) when planning was done within the framework of a single country situation. Planning on the basis of "achieved level" remained, because the State Planning Committee of the USSR continued to plan for an expansion in the area of land under irrigation, a policy which further exacerbated the problem of water deficits. A solution could probably have been found through better organized water use and water distribution, and a reduction of non-productive water consumption and losses, but for this purpose it would have been necessary to raise awareness and make investments into the O&M of the irrigation and drainage systems at all levels, starting from river basin, through the systems of main and inter-farm canals, down to the on-farm systems. However, the beginning of the 1980s saw the first signs that governments were paying attention to the problems of managing the large river basins (Amudarya and Syrdarya) in Central Asia. The Ministry of Water Resources of the USSR together with five republican ministries initiated a revision for the schemes of integrated water resource use in these river basins.

Nevertheless, most funds were used for rehabilitating the inter-farm irrigation and drainage networks, because O&M fell within the responsibility of the water authorities. For example, in Uzbekistan the respective funds allocated for the O&M of inter-farm and on-farm irrigation networks were extremely disproportionate. More than 70 rubles per irrigated hectare were annually allocated to water management organizations (responsible for 7.6 m/ha of inter-farm canals on average) and only 16.5 rubles per irrigated hectare were allocated to collective farms or state farms (responsible for the O&M of 39.7 m/ha of on-farm canals on average). Accordingly, over a five-year period (1980–1985), fixed assets at the inter-farm level increased by 36% but they only increased by 11.34% at the on-farm level. During the same period, the number of service staff at the inter-farm level increased by 20%, but the numbers of on-farm level service staff remained unchanged. The number of inter-farm and on-farm irrigation networks equipped with water-measuring structures increased by 93% and 12% respectively. For water-distributing structures, these figures were 94% and 32% respectively (Dukhovny et al. 1989).

The USSR Long-Term Land Reclamation Program adopted in 1984 was extremely topical. Based on this program, the Council of Ministers of the USSR issued the resolution "On Measures for Providing High-Productive Use of Irrigated Land in Collective Farms, State Farms and Other Agricultural Enterprises". A main feature of this resolution was its awareness of the nationwide tasks necessary for the radical reform of the system of O&M of the whole irrigation and drainage (I&D) network, starting with the on-farm I&D network. It was therefore necessary to develop measures setting out the joint

responsibility of water and agricultural organizations for implementing the major task detailed in this resolution, which was to rehabilitate all obsolete irrigation and drainage systems by 1990.

However, these administrative measures and the shifting of responsibility for O&M to water organizations didn't solve all existing problems. The water ministries of the republics were not in a hurry to accept this responsibility. Water organizations only undertook to carry out O&M of on-farm pumping stations and tubewell drainage systems as well as accepting the commitment to introduce advanced irrigation methods.

The key issue, the financial basis for this new organization of responsibilities, was not resolved. The introduction of water charges was supposed to be basis for a transition process in which the water organizations would be come self-financing. However, this proposal had only resulted in pilot projects, in the course of which the O&M budget was distributed among farms, making it seem as if they were being paid for services rendered. This was too far from reality, and these pilot projects ended in a fiasco because this kind of system did not work for either the agricultural bodies nor the water organizations.

It was necessary to create a kind of financially efficient system and establish proper economic relations between agricultural producers and the water organizations that serviced them. SIC-ICWC proposed a system based on contract relations between farms and water organizations and based on tariffs that take into consideration local features related, for example, to specific O&M aspects of on-farm irrigation and drainage systems.

It was assumed that the district water administrations and the local water construction organizations would be turned into operational enterprises that would sign contracts with farms for technical servicing and assistance. The cost of O&M work, charged in the form of a two tier tariff, had to be established by taking into consideration the local land reclamation conditions and the parameters of the I&D networks. Part of the tariff was taken as a percentage of land productivity, and an additional charge was collected for each cubic meter of water delivered. It was also assumed that payments to each of the operational organizations would be related to the output of these organizations. The district agricultural administrations would evaluate the services and specify a payment plus any sanctions (or bonuses in the case of reducing operating costs, saving water, etc.).

A key reason for declining productivity indicators for land and water use in this period, and accordingly reduced GNP in the agricultural sector, was that the pricing system for agricultural output did not take into account the cost of land and water resources. One may also object that the federal government considerably subsidized the water and agricultural sector. However, an optimal combination of sectoral and territorial planning – and the interests of the national economy and of the regions – can only be achieved by accounting for the cost of land and water resources. A lack of accounting for the cost of land and water in the irrigated agriculture sector resulted in a situation where the residual income from cotton production and other agricultural products (generated in the form of a turnover tax) was invested in product manufacturing rather than reinvested in cotton growing in the region.

Undoubtedly maximizing the manufacturing of finished products could be economically productive, but unfortunately until 1985 only 10–15% of the agricultural output of Central Asia was processed in the region. Only in the twilight of the Soviet period did the government begin creating the processing enterprises and small-scale businesses necessary to provide more employment opportunities for the local population.

Until 1975, the growth in the productivity of irrigated land contributed to positive economic trends in Uzbekistan compared to other regions and, as a result, lower costs per unit increment of agricultural output of 0.7 rubles per ruble compared with an average of 1.6 rubles per ruble in the country as a whole (Figure 3.28). However, by 1985, the situation had drastically changed and the efficiency of producing a unit increment in agricultural output had fallen by a factor of 2, and the costs had increased from 1.6 to 2.9 rubles per ruble on average in the country as a whole. In Central Asia, costs increased from 0.7 rubles per ruble to 3.8 rubles per ruble in the same period.

There are two reasons for this result. First, it can be attributed in part to the considerable differences in the productivity per irrigated hectare between various areas of Central Asia. This ranged from 2100 rubles/ha in the Surkhandarya Province of Uzbekistan, South Kazakhstan and some regions of Turkmenistan to 600 rubles/ha in Karakalpakstan and Dashhowuz Province of Turkmenistan. Moreover, the potential productivity of irrigated land in Central Asia ranged from 2600–3200 rubles/ha in a zone of fine fiber cotton cultivation to 1600–2000 rubles/ha in a zone of medium fiber cotton cultivation (under rotation of cotton and alfalfa); from 1500–1800 rubles/ha for land under forage crop rotation (corn and alfalfa) to 3000–3500 rubles/ha for land used for vegetable cultivation, orchards and vineyards. Analysis shows that a scientific approach to the use of irrigated land in the region could almost double productivity. The main reasons for low crop yields are not only a failure of farming techniques and bad organization but also poor hydro-geological conditions, non-leveled micro-reliefs and unfavorable soil properties. As a result, cotton yields vary from 1.5–5.1 tons/ha even on the best fields (and are about 3 tons/ha on average) (Central Statistical Office of the USSR 1984).

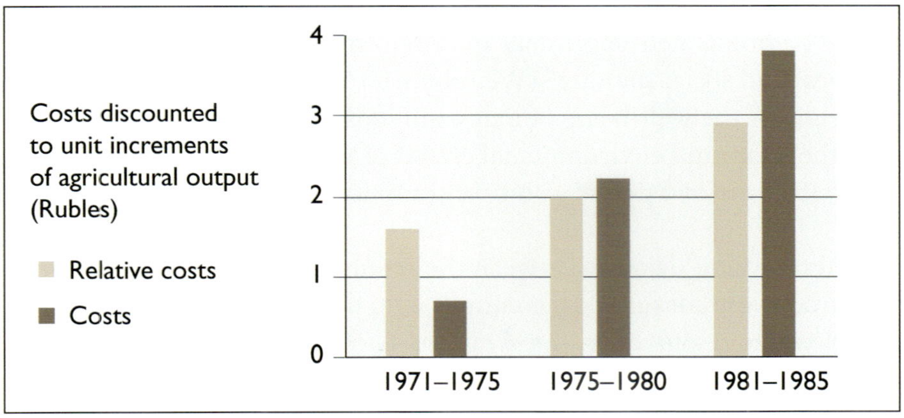

Figure 3.28 Trends in relative costs of irrigation productivity (source: SIC-ICWC).

The second serious reason for the rise in the cost per unit increment of agricultural output was the growing water deficit, especially in some areas in the lower reaches of the rivers in dry years. However, it should be noted that these water deficits related to improper water resource management in certain zones (ill-timed water delivery and inadequate water distribution) rather than an absolute water shortage.

Increasing water deficits in the river basins accelerated the rate at which the Aral Sea water level was falling. Between 1982 and 1986, only 4 km^3 of water flowed into the sea (Figure 3.29). This was also a manifestation of the contradictions that became more strained in the region after disintegration of the Soviet Union, namely those resulting from the need to redistribute water resources for the natural environment and the need for water for socioeconomic development as well as those arising from the need to meet the demands of both upstream and downstream users. However, initially, insufficient consideration was given to the Aral Sea (both in socioeconomic and environmental terms) as it was not regarded as being important when compared with the efficiency of irrigated agriculture or with the social problems in the region.

In 1982, one of the authors of this book tried, during a personal conversation, to alert the Minister of Water Resources to the fact that water was not flowing into the Aral Sea. The minister said: "I have my own headaches because of water sharing among the republics. The Aral Sea is a only problem for the state committee of natural resources." To this minister's credit, it should be noted that less than one year later he had changed his opinion, but the options were limited by the existing system of finance. It was impossible to receive funds for nature protection from the State Planning Committee, as at that time both Kazakhstan and Karakalpakstan were interested in developing their rice growing sector in the lower reaches of the rivers.

One cannot but agree with Weinthal (2002) who has written that the Soviet federal government tried to smooth over the contradictions in the water resource management regime. It acted as the mediator in conflict situations and had considerable influence and capabilities in suggesting solutions that were either correct in essence or ostensibly mutually acceptable. Moscow "orchestrated" the collective activity of the republics in the region, keeping stability at the local level and stimulating co-operation in water sharing. It "demonstrated hegemony for environmental exploitation while providing a collective form of political and social stability" (Weinthal 2002). As Weinthal acknowledges, the dilemma was whether to provide for the well-being of many millions of people by "taking" from the Aral Sea or whether to prevent the social and environmental effects of this infringement. Preference was obviously given to irrigated agriculture (and public water consumption) at the time.

Moscow was also settling disputes between the republics related to water sharing. In the federal country, water resources were considered a common state property. Weinthal confirms that "the Soviet authorities managed water resources in the Aral Sea basin in an integrated and quite co-ordinated manner". One cannot say that decisions on water sharing ignored the opinions and proposals of the republics. The schemes of integrated water resource use and protection in the Syrdarya and Amudarya river basins (see, for example, Sredazgiprovodkhlopok Institute 1974) were approved by the State Planning Committee following joint submissions from the Ministry of Water Resources and the republic

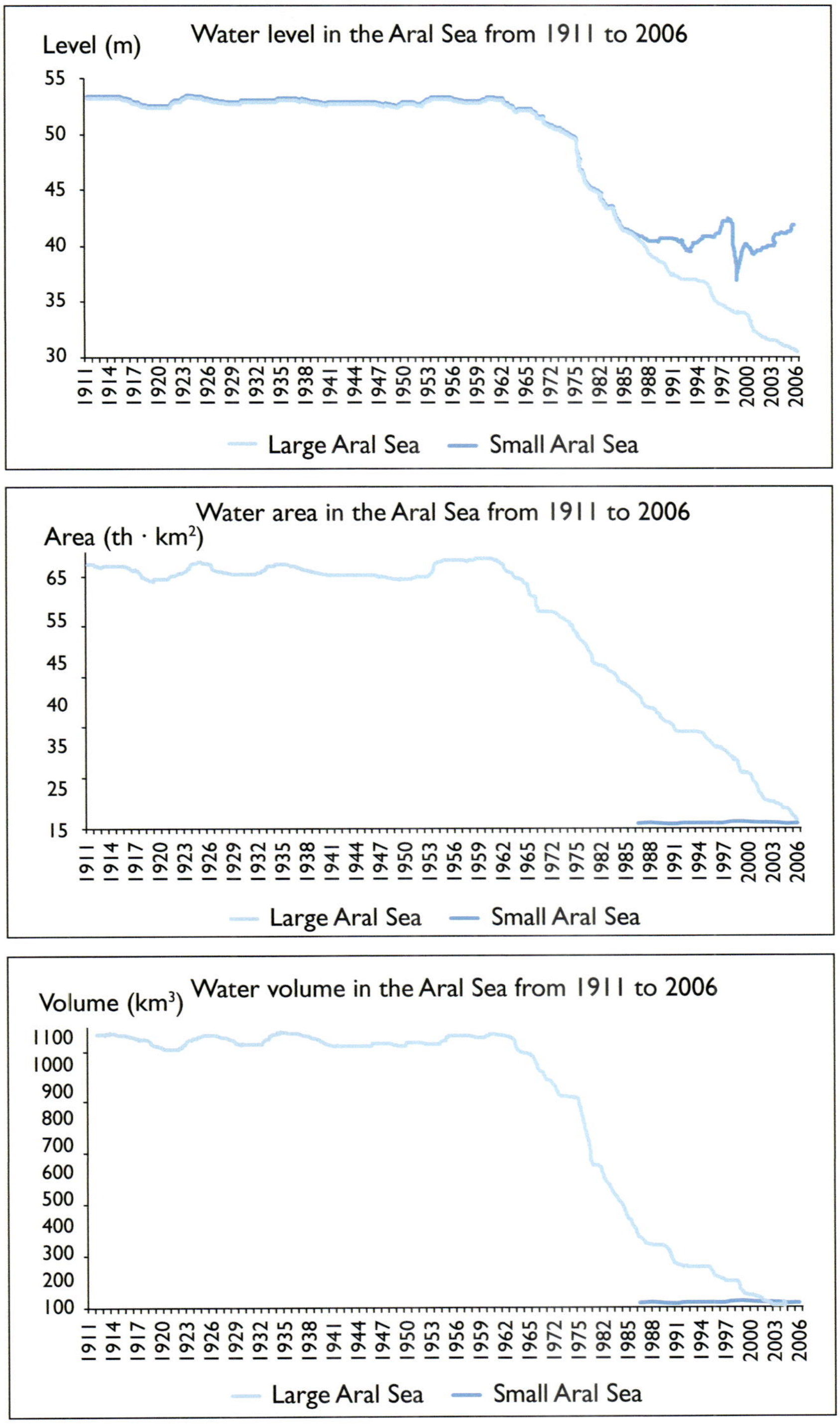

Figure 3.29 Main indicators showing the consequences of the Aral Sea disaster.

governments, and each year the Ministry of Water Resources approved the plans for water sharing, which covered the size of irrigated areas, water use limits for different economic sectors and the water release regimes. The water-sharing process became especially strained in dry years (in, for example 1974, 1975 and 1982), when a headquarters was established for water sharing, comprising the ministers of water resources from all republics and headed by the Deputy Minister of the Ministry of Water Resources. This body set water limits depending on annual water availability predicted by the Hydrometeorological Service and issued a special governmental decree on water shortage control. Considerable funds were allocated under this decree (more than 340 million rubles, for example, in 1974 and 1975) for drilling additional water supply tubewells, for adding lining into canal sections with the largest seepage losses, and for other measures.

Regarding an awareness of the inevitable drying up of the Aral Sea, one can say that the scientific community, including water and environment specialists in Central Asia, were aware of these problems in an early stage. They were certainly aware of the problems before the publication of books and articles by foreign (that is, non-Russian) publishers in which the phenomenon was characterized as a manifestation of the decay of the social regime. However, it stands to reason that the socialist regime committed many mistakes or even sins, including neglecting environmental protection. Glantz (1998) stigmatizes the Soviet government as being blind to the problem of the Aral Sea, though notes that: “In reality there were adverse impacts on those in Moscow who made decisions that led to the declining Aral Sea levels.”

It is necessary to admit frankly that large-scale irrigation in Central Asia and the pursuit of increasing the area of land under irrigation as a main goal of water development had to result in the tragedy of the shrinking Aral Sea. However, as early as the 1960s, there was discussion about the territorial redistribution of water resources in the USSR. A wide range of conferences were devoted to this issue. In 1962, the need to transfer extra volumes of water to Central Asia was expressed at an All-Union Conference on Integrated Land and Water Resource Use in Central Asia that was held in Tashkent. This idea was later supported at a conference devoted to the action plan for integrated water resource use in the Ob river basin (Novosibirsk, 1965) and at the meeting of the extended expert panel of the State Committee on Science and Engineering Research of the USSR (Moscow, 1967) and was followed up by various governmental resolutions.

Although many scientific papers of the time provide sound proof of the Aral Sea desiccation and present proposals for saving the sea (see Sadykov 1975 and others), the fact is that the USSR government, like all other governments, only began to see the environmental risks at the beginning of 1980s, when nature conservation ideas and approaches were spreading all over the world (the Dublin Conference, Rio Summit, etc.). That said, the first actions were already being undertaken at the beginning of the 1970s due to the rising concern about the water resource imbalance in Central Asia.

In 1968, the plenary session of the Central Committee of Communist Party of the USSR commissioned the State Committee of Science and Technology (CSST), the Ministry of Water Resources, Ministry of Agriculture, Ministry of Energy, the Academy of Sciences, and the Academy

of Agricultural Sciences together with appropriate organizations in the republics to develop a plan covering land reclamation, river runoff regulation and water transfer. The Souzgiprovodkhoz Institute prepared a feasibility report covering these aspects for the period 1971–1975. This feasibility report gave proof that there would be a need to transfer part of the Siberian rivers runoff to the Aral Sea basin by around 1995–2000. The Central Committee of the Communist Party of the USSR and the Council of Ministers of the USSR later approved this feasibility report (under resolution 612 of June 24, 1970).

In 1973, on the initiative of the Institute of Geography of the Academy of Sciences of the USSR, a commission under the leadership of Academician I.P. Gerasimov was established for reviewing these documents. In 1974, a similar governmental commission under the leadership of A.A. Borovoy, the deputy chairman of the State Construction Committee, was also established. Professor Dukhovny was a member of both these commissions. Both commissions confirmed the conclusions, contained in the schemes of integrated water resource use and protection in the Amudarya river basin (1967), Syrdarya river basin (1968) and Aral Sea basin (1972), that available water resources would be completely exhausted by 1995–2000. They also agreed to the need for transferring an extra 20–30 km^3/year into the region for further development and to prevent further shrinking of the Aral Sea. Thus, the thinking at that time was aimed at searching for alternative supplies that could provide extra water for the region, without neglecting two other aspects – the preservation of the Aral Sea (or rather the coastal and delta areas) and water conservation for the future development of the region.

The resolution of the government committee led by Borovoy was approved by the cross-sectoral scientific and technical session under the aegis of CSST (Moscow, 1977) in December 1978. The Central Committee of Communist Party of the USSR and the Council of Ministers of the USSR issued the resolution "On Executing Scientific-Research and Design Works Related to the Issues of Transferring a Part of Runoff of Northern and Siberian Rivers to South Regions of the Country".

All government commissions confirmed the need to initiate work to select a final option for a canal route to import water into the region. Two leading institutes of the Academy of Sciences – the Institute of Water Problems (in a study by Voropaev, Bostanzhoglo and Dunin-Barkovsky) and the Institute of Geography (a study by Vendrov and Kuznetsov) – studied numerous options for transferring water from Siberia (from the Irtysh, Ob and Yenisei rivers) and even from the Pechora River via the Volga. As a result of these large-scale studies carried out by the Academy of Science and the Souzgiprovodkhoz Design Institute (a study by Grischenko and Gerardi), with support from more than 100 different organizations, the necessary data were collected for developing options for the Sib-Aral Main Canal. The feasibility study for transferring part of the Siberian rivers runoff to Western Siberia, the Urals, Central Asia and Kazakhstan was discussed at numerous conferences and workshops held in Tashkent, Moscow, Novosibirsk and Nukus. The State Committee of Science and Technics of the USSR, the Academy of Sciences together with the governments of interested republics submitted this feasibility study for judgment to the State Commission of Experts under the State Planning Committee.

In June 1983, the State Planning Committee approved the resolution of the State Commission of Experts on the feasibility study and recommended to the Ministry of Water Resources of the USSR that it should proceed to design the Sib-Aral Main Canal with a capacity for transferring 27.2 km^3 of water annually. The Presidium of the Council of Ministers of the USSR reviewed this issue on January 31, 1984. At this session some questions arose, and to settle these matters seven teams of experts were established under the leadership of I.S. Shatilov, A.V. Ratkovich (two teams), U.I. Bokserman, U.M. Andrianov, I.P. Aydarov and A.V. Mikhaylov. Their work was evaluated by the State Economic Committee under the State Planning Committee of the USSR, which issued a special resolution on June 13, 1984 (key statements are given in the final section of this book). At the sessions held on September 15, 1983 and on May 23, 1984, the Political Bureau of the Central Committee of the Communist Party of the USSR, after reviewing the Long-Term Land Reclamation Program in the USSR, made a positive decision in respect of this territorial redistribution of water resources.

The selected route of the Sib-Aral Main Canal (Figure 3.30) started from Belogorye on the Ob River and ran along the left bank of the Irtysh River towards Tobolsk (the Tobolsk hydro scheme), and then along the right bank of the Tobol River across the watershed and through the Turgay Saddle to come to the Syrdarya near Djusaly. After this, it crossed the territory between the Syrdarya and Amudarya and, at the end of the 2250-km route, reached the Amudarya at a site between Tuyamuyun and Takhiatash. The carrying capacity at the head of the canal amounts to 1150 m^3/sec. The water needed to be lifted by a pumping cascade consisting of seven pump stations with a total lifting height of 110 m. A system of waterworks was planned all along the canal route (the Tobolsk hydro scheme on the Irtysh River, pump stations, check dam structures, spillway structures, aqueducts, outlets into main canals and bridges). The Tengiz Reservoir was planned in the North Pre-Aral (coastal) area to provide a smooth water supply at the head section of the canal during the peak irrigation water consumption.

This project was quite feasible from the point of view of available technical capability, given that the designed canal was only two times longer than the Karakum Canal and the capacity of the pump stations was only three times higher than the capacity of the Karshi Pumping Cascade. With an estimated total project cost of about 20 billion rubles and a project implementation period of 15 years, the annual scope of work was commensurable with that which was jointly being executed by the Glavsredazirsovkhozstroy and the Ministry of Water Resources of Uzbekistan at that time.

The project was being planned under quite difficult conditions, and was beset by delays and was accompanied by numerous doubts and conflicts of interest. It would seem that water was needed for the south of the country (not only Central Asia but also part of Russia and the north of Kazakhstan), but the period of "great funerals" and frequent changes of power in the Kremlin could not but affect the project's progress.

During the rule of Andropov, Moscow had declared its support to the project, and even Gorbachev – the future head of state and at that time the secretary of the Central Committee of the Communist Party of the USSR responsible for the agricultural sector – spoke at a meeting held

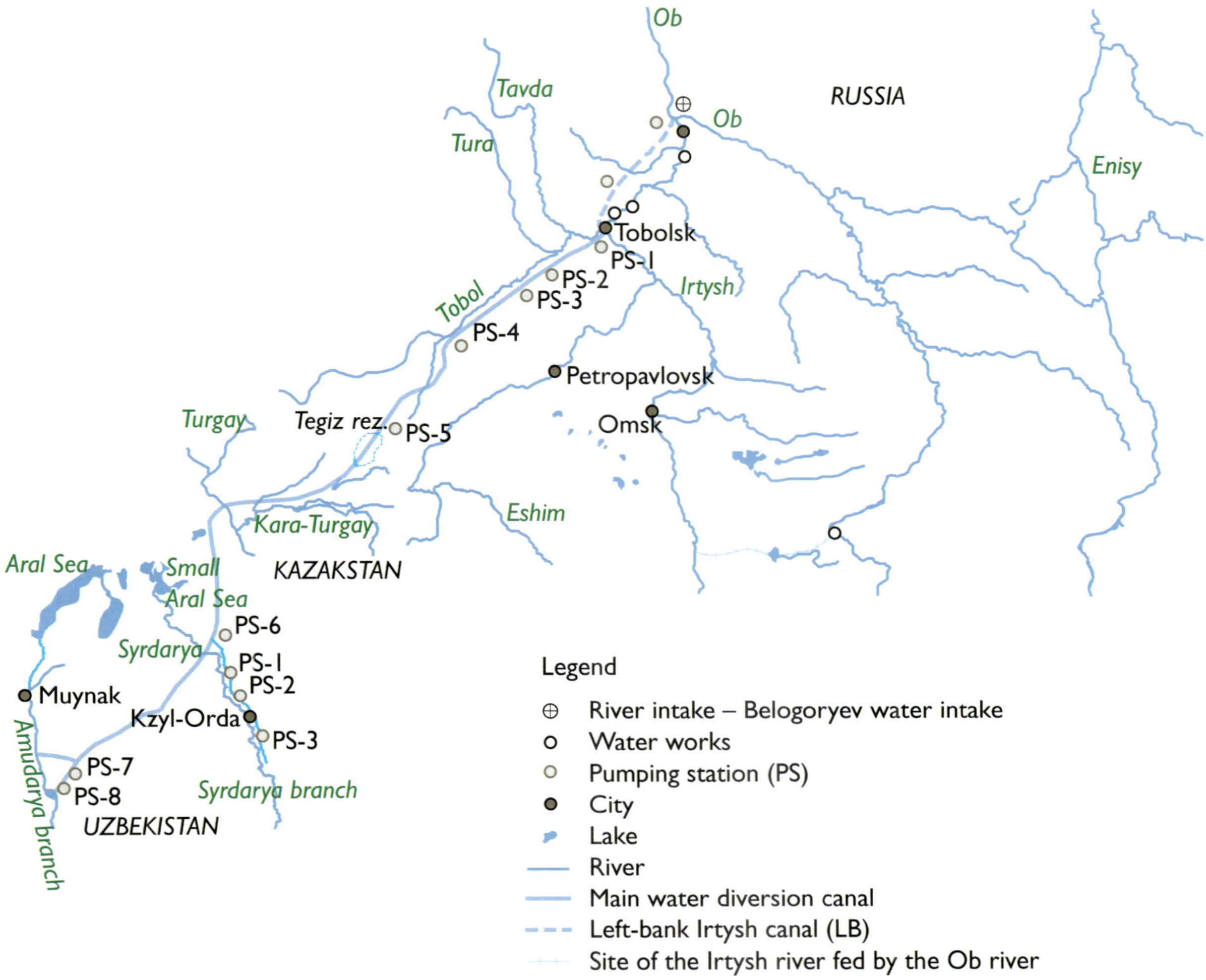

Figure 3.30 Sketch map of the water transfer scheme from the Irtysh and Obrivers to the Aral and Caspian Sea basins (after Gerasimov and Gindin 1976).

at the Giprovodkhoz Institute (in February 1985) where this project was presented, where he said "This project is quite needed from a political point of view. The steering wheel of Central Asia will always be in the hands of Russia." Even at the time of Gorbachev's rule, V.P. Nikonov (the secretary of the Central Committee of the Communist Party of the USSR) and N.K. Baybakov (the chairman of the State Planning Committee), as well as many other leaders of the state, actively supported this project.

However, the project was also attracting sharp criticism, under the slogan "anti-environmental mania for grandiose scale projects", and protest was mounting in the mass media supported by some literary scholars and scientists who were actually far removed from both the water sector and the environmental protection sector. The writer S. Zalygin, geologist A. Yanshin, chemist B. Laskorin and journalist V. Reznichenko competed in their haste to bring out often distorted facts and figures, and rushed to vulgarize a project which, from the perspective of that particular time, was an accepted plan of land reclamation and water resource management.

This campaign can be reviewed from both the point of view of world trends and the changing situation in the country. People started to look in a new way at the global struggle for survival and Soviet society experienced delay on the way. In the three social spheres of socio-economy, ecology and ethics, for a long time the Soviet Union gave preference to the first, victoriously neglected the second, and boasted of the third (the ostentatious "iron" moral). Later on, after economic and social decline, people felt a loss of ideals and social restrictions, but at the same time their eyes were opened to environmental issues. The countrywide electrification that was the basis of the Soviet socialist society resulted in the Chernobyl disaster and the loss of millions of hectares of floodplains. The use of agricultural chemicals contaminated soil and water, and even had adverse impacts on people's health. It is therefore not surprising that land reclamation work also became to be considered "sinful". Nevertheless, the unbridled campaign in the mass media portrayed land reclamation activities as a key source of all troubles in the country. During the last plenary session of the Writers' Union of the USSR, the problems of land reclamation were discussed more intensively than during the sessions of the Academy of Agricultural Sciences or at the panel of experts of the Ministry of Water Resources. One writer even initiated a discussion on the candidature of a future minister, and maybe he still is persona non grata for writers today.

In general, during the last three years of the USSR's existence the water sector was presented as the terrible monster Karabas Barabas[22], who suppresses a whole state system and devours billions of cubic meters of water and billions of rubles, or as a terrible dragon able to destroy nature and all the world. In the course of an election campaign, each candidate (whether a great physicist or an ordinary worker) presented the land reclamation sector as the number one enemy of the state and a chief embezzler of state funds.

Thanks to the writers, the water sector came to the attention of the public. However, it is necessary to find some truth in this sea of emotions. In fact, in the process of improving land productivity, additional amounts of water needed to be diverted from rivers. However, water-saving practices both at the field level and in the water conveying systems can minimize the adverse effects. Therefore, the amount of water supplied per unit area for crop irrigation is an indicator of the correctness of our irrigation practice and of our attitude to nature. (The key question of our time is whether it is possible to further reduce irrigation water consumption?)

Comparing the Soviet anti-ecological position with the situation in the USA, which became possible after a meeting of Soviet and US experts in Bloomington, Indiana (Dukhovny 1991), is quite instructive. It was to their credit that most American scientists at this meeting made an objective assessment of the situation in their country when compared with the Aral Sea tragedy. Daniel Willard, the

[22]A malevolent character in a story for children written by Tolstoy. The story was a Russian version of the Adventures of Pinocchio.

author of the famous book *Impacts of irrigation on water quality problems*, made a presentation on the regional crisis in California in which he characterized the situation as being quite similar to the situation in the Aral coastal area. The San Joaquin, Sacramento, and Colorado river valleys were home to 24 million people at the time. It was an economically developed region (with an average per capita annual income of US$10,600), with large-scale crop irrigation, under conditions of a water deficit. By the time of Daniel Willard's presentation, all available water resources, including water transferred in from other regions (potable water was supplied to Los Angeles from sources located 1500 miles away) were practically depleted. The drastic decrease in the volume of fresh water entering San Francisco Bay had resulted in the deterioration of its water quality and sharp drop (by a factor of 3.5) in fish productivity.

Frank Konty, a professor from Oregon University, devoted his presentation to comparing the Aral Sea and the Mono Lake in California. The fate of these water bodies share many common traits, although the water level of the Aral Sea started to fall in 1961 while the level of Mono Lake had been falling since 1941. The depth of the Aral Sea had decreased by 14 m and the depth of the Mono Lake by 12 m. The public began to get concerned about the Aral Sea in 1986 (Mono Lake in 1980), and the government of the USSR issued its resolution on the Aral Sea in 1988 (the government of the USA in 1985). In both cases, these disasters were caused by a drastic increase in water withdrawal for public needs.

Michael Rozengurt reported on the problems facing San Francisco Bay. He compared the degraded environment of the bay with the ecological processes in the Sea of Azov, Caspian Sea and Aral Sea. During the previous 35 years, the volume of fresh water flowing into the delta of the Sacramento River and San Francisco Bay had fallen by 50%. It could be assumed that, in the future, fresh water inflow will reduce further and, after completion of the San Luis Drainage Canal, the disposal of brackish drainage water with a high chloride and hydrocarbon content will lead to further degradation of water quality in the bay. Referring to world experience (including the Soviet experience), Rozengurt stressed that diverting more than 25% of a river's natural runoff can lead to catastrophic effects. In this case, salinization processes in the river deltas, intrusion of sea water into the subterranean water-bearing formations, and changes in directions of sea currents and origin of storms had been observed. As a result, the biological productivity, primarily the fish productivity, of this body of water had been sharply reduced.

The only presentation by an American scientist devoted to the problems of the Aral Sea was made by Professor Micklin. His ideas coincided with SANIIRI's thinking in many aspects. For example, Micklin considered that it would be impossible to preserve this sea within its present boundaries. In his opinion, it would be necessary to restrict the area of water and to develop a single program of water saving in the Aral Sea basin.

However, circumstances within the country strengthened the opposition to the Siberian rivers runoff redistribution project. The financial and economic potential of the country had considerably decreased. The country was becoming less manageable under the rule of Gorbachev. There were arguments by Aganbegyan, a leading economist of the time, that the project feasibility study had been

insufficiently elaborated (although these were arguably an attempt to justify the abandonment of the project given the lack of money in the country). In addition, the "Uzbek cotton criminal case" gave rise to further anti-Central Asian sentiments.

This was the context in which the government of the USSR issued the resolution "On the cessation of work for the river runoff redistribution into Central Asia" in August 1986. According to this resolution, to assist the Central Asian republics and Kazakhstan, the State Planning Committee together with the Academy of Sciences, SCST, and the Academy of Agricultural Sciences had to develop a strategy of socioeconomic development and rehabilitation of the environment in the region based on their own water resources. However, this decision was only an "excuse", because in 1988 the State Planning Committee transferred this matter to the Ministry of Water Resources that already was "at death's door" at that time. The strategy was ready by the end of 1989, but nobody reviewed it or a similar concept note developed by the Academy of Sciences. Nevertheless, we will consider these two documents later chapter in this book.

The assessment made by N. Ryzskov, the former chairman of the USSR government, at the jubilee celebrations of P. Poladzade, the former leader of the Ministry of Water Resources, is quite interesting: "We, leaders of the government and members of the political bureau of the Central Committee of the Communist Party of the USSR, are guilty [and have let down] the water sector's employees. When your land reclamation activities and Siberian rivers runoff redistribution project were attacked, we overlooked this attack on the Soviet state, [and did not fight] when they executed a political order."

However, undoubtedly the economic decline in the efficiency of irrigated agriculture, and of land reclamation activities as a whole, contributed to the decision to cancel the river runoff redistribution project. Moreover, the socioeconomic feasibility of such a project, as well as its technical and environmental feasibility, would add arguments to any discussion of the project's merits in the present time frame.

References

1. Aminova, R.Kh. Agrarian policy of the Soviet Power in Uzbekistan (1917–1920), Publishing House: Fan. Tashkent, p. 314. (in Russian)
2. Asarin, A.E. 1973. Components of the Aral Sea water balance and their impacts on long-term fluctuation of its water level. Journal "Water Resources" No 5. (in Russian)
3. Avakyan, A.B., Butorin, N.V. and Vendrov, S.L. 1976. Impacts of water reservoirs on the environment. Proceedings of the 23th International Geographic Congress, Publishing house: "Gydrometizdat", Leningrad. (in Russian)
4. Borowets, S.A., Bostanzhoglo, A.A., Kornakov, G.I. and Rukhalis, I.M. 1969. On rational water resources use and protection in Central Asia. Transaction of Sredazgiprovodkhlopok, Tashkent. (in Russian)

5. Chelguzov, N.M. 1934. Problems of irrigation in Uzbekistan. Proceedings of the 1st Conference of Studying Productive Forces in Uzbekistan. Publishing House of the Academy of Sciences of the USSR, Leningrad, pp. 78–80. (in Russian)
6. Dukhovny, V.A. 1973. Irrigation and developing of the Golodnaya Steppe. Moscow, Publishing House: Kolos, p. 239. (in Russian)
7. Dukhovny, V.A. 1974. On the issue of improving integrated desalinization reclamations. Journal "The Cotton Business" No 11, Publishing House: Kolos, Moscow, pp. 35–38. (in Russian)
8. Dukhovny, V.A. et al. 1976. Subjugation of the Golodnaya Steppe. Publishing House "Uzbekistan", Tashkent, p. 310. (in Russian)
9. Dukhovny, V.A. 1976. High cotton yields under conditions of water shortage – the valuable experience of scientists and practitioners in Central Asia. Herald of the Academy of Agricultural Sciences No 5. Moscow, pp. 92–99. (in Russian)
10. Dukhovny, V.A. and Litvak, L.S. 1976. Impacts of irrigation on Syrdarya flow regime and water quality. Irrigation of arid lands in developing countries – the environmental problem and effects. Alexandria, pp. 216–236.
11. Dukhovny, V.A. Belotserkovsky K.I. and Bocharin, A.V. 1977. From maintenance of irrigation systems towards management of water complexes. Journal "Hydrotechnics & Land Reclamation" No 3, Moscow, Publishing House: Kolos, pp. 13–23. (in Russian)
12. Dukhovny, V.A. Editor-in-Chief. 1981. Encyclopedia: Irrigation of Uzbekistan. Volume IV, Publisher: "Fan." Tashkent, pp. 81–93. (in Russian)
13. Dukhovny, V.A. and Vilenchik, V.B. et al. 1985. Scientific-technical progress and land reclamation in Central Asia. Tashkent. Publishing House: Mekhnat. p. 163. (in Russian)
14. Dukhovny, V.A. 1986. Maintenance of the on-farm networks at the new stage. Journal "Land Reclamation & Crop Yield" No. 1, Publishing house: "Kolos". Moscow, pp. 31–343. (in Russian)
15. Dukhovny, V.A., Avakyan, I.S. and Mikhaylov, V.V. 1989. Land reclamation, water sector, and socio-economic problems in Central Asia. Collected papers "Land Reclamation and the Water Sector" No. 9, p. 4. (in Russian)
16. Dukhovny, V.A. 1991. What has the Bloomington meeting shown? Journal "Land reclamation and Water Sector", No. 5, pp. 58–59. (in Russian)
17. Explanatory dictionary on water management of Turkmenistan (edited by G. Hojamyradov), Ashgabat, 2007, p. 174. (in Russian)
18. Gerasimov, I.P. and Gindin, A.M. 1976. Problems of transferring Siberian Rivers' water to desert plains of Central Asia. Proceedings of the 23th International Geographic Congress. Symposium "A Human Being & the Environment", Moscow.
19. Gerasimov, I. and Kovda, V. 1952a. "The Main Turkmen Canal", Moscow. (in Russian)
20. Gubarevich-Razdolsky, A. 1905. An economic essay about Bukhara and Tunis. Publishing House of Kirshbaum. St. Petersburg, p. 201. (in Russian)
21. Gumilev, l. 1984. Russian and Ancient Steppe, Moscow. Science, p. 312, on Russian.
22. Gumelev, L. 1990. Ethnogenesis and the Biosphere of Earth. Leningrad. (in Russian)
23. Journal "Bulletin of Irrigation" No. 3, 1923. p. 85, No. 9, 1924, p. 67. (in Russian)
24. Kadyrov, A.A. 2005. Water & Ethics. GWP, Tashkent, p. 103. (in Russian)
25. Khodjaev, F. About the revolution history in Bukhara and national demarcation in Central Asia. Collected Works, Volume 1, p. 289. (in Russian)

26. Kholdjuraev, Kh. 2003. Irrigation civilization. Publisher: "Umed", Khudjant, p. 510. (in Russian)
27. Kilchevsky, V.A. 1927. Districts of water and land reclamation associations. Journal "Bulletin of Irrigation" No. 1, pp. 27–35. (in Russian)
28. Krasnopolsky, A. 1915. Cotton-growing in Khiva. Chief Directorate of Land Management & Agriculture, the Cotton Committee, St. Petersburg, Printing Office of Kirshbaum, p. 20. (in Russian)
29. Kobori, I. and Glantz, M. 1998. Central Eurasian Water Crisis. (in Russian)
30. Kostyakov, A.N. 1956, Basic of irrigation. Nauka Moscow, p. 638. (in Russian)
31. Kostyakov, A.N. 1919. Chief elements of developing the irrigation systems and their study. Moscow, Printing Office of Kushnerev, p. 434. (in Russian)
32. Mamedov, A.M. 1965. Russian scientists and developing irrigation in Central Asia, Tashkent, pp. 78–84. (in Russian)
33. Martha Brill Olcott. 2005. Central Asia's second chance, Washington DC.
34. Masalsky, V.I. 1913. Turkistan Krai. St. Petersburg, Publisher: A.F. Devren. p. 861. (in Russian)
35. Matley, 1994. Agricultural development in Central Asia 1865–1963, 130 years of Russian dominance, Dune University press, pp. 294–295.
36. Glantz, M. 1998. Creeping environmental problems in the Aral Sea basin", UNU Press, Tokyo – New York – Paris, p. 35.
37. Micklin, Ph. P. 1991, The water management crisis in Soviet Central Asia, No. 905, Carl Beck Papers, Center for Russian and East European studies, University of Pittsburg.
38. Central Statistic Office of the USSR. 1984. National Economics of the USSR in 1983. Moscow. (in Russian)
39. Palen, 1911. Report on social – economic survey of Turkistan, Tashkent. (in Russian)
40. Pokrovsky, S. 1926. Bulletin of Irrigation, N 7. Published in Central Asia, p. 139. (in Russian)
41. Predtechensky, A.A. 1921. Agriculture and Objects of Irrigation in the Zeravshan Valley. (in Russian)
42. Proceedings of report "Data of available land resources suitable for irrigation in Turkestan", 1923, Moscow, p. 64.
43. Proceedings of the IRTUR, p. 239, 1921, Moscow. (in Russian)
44. Rabochev, I.S. 1973. Improving integrated land reclamations for soil desalinization. Journal "The Cotton Business" No. 11, Publishing House: Kolos, Moscowl. (in Russian)
45. Rauner, S.N. 1887. The artificial irrigation of lands, St. Petersburg, p. 156. (in Russian)
46. Rizenkampf, G.K. 1921. Problems of irrigation in Turkistan. Proceedings of the Central Asian Water Management Administration, Moscow, p. 144. (in Russian)
47. Sadykov, A.S. Editor-in-Chief. 1975. Encyclopedia: Irrigation of Uzbekistan. Volume I, Publisher: "Fan." Tashkent, p. 352. (in Russian)
48. Sadykov A.S. Editor-in-Chief. 1975. Irrigation of Uzbekistan, Volume 2, Tashkent, "Fan" p. 360. (in Russian)
49. Sarkisov, M.M. 1992. Irrigation in South Turkmenistan, Publishing House: Geoinformmark, Moscow, p. 134. (in Russian)
50. Savitsky A. The Land Ownership in Turkestan, Tashkent, 1963, p. 11. (in Russian)
51. Schutter, J., Dukhovny, V. and Tuchin, A. The South Prearalie – new perspectives. Tashkent, 2000, the SIC ICWC, p. 147.
52. Shastal, I. 1923. Land reclamation associations and their significance for Turkistan, Tashkent. (in Russian)
53. Solovyev, S.M. 1989. Readings and stories about Russian history, p. 768, Moscow. (in Russian)
54. The History of Uzbekistan. Volume 1, Tashkent, Fan, p. 176. (in Russian)

55. Gerasimov, I.N., Kovda, V.A. (eds.). 1952b. The Main Turkmen Canal. Moscow, Nauka. (in Russian)
56. Sredazgiprovodkhlopok Institute. 1974 (revised in 1987). The Scheme of Integrated Water Resource Use and Protection in the Amudarya River Basin. (in Russian)
57. Trombachev, S.M. 1924. On the issue of rehabilitating the water sector in Turkistan, Publisher: Central Asian Water Management Administration, Moscow, pp. 3–11. (in Russian)
58. Tzinzerling, V.V. 1927. Irrigation in the Amudarya Basin. Publisher: Central Asian Water Management Administration, Moscow, p. 736. (in Russian)
59. Weidel, H. Dukovny, V. and Rubin, M. 2004. Economical assessment of joint and local measures for the reduction of socio-economic damage in the Aral Sea coastal area. Vienna-Tashkent. SIC ICWC, p. 143.
60. Weinthal, E. 2002. State making and environmental cooperation, The MIT Press, Cambridge, London, pp. 78–79.

The Sudoche Wetlands in the Amu Darya Delta 5 years after restoration

4

Water for Independent States

– Apple of Discord or Subject for Collaboration

4

Water for Independent States – Apple of Discord or Subject for Collaboration

The Soviet state ceased to exist on December 8, 1991, 74 years since the October revolution and after 69 years of official existence as the union state that was created in 1922. This event was announced in the Belovezhsk Agreement, signed by the heads of Russia, Belarus and Ukraine and ratified by the parliaments of these countries a few days later. Although the leaders of the Central Asian republics did not attend the Belovezhsk meeting, they got the message and readily assumed power. The newly independent states had a good opportunity to develop their own strategy for using their natural and economic potential and their labor force, but it would take time for each republic to identify and develop their own strategies. They understood that it would be impossible to manage the water resources of transboundary rivers separately, which led them to sign a protocol to establish the Interstate Commission for Water Co-ordination (ICWC) and the Agreement on Co-operation in the Field of Joint Water Resources Management and Conservation of Interstate Sources (October 12, 1991). A variety of treaties and establishment of more interstate organization followed thereafter. As far as actual water management regimes were concerned former Soviet State agreements remained in place to a large extend.

It soon became apparent that a large scale program for technical assistance and investment would be required to continue the operation of irrigation works, which had greatly suffered during the last decades of the communist period. Many donors took an interest in this "region of disaster" and formulated the Aral Sea Basin Program one. The implementation of this program (with an initial cost of US$41 million, later revised to US$60 million) may be assessed in different ways. Its effectiveness was variable, but it did play a significant role. First, it consolidated the regional efforts aimed at solving some key problems and, second, it assisted in the search for solutions that were later developed in the course of follow-up activities by both national and regional organizations. Not all projects funded under the ASBP-1 were successfully organized, but some of them had merit which should be noted. The chapter extendedly deals with issues related to the cooperation between donors and international financing agencies and the first steps made on the way to modernization of the sector.

The further growth towards independence showed considerable variation in the political and economic approaches adopted by the five states and led to different transformation processes with a differing interpretation and implementation of adopted decisions and agreements depending on the actual economic and geopolitical conditions in each state. Both Kyrgistan and Tadjikistan for example started to increase exploitation of their relatively cheap hydropower at the cost of the irrigation water requirements downstream. The chapter presents an analysis of potential and output of the economies of the five Central Asian States. A key role is played by the liberalization policies of the different countries with land ownership as a central variable. Although the Aral Sea Basin is often characterized as one of the basins in the world with very scarce water resources, this does not really correspond to the facts because the available water in principle, should be sufficient. The water resources per capita still considerably exceed UN indicators for extreme water deficit even under current population growth rates. The key towards the future is not so much in quantity of resources in the region, but in how the, now independent, countries will cooperate to share, which is also the lead element in all literature that is produced about the development of future Central Asia.

4.1 The First Steps – New Expectations

The Soviet state ceased to exist on December 8, 1991, 74 years since the October revolution and after 69 years of official existence as the union state was created in 1922. This was announced in the Belovezhsk Agreement, signed by the heads of Russia, Belarus and Ukraine and ratified by the parliaments of these countries two-to-four days later. Three elbow-benders[1] gathered for hunting in the Belovezhsk Pushcha Reserve[2] and, in the course, of their trip liquidated this vast country. After the putsch in August 1991, the leaders of the Soviet state had practically lost power and Gorbachev and his immediate retinue could not keep this country in their irresolute hands.

Although the leaders of the Central Asian republics didn't attend the Belovezhsk meeting, they heard its message and were quite ready to assume power. Centrifugal political aspirations were already apparent in Central Asia in 1990. On March 24, 1990, Islam Karimov, at that time the First Secretary of the Central Committee of the Communist Party of Uzbekistan, was elected President of the Republic of Uzbekistan (though this, at the time, remained within the political framework of the USSR) at the session of the Supreme Council of the Uzbek SSR. On August 24, 1990, the Supreme Council of Tajikistan adopted a declaration "on the sovereignty of Tajikistan". In October 1990, at the sessions of the Supreme Councils of Kyrgyzstan and Turkmenistan, Askar Akaev the former president of the National Academy of Science, was elected president of Kyrgyzstan, and Saparmurat Niyazov, who had a similar role to Karimov as the First Secretary of the Central Committee of the Communist Party of Turkmenistan, was elected president of Turkmenistan.

The parade of sovereignties started right after the putsch in August. Uzbekistan was again the first to make a move. On August 31, 1991, the republic declared its independence, and on December 29, 1991, Islam Karimov was elected the first president of this country by a majority of votes (84%). On September 9, 1991, the Supreme Council of Tajikistan adopted a resolution declaring the state independence of the Republic of Tajikistan". Similar resolutions were adopted by the Supreme Council of Kyrgyzstan on October 12, 1991, by the Supreme Council of Turkmenistan on October 27, 1991, and by the Supreme Council of Kazakhstan on December 1, 1991. Thus, the geopolitical unit with a single hydrographic network in Central Asia had turned into separate state entities with all subsequent implications.

This political transformation immediately created a platform for the reappraisal of all approaches to managing water resources and irrigated agriculture. This event was expected but, as the last chapter showed, it had not been prepared for sufficiently. With federal governance of water resources and land reclamation activities, there was a "regular arbiter". The individual republics had developed a habit of

[1]People who like to drink a lot of alcohol in a jolly crowd.

[2]The Belovezhsk Pushcha Reserve is in Belarus. Part of the oldest existing European forest, it is the sanctuary of the almost extinct European bison.

acting circumspectly – of asking "Oh, my God! What will Princess Marya Aleksevna say?"[3] Now each country had to develop its own policy, but taking into account the interests of its neighbors.

Some legacies of the Soviet epoch undoubtedly had a positive influence upon the present and future development of this region. These are some of the positive features.

- As far back as the 1960s the so-called integrated approach to the development of virgin lands was adopted, first in the Golodnaya Steppe, and then in the Karshi Steppe, Jizak Steppe, Syrkhan-Sherabad Steppe, Kyzylkum Steppe, Yavan-Obykiik Steppe and other regions, where irrigation and drainage systems were being built at the same time as new settlements, factories, roads, communication lines, etc. These projects demonstrated to water professionals and economists the advantages of this advanced method of operating and maintaining irrigation systems. These developments actually took place long before the modern concepts of integrated water resource management (IWRM) were introduced all over the world.
- The high level of water education and scientific research established a sound base for building up water resource management expertise.
- Through water professionals in different republics of the former USSR taking part in joint activities during the Soviet period under integrated supervision and according to uniform standards, rules, methods and approaches, the appropriate conditions were established for the future co-operation of generations of water specialists. This provided the inspiration to retain the type of co-ordinated actions developed in the Soviet time.
- In the last six-to-eight years before the collapse of the USSR, the Soviet government had given much consideration towards improving the socioeconomic and environmental situation in the Aral Sea basin. It approved the State Aral Sea Program (1986), established two BVOs (river basin management organizations), and allocated considerable finance to implement various projects in the field of water supply and social rehabilitation.

In the water sector, this created the necessary prerequisites for a smooth transition from one political system towards another or, as one might say, "going from imperfect socialism towards initial capitalism, under different paces and methods of reform in the various countries of the region".

The newly independent states now had a good opportunity to develop their own strategy for using their natural and economic potential and their labor force without having to watch for the reaction from Moscow. It would take time for each republic to develop their own strategies. However, the management and especially use of water resources drawn from international or transboundary rivers (different terms do not change the fact of the matter) would require the preservation of their unity. There was a deep insight that it would be impossible to manage the water resources of transboundary rivers separately, and this led the leaders of water departments to sign a protocol to establish the Interstate Co-ordination Water Commission (ICWC) and the Agreement on Co-operation in the Field of Joint Water Resources Management and Conservation of Interstate Sources (October 12, 1991). The heads

[3]This famous aphorism from the comedy *Wit Works Woe* by A. Griboedov is a synonym for dependence on another's opinion.

of the water departments of the Central Asian republics and Kazakhstan also adopted a declaration in which, without assistance from Moscow and as real professionals, they evaluated the difficult situation arising from the growing water deficit and environmental tensions in the Aral Sea basin. Drawing on the historical unity of the Central Asian nations and their equal rights and duties in respect of rational water use in the region, they declared their recognition of the exceptional features of the closed hydrographic Aral Sea basin, and thereby asserted the common interests of all Central Asian republics in the mutually beneficial use of the water resources of this region. Acknowledging the integrity of the common watershed that is drained by the Aral Sea (which was now drying up), the declaration asserted the need for developing the principles of fair regulation of water consumption taking into consideration the interests of all nations of the region. The parties to this agreement made a commitment to avoid unilateral actions that could adversely affect neighboring republics.

Intense work by representatives from all these countries allowed to the draft an agreement to be co-ordinated with all national governments in just four months. On February 18, 1992, the ministers of water resources from all Central Asian countries – N. Kipshakbaev (Kazakhstan), M. Zulpuev (Kyrgyzstan), A. Norov (Tajikistan), A. Ilamanov (Turkmenistan), and R. Ginyatullin (Uzbekistan) – signed this basic agreement. It had an enormous significance for supporting co-ordinated actions by the five countries in the field of joint use and conservation of water resources in the transboundary rivers.

This framework agreement contained some important regulations that laid the basis for the future co-operation:

- it recognized the community and unity of the region's water resources, established that the parties have equal rights of use and responsibility for ensuring rational use and protection;
- each party to the agreement was obliged to prevent actions on its territory which could infringe on and cause damage to the interests of the other parties, including deviation from agreed values of water discharges and pollution of water sources;
- the parties decided to establish, on parity conditions, an Interstate Co-ordination Water Commission (with one member from each country) to deal with problems of the regulation, rational use and protection of water resources from interstate sources – the commission membership would include authorities of water management agencies, and quarterly meetings were envisaged with special meetings as required on the parties' initiative;
- the parties agreed to information exchange on scientific and technical progress in water economy, integrated use and protection of water resources, and they agreed to conduct joint research for the scientific and technical support of all problems and to develop expertise in water-related projects;
- agreements could only be changed or supplemented by way of joint consideration of all parties to the initial agreement on the basis of consensus;
- the ICWC and its executive bodies would strictly observe the release regimes and water use limits established by the rules and schemes of the former Soviet governments;
- democratic procedures would be adopted for ICWC activities.

It was very important that the basin water management organizations – the Amudarya BVO and the Syrdarya BVO – that were established by a resolution of the Soviet government four years before the creation of the ICWC were already standing on their own feet and thus could provide the organizational basis for the ICWC (this was reflected in article 9 of the agreement). This agreement stated that "the Syrdarya and Amudarya basin water management organizations shall act as the executive and interdepartmental organs of the Interstate Co-ordination Water Commission, and shall function on the condition that all structures and facilities on the rivers and the water services belong to the corresponding republic and are deemed transferred for temporary use without the right of transfer and redemption".

At the first meeting of the ICWC at the end of 1992, the statute of this commission was approved together with regulations for the use of the Amudarya and Syrdarya water resources. The Scientific-Information Center (SIC) was assigned as a third executive body of the ICWC. This initially functioned within the framework of SANIIRI, but since 1996 it had been a self-sufficient body. When, at a heads of state meeting in January 1993, it was decide to establish the Interstate Council on the Aral Sea (ICAS), the ICWC members immediately started preparations for the next meeting of the heads of state to be held at the end of March in Kyzyl-Orda City. This meeting officially proposed to establish the International Fund for the Aral Sea (IFAS) and the Interstate Council on Aral Sea Basin Problems (ICAS). The first body was tasked with attracting funds from regional countries and international donors (playing a similar role to the Rockefeller Foundation, Carter Foundation, etc.). The second body was established for developing the action plans and implementation schedules for those plans. Participants at this meeting insisted on an exact delineation of the functions of these bodies. IFAS would be a voluntary association of all regional countries and international donors, but ICAS would be an interstate body vested with specific powers by its founders and responsible to the peoples of this region for solving the Aral Sea basin problems. Regional monitoring of all water resources and water-related natural resources was planned as a first priority.

On March 26, 1993, in Nukus, the heads of the Central Asian states together with the representatives of the Russian Federation signed a very important agreement entitled "On Joint Actions for Mitigating Effects of the Aral Sea Crisis; Rehabilitating the Environment and Enhancing Socioeconomic Development in the Aral Sea Basin". The Interstate Council on the Aral Sea Basin (ICAS) with parity representation of all regional states, the Executive Committee of ICAS (EC-ICAS), a secretariat for ICAS and the Interstate Commission for Socioeconomic Development and Scientific, Technical and Ecological Co-operation (later renamed as the Sustainable Development Commission) were all established through this agreement (see Figure 4.1). The inclusion of the ICWC on the list of interstate bodies of the Aral Sea basin provided legal status to the signatures of water ministers from the five Central Asian countries and allowed them to develop co-operation in the field of water resources management in a quite successful manner. This enabled the development of specific procedures for establishing water use limits, operational monitoring procedures and the means to implement co-ordinated actions of the commission and its executive bodies.

The ICWC and its executive bodies were the co-authors of the Action Program for Improving the Socioeconomic and Environmental Situation in the Aral Sea Basin (The Aral Sea Basin Program). This significant document was approved by the heads of state on January 11, 1994 and presented to the

ARAL SEA BASIN
REGIONAL ORGANIZATION CHART

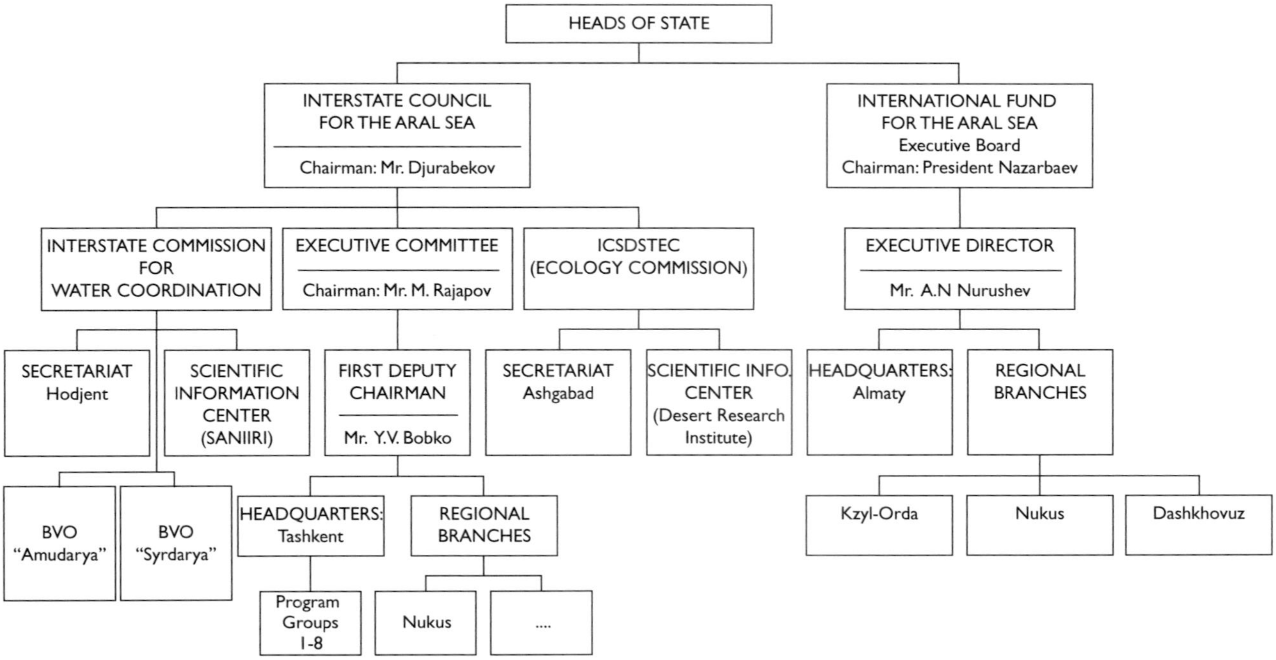

Figure 4.1 Aral Sea basin regional organization chart (after SIC-ICWC).

donor community at a donors' meeting that was jointly held by the World Bank and the UN in Paris on June 24, 1994.

These actions of the leaders and governments of Central Asian countries were well received by the international community. Guy Le Moigne (1994), a senior adviser at the World Bank, wrote in his paper: "After acquiring their independence in 1991, the countries located in this basin were aware of water problems and therefore were conscious of the need to co-operate to tackle these problems." The heads of the Central Asian states organized several summits (in Almaty in 1992, in Kyzyl-Orda in 1993 and in Nukus in 1994) to discuss issues arising from co-operation in the field of joint use and protection of transboundary water resources. The international financial institutions and the donor community were also concerned by the gravity of the problems. With varying success, they attempted to support the development projects that required regional co-operation for stabilizing the Aral Sea water level and rehabilitating the adjacent disaster area.

In September 1995, the heads of state organized the UN Conference on Sustainable Development in the Aral Sea Basin and signed the Nukus Declaration of Central Asian states and international organizations. In this declaration, they recognized the validity of the agreements that they had already made and the other instruments that regulated their mutual relations in the field of water resource management in the Aral Sea basin and made a commitment to ensure their strict observance. The declaration also made an appeal to the international community to assist Central Asian countries in solving the problems related to sustainable development and improving the environment in the region.

The text of this document was signed by the heads of all five states – the President of Turkmenistan, S. Niyazov, didn't attend the conference, but signed the document later – and retains its significance today, although at present, not all signatories strictly adhere to its principles. These are the declaration's main commitments.

1.4. We reaffirm our commitment to comprehensive co-operation at a regional level on the basis of mutual respect, neighborly relationships and our aspiration to work for the sake of overcoming the consequences of the environmental crisis in the Aral Sea basin and its impacts on nature and human beings.

1.5. We declare our adherence to the principles of sustainable development and consider that to realize this it is necessary to undertake the following: i) to recognize the significant value of water, land and biological resources as the basis for sustainable development; ii) to shift towards a more balanced and scientifically substantiated system of agriculture and forestry; iii) to enhance irrigation efficiency by means of developing economically driven methods of water resource use and applying advanced technologies for irrigation and conservancy; and iv) to create incentives for long-term reform of land and water resource use.

1.7. We recognize the need to improve the integrated system for managing natural resources in our region by means of the following measures: i) establishing a regional system of monitoring of the environment and especially the status of water resources; ii) creating a regional information sharing system related to monitoring of the environment; and iii) harmonization of environmental standards and environment-related legislation.

It is significant that already by this time all heads of state agreed (in paragraph 1.9) to completely adhere to relevant international conventions, including the UN Convention to Combat Desertification, the UN Framework Convention on Climate Change, and the Convention on the Protection and Use of Transboundary Watercourses and International Lakes (the UNECE Water Convention). Unfortunately, the UNECE Water Convention was only ratified by Kazakhstan and Uzbekistan. Moreover, later, two countries in the region (Kyrgyzstan and Tajikistan) demonstrated complete unwillingness to sign this convention and even opposed the use of the term "transboundary watercourses".

A characteristic feature of this period, in relation to creating a legal foundation for regional co-operation, was the openness and unity in the interactions between the water organizations and political bodies, as well as in dialogue with the World Bank representatives, who encouraged all parties to work rapidly towards achieving regional consensus and approval for the Aral Sea Basin Program (ASBP).

Special emphasis should be put on the outcome of the heads of Central Asian states summit that was held in January 1994 in Nukus. A concept for improving the socioeconomic and environmental situation in the Aral Sea basin was adopted at this summit with the following priorities:

- safe water supply to the settlements in the Aral Sea disaster area and developing a network of public health and child care institutions;
- control of chemical contamination of the surface and ground waters by adopting new biological protection technologies in the region;
- restoration of wetlands and wetland ecosystems in the Amudarya and Syrdarya deltas;
- improving water resource management at the interstate level;
- preparing a set of legal instruments and procedures for water resource management by drawing on examples available in international practice;
- developing a Syrdarya Basin Automatic Control System (BACS) and progressing with the elaboration and introduction of a Amudarya BACS;
- inclusion of activities related to monitoring of surface and ground water quality into the operational program of BWOs and assessment of water resource management in the river deltas;
- improvement of forecasting methods of annual water availability, development of a hydrometric observation network and a system of satellite-based visual observations of snow and glacier fields;
- joint monitoring of natural events affecting the quantity and quality of water resources;
- establishing a common policy for joint activities related to water conservation and the phased reduction in unit water consumption in the agricultural sector for the benefit of regional sustainable development.

This document actually represented the opinion of the governments of all Central Asian states with respect to the necessity to preserve the Aral Sea as a natural object and to improve the socioeconomic and environmental situation in the regions adjacent to the Aral Sea and subject to the disaster.

The measures that had been implemented since 1989 were aimed at the improvement of the supply of water into the deltas and restoration of their system of lakes, which had substantially improved the microclimate, as well as to initiate the recreation of sustainable and active ecosystems in the deltas and on the exposed sea bed. These activities had helped to rehabilitate and increase the productivity of fishery activities and muskrat breeding, to recreate breeding grounds for migratory birds, to reduce dust and salt transfer from the dried sea bed, and to recover much local flora and fauna. There were also many plans to recreate active and controlled ecosystems to provide a sustainable rehabilitation of the disturbed natural equilibrium in the region adjacent to the Aral Sea.

Activities related to creating manmade ecosystems in the deltas and on the dried sea bed were considered the top priority from the point of view of conservation measures within the region directly adjacent to the Aral Sea. Based on available water resources, these included the following works: i) creation of a system of regulated water bodies in the Amudarya delta and the water infrastructure for managing the Northern Small Sea in the Syrdarya delta; ii) tree planting for the fixation of sand blown over the region, which threatened infrastructure and human habitats; iii) disposal of drainage water into the Aral Sea via the "dispersed" zone; and iv) construction of so-called polder systems[4] on the dried sea bed.

[4]This refers to ideas to develop large water bodies with controlled water management systems in the delta of the Amudarya.

To support this program with action on the ground, it was intended to implement some water supply and sanitary projects under the IFAS umbrella, using the funds of Central Asian states and international donors. Therefore the ASBP-1, consisting of 9 key programs and 19 projects (Table 4.1), was prepared and presented to the international donors. At the donors' meeting in Paris (held on 23–24 June 1994), this program was adopted for implementation and the World Bank was appointed as the agency

Table 4.1 Progress of implementation of the ASBP-1 as of January 1, 1997 [US$ .000] (World Bank data).

Name of program	Estimated cost	Total allocation	Committed	Outcome
Program 1.1 Regional water resources strategy	8,000	1,745	1,445	Principal provisions of the regional strategy
Programs 1.2; 1.3 Improvement of water management efficiency; and Dam safety and reservoir management	2,000	300	–	
Program 2 Hydrometereological network and Regional Environmental Information System	3,000	–	–	Proposals for development
Program 3.1 Water quality management	7,500	675	675	Remote sensing technology
Program 3.2 Uzbekistan drainage program	5,000	1,750	1,750	Principal positions WARMIS
Program 4 Wetlands restoration and Syrdarya Control	6,700	3,480	1,640	Feasibility study for Lake Sudoche and the Small Northern Sea
Program 5 Clean water & health	13,000	5,940	2,930	Design water supply system in Nukus, Dashhowuz, Kyzyl-Orda
Program 6 Upper watershed management	3,000	600	500	
Program 7 Automation of water infrastructure	1,500	300	100	Prefeasibility study for the automatization system
Program 8 Capacity building	9,100	5,700	5,700	
Planned	60,800	22,250	16,780	

responsible for co-ordinating and monitoring this initiative. The UN also offered its services to help co-ordinate this program, citing its experience in implementing the Mekong River Basin Project, but Central Asian countries eventually preferred to use the World Bank. This decision may have been ill advised to some extent.

The implementation of this program (with an initial cost of US$41 million, later revised to US$60 million) may be assessed in different ways. Its effectiveness was variable, but it did play a significant role. First, it consolidated the regional efforts aimed at solving some key problems and, secondly, it assisted in the search for solutions that were later developed in the course of follow-up activities by both national and regional organizations. Not all projects funded under the ASBP-1 were successfully organized, but some of them had merit which should be noted.

The key findings of Sub-Program 1.1 were presented in the report *Developing the Basic Provisions of a Regional Water Strategy for the Aral Sea Basin*. This was prepared with participation of the World Bank (led by Professor Y. Kindler, task manager), and was the outcome of the activities of five national groups and a regional group consisting of representatives from the five Central Asian countries and regional organizations. Until now, this is the only document that was agreed to by representatives of all countries and, at the same time, was a crucial instrument for further regional development. Another important project was the TACIS project – Water Resource Management and Agricultural Production (WARMAP) in the Central Asian Republics. Both projects – Sub-Program 1.1 and TACIS – were mainly implemented by local experts in collaboration with foreign consultants, and enabled water professionals from the five Central Asian republics to exchange views with their regional and foreign colleagues in the course of roundtable discussions and seminars. Based on their analysis of local and western technologies and the procedures employed in water resource management, they prepared reports with sound recommendations that could be put into practice. The most significant components of the Sub-Program 1.1 and TACIS project were: i) a water use and farm management survey (WUFMAS), under which field surveys were conducted in sample farms selected in different regions; ii) the Water Resource Management Information System (WARMIS); and iii) a legal framework for the regional water strategy. The joint work in these projects created a framework which allowed efficient co-operation between representatives of the different countries and organizations and created a common basis for future development.

The projects financed by USAID (on improved management of critical natural resources, including energy, and others) were less effective due to the lower level of involvement of local initiatives and knowledge and an insufficient focus on practical results. At the same time, several projects with much smaller budgets (US$ 0.2–1.5 million) compared to the USAID projects were implemented by local specialists (SIC and BVOs) with the participation of the national ministries of water, agriculture, ecology, etc. and with the assistance of various sponsors, including CIDA, SDC, NATO, INCO-Copernicus and others. There were several advantages to this local approach, including:

- direct relationships with the executive agency that participates in the preparation of the projects;

- highly efficient investments, because of the low labor cost for very experienced local personnel;
- practical assistance to local specialists, rather than trying to transfer western theoretical knowledge that was not regionally adapted;
- focusing on the principal targets of special interest for the region;
- the representatives of different countries have an opportunity to come to a common point of view and make mutual commitments in the process of joint work in these projects.

One should note here projects such as Developing the Hydrometeorological Service, Water supply in the lower reaches of the Amudarya and Syrdarya rivers, and Preserving the deltas of the Amudarya and Syrdarya rivers. In subsequent years, these projects were supported by:

- establishing the United Land and Water Resources Information System (a project financed by the Swiss Agency for Development and Co-operation (SDC)), which still is successfully functioning and at present being further developed under the aegis of the ICWC (www.cawater-info.net);
- numerous water supply projects in Kazakhstan, Tajikistan and Uzbekistan, which have contributed to solving problems related to potable water supply in these regions;
- the wetlands restoration project in the Amudarya delta, where the pilot project on Lake Sudoche provided a sustainable water supply into the western part of the Amudarya delta and proved the feasibility of maintaining such wetlands using scanty water resources and with limited investment;
- the Northern Sea and stabilizing water supplies into the Syrdarya delta (SYRNAS) project – a joint project of the World Bank and the State Water Resources Committee of Kazakhstan – which resulted in the creation of a separate Northern Sea and improved the carrying capacity of the downstream section of the Syrdarya riverbed.

Efforts to establish a legal base for interstate water relations were initiated in the framework of the WARMAP project, in which the participation of Italian and Israeli specialists merits special attention. The water management bodies, encouraged by international consultants, donors and financial institutions (through the TACIS project, World Bank, etc.), came to realize that the existing framework agreement had to be enhanced with a set of detailed agreements that would regulate water-related activities at the interstate level. The experience in the first three-to-four years following the end of the Soviet Union (1992–1995) showed that the ICWC had quite efficiently solved current tactical issues and was seeking to be an initiator of new ideas for achieving the strategic goals of the region.

The significance of strengthening the legal base became obvious in 1994, when the interstate water management bodies displayed indecisiveness and there was no agreement between the national governments on how to solve many operational issues related to water management. The minutes of the ICWC meetings already attracted the attention of all ICWC members and national governments. It was clear that only with their help and with the full co-operation of the riparian countries, could the complex activities needed to solve water-related problems of the ICWC executive bodies (see Table 4.1) become a reality.

The memorandum that summarized the results of the working visit of water ministers and members of the ICWC to the European Union (Belgium, Italy and Germany) in December 1995 is worth quoting:

By understanding our responsibility for a way out of the crisis situation in the Aral Sea basin, which is unprecedented in human history in scale and complexity, we declare:

- *that the ICWC member countries share a common awareness of the need for joint efforts to improve water resource management in the basin for the benefit of our nations, with the Pre-Aral region being recognized as a separate water consumer;*
- *based on our previous agreements and taking into account the joint works in the framework of Sub-Program 1.1 and WARMAP, the ICWC member countries clearly realize the ways to tackle these problems and are ready to implement joint and synchronous measures both at the interstate and national level that should promote creation of mechanisms for harmonious water resource management and use;*
- *the ICWC places great hopes on the elaboration of a set of legal and standardizing documents as a rules of play for all participants in water resource management and water resource use. This development should have an evolutionary nature and be based on "interim milestones" which take into account national interests and regional requirements as building blocks towards sustainable consensus. The only precondition is the readiness to make concessions to each other under parallel appraisal of national and regional aspects.*

The participants of a workshop organized to discuss the outcomes of this working visit came to the conclusion that "the development of the legal and institutional base for water resource management at the regional level is the main task of our future joint activities. At the same time, environmental aspects should be also taken into consideration in regional agreements[5]." To do this, the following interstate agreements had to be developed:

- an agreement enhancing ICWC rights and liability;
- an agreement on the use of water resources under current conditions;
- an agreement on ensuring the environmental sustainability in the region, on water resource protection and on rules of monitoring water and environmental quality;
- an agreement on joint planning and management of water resources.

In addition, it was deemed necessary to enhance or revise the mandates of all ICWC organizations with the purpose of providing compliance with the provisions of international water law. The basic aim of these measures consists in enhancing the ICWC's power and mandate and transforming this commission into a regional water authority. While revising the institutional framework, it was necessary to enhance the participation of all countries in the activities of the executive bodies of the ICWC and to establish branch offices in each country.

One of the more important achievements of this period was the practice of preparing joint documents and recommendations with the participation of national thematic groups. A typical example is

[5]Archive of SIC-ICWC-unofficial papers in the collection of the authors.

the work to prepare programs 1.1, 2.2, 3.1 and others. When preparing the document entitled Principle Provisions of the Regional Water Strategy, each country formed its own national working group with two water experts, a power engineering specialist, environmentalist and economist. The five heads of the national working groups were members of the regional working group, together with representatives of the EC-ICAS, SIC-ICWC and both the BVOs. The national working groups (NWG) each prepared their water visions, which were then integrated into a regional vision document. The documents were first co-ordinated and combined by the members of the regional working group (RWG) and then by all the members of the NWG. Complete consensus was usually achieved as a result of this joint working process (see Figure 4.2). It should be noted that prior to 1997, there was a greater willingness to reach mutual understanding and stronger aspirations to reach the compromises needed for interstate cooperation. Later on, growing national ambitions and an adherence to a doctrine of "absolute national sovereignty" started to affect the joint decision-making process more and more.

In October 1997, the executive committee of the International Fund for Saving the Aral Sea (EC-IFAS) organized a donors' meeting to review the implementation of the ASBP-1. The situation with regard to donor aid was found to be much more encouraging than had been expected during the first meeting of donors in June 1994.

The formation of IFAS, ICAS and two regional commissions (ICWC and Sustainable Development Commission) – and the preparation of their legal mandates – were substantial outcomes of the activities undertaken in the first years of the newly established regional structures. The support of the international community and the overall co-ordinating role of the World Bank was crucial at this stage. In the beginning, these different regional bodies were established in parallel but using a quite logical framework (see Figure 4.3). If IFAS could mobilize funds from regional countries and donors as planned, then ICAS could proceed with its work. However, IFAS could not mobilize these funds according to plan, owing to the fact that each country specified its own scope of work within its own national share of funding and allocated its funds to these works only. All this happened almost

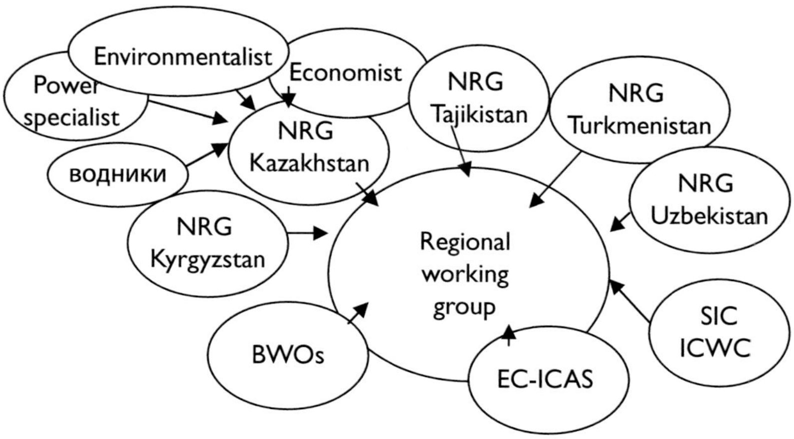

Figure 4.2 Flow chart showing the process of preparing the interstate documents (after SIC-ICWC).

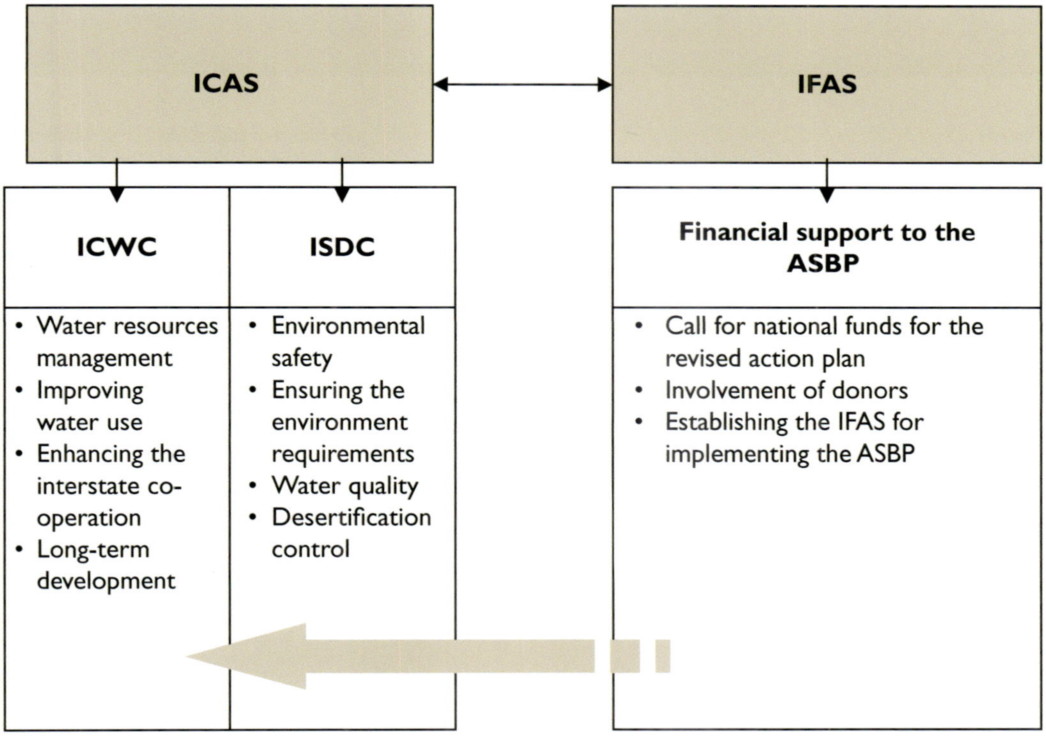

Figure 4.3 Organizational principles governing the interrelationships between ICAS and IFAS in the initial phase (after SIC-ICWC).

completely without participation of IFAS, which merely informed the executive directorate (ED) about these actions. Finally, and step by step, the IFAS (as represented by this executive directorate) turned itself into a Kazakh organization.

Nevertheless, the EC-ICAS, ICWC and SDC carried out a considerable scope of work (including institutional arrangements, procurement of computers and software and introduction of new technologies, establishing new ties with international agencies, etc.), which finally resulted in some actual progress towards the implementation of the ASBP-1. One of the advantageous outcomes of this activity was an agreement with the World Bank on submitting a new program to donors (for the extension of work initiated under the ASBP-1), that was aimed at attracting loans and investments of some US$470 million in total. Table 4.3 shows the scope of financing under this program (according to a World Bank report).

Figure 4.4 shows that after 1996 the World Bank paid less attention to regional water and environmental problems in Central Asia. In their review [38], the World Bank's experts admit the problems in transboundary water management but point out right away that "although the World Bank encourages solving the Aral Sea issues, it did not accept that the bank's participation in improving transboundary basin management should be a key part of the World Bank's policy for assisting the region and countries". Another equally remarkable statement is that the World Bank ascertains that

Table 4.2 Agenda of the ICWC meetings devoted to organizational and legal aspects.

Issues discussed	Reference number for ICWC minutes of meeting	Final decision made? (yes/no)
Ensuring taxation and customs supervision and border control on preferential terms for the interstate water organizations as provided by articles 9 and 10 of the action program signed by the EC-ICAS members	Minutes 8 (1994), 12, 13	no
Establishing regular information exchange between the ICWC executive bodies and national water management organizations. Developing the information system and regulations for its operations	Minutes 10 (1995), 12, 16, 19, 22, 24, 25, 26, 30	for most part
Ensuring environmental (minimum) base flows through the Syrdarya and Amudarya rivers	Minutes 12 (1995), 33, 34	no
Preparing and co-ordinating the main provisions of the regional water strategy	Minutes 12 (1995), 13	yes
Maintenance of riverbeds and flood control	Minutes 13 (1995)	partly
Allocating funds for O&M of water infrastructure on transboundary watercourses	Minutes 13 (1996), 18, 20, 22	partly
Lack of a mechanism for the examination of interstate projects	Minutes 15 (1996)	no
Lack of regulations for cooperation with donors	Minutes 15 (1996), 31, 32, 33, 34	partly
Unsatisfactory links with hydrometeorological services and the insufficient accuracy of hydrological forecasts	Minutes 16	partly
Strengthening activities in the framework of adopted agreements	Minutes 16, 18	partly
Elaborating regional and national criteria for water consumption in the region	Minutes 17	no
The necessity of involving prime ministers in ICWC activities	Minutes 17	no
The necessity of creating a water and energy consortium	Minutes 17	no

"the operating condition of water infrastructure has deteriorated drastically since 1990. [Prior to 1990] water infrastructure was well-developed and there was quite good access to it". However, the review noted that the World Bank will "support a dialog on water resource management between countries in the Aral Sea basin through IFAS and EC-IFAS". In fact, the World Bank switched its attention and

Table 4.3 The Aral Sea Basin Program Phase 2, in million US$ (World Bank and IFAS, 1997).

Sub-program	Total financing	Sources		Note
		Grant	Investments	
Program 1	11.0	10.0	1.0	Executed through GEF, WEAMP
Program 2	27.5	5.0	22.5	Partly implemented through SDC, USAID and EU
Program 3	141.5	12.5	129.0	By GEF, Netherlands and World Bank, but not completed
Program 4	83.0	6.7	76.3	Partly implemented by Netherlands, GEF and mainly the WB
Program 5	173.3	13.0	160.3	By World Bank and SDC. Loans were much exceeded
Program 6	8.0	3.0	5.0	Not implemented
Program 7	25.0	1.5	23.5	By CIDA and SDC with some participation of USAID
Total	469.3	51.7	417.6	

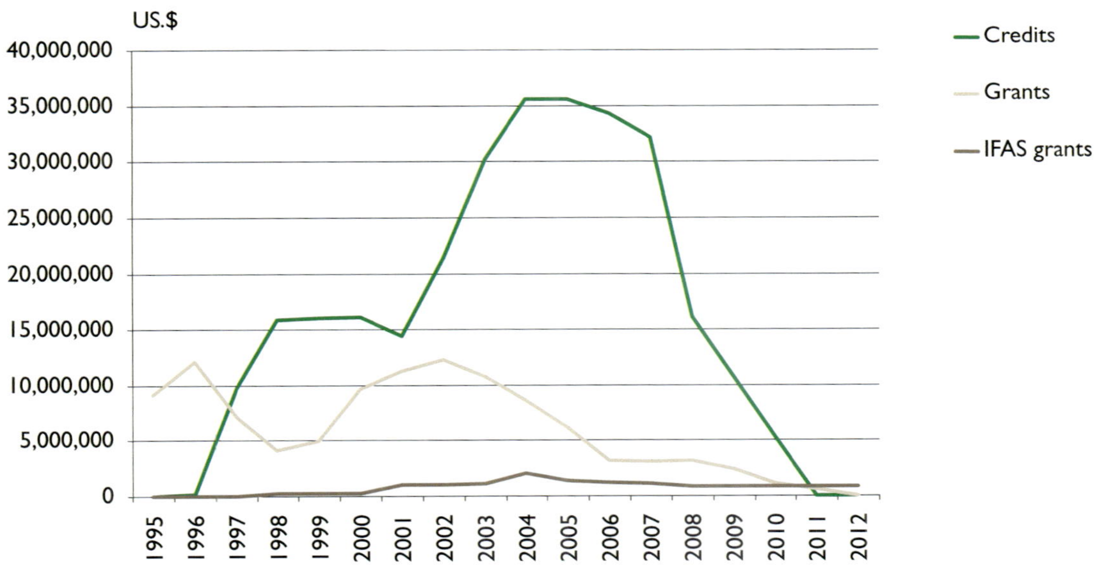

Figure 4.4 World Bank investment program in Central Asia between 1995 and 2012.

Table 4.4 Trends in the financing of IFAS programs, US$.

	1995	1996	1997	1998	1999	2000	2001	2002	2003
CREDITS	0	200,000	9,742,040	15,860,000	16,000,000	16,093,929	14,400,000	21,400,000	30,225,000
GRANTS	9,114,000	12,104,095	7,136,722	4,095,800	4,974,824	9,629,228	11,287,170	12,290,686	10,785,384
GRANTS OF IFAS	0	0	0	250,000	250,000	250,000	1,059,000	1,059,000	1,123,654
	2004	**2005**	**2006**	**2007**	**2008**	**2009**	**2010**	**2011**	**2012**
CREDITS	35,550,000	35,550,000	34,250,000	32,150,000	16,150,000	10,650,000	5,325,000	0	0
GRANTS	8,627,500	6,159,510	3,125,000	3,069,234	3,170,000	2,426,000	1,100,000	600,000	0
GRANTS OF IFAS	2,056,294	1,406,406	1,168,200	1,127,000	877,000	877,000	877,000	877,000	869,000

most of its funding to credit agreements, but with a big time lag in implementation (see Table 4.4). A team of UN and IFAS experts have published a general assessment of donor inputs in the 1995–2005 period (UN-IFAS 2008).

One of the typical characteristics of the activities financed by the World Bank and some other donors was poor project preparation, which resulted in delays in implementation and sometimes saw successive replacement of task managers, with responsibility passing between two or three different people. Let us consider the progress of the Drainage, Irrigation and Wetlands Improvement Project in Uzbekistan. This project, under the title "The Right-Bank Collector Drain," had been launched in the Soviet time, when work worth hundreds of millions of rubles was begun. After independence, the project was frozen and it was only submitted to the World Bank in 1995 as Sub-Project 4.3 in the ASBP framework. The project took six years to pass through all World Bank procedures and detailed design and implementation was only started in 2001. Phase I of the construction of this drainage canal was finally completed in 2009. This is very illustrative of the complexity of the bureaucratic procedures faced by Central Asian countries when seeking finance for projects from the World Bank and other international financial institutions.

4.2 Independence – A New Policy and Economy

The Soviet political and economic systems tried to support equal social and economic development in the regional republics, although it was clear that Kazakhstan and Uzbekistan were in leading positions for industrial development and for cultural, scientific and educational potential.

The State Planning Committee always skillfully combined key national aspirations with the interests of the federal state in the process of planning. As a result, all republics had sufficient supplies of water, gas and energy. The economic capacity created in Central Asia during the Soviet time was based on co-operation, as well as joint transport and power communications developed over 70 years, and this have closely linked the countries and the people of the region.

Independence provided new opportunities for development, but at the same time deprived the new countries of their former economic base and ties. These had to be created over again, and this meant forming simultaneously a political base, an economic foundation, a legal framework and a system of transnational relations. It was also necessary to take into consideration "the oriental image" of the political elites which come into power and deal with all the ensuing consequences. These elites were seeking to identify their national roots and traits, while not hiding their ambitions nor claiming supremacy over their neighbors. Erika Weinthal (2002) was right in saying that all five states rapidly ran away from the communist system and, although four of them proclaimed their status as a republic, their political structures, aspirations and ideals were both quite different from any well-known political model and quite different from each other. If, at the beginning, some countries tried to adapt a "Turkish model", other states started to introduce a "Western model" that, in essence, proved a rather far-fetched

Table 4.5 Comparative socioeconomic Indicators for Central Asian countries in the 1990s [US$ or specific](SIC ICWC).

Indicator	Kazakhstan*	Kyrgyzstan*	Tajikistan	Turkmenistan	Uzbekistan
Population, '000 people	2474.1	2364	5359	3346	20,606
GDP, US$ millions	3851.3	1874.2	4836	3334	20,213
Annual growth 1980–1990, %	206	229	137	112	143
GNP per capita, US$	1557	793	902	996	981
Industrial output, US$ millions	560	449	1348	533	4851
idem per capita, US$	240	190	252	159	235
Agricultural output, US$ millions	1093	533	1733	1500	8894
idem per capita, US$	468	225	324	448	432
Irrigated area per capita, ha	0.315	0.1741	0.1266	0.387	0.2031
Water supply per capita, '000 m3	4803	1994	2343	6603	3041
Power production kwh/capita,	441	5597	3376	4353	2728
Access to water sources, %	65	76	48	39	56

*For Kazakhstan and Kyrgyzstan within the Aral Sea basin (national statistics).

model. Some very distinctive features of the models adopted by these states were egoism, a domination of clannish and group interests, nationalism and a complete lack of understanding how to swim in the decentralized and poorly governed "sea of a volatile market".

Many consultants and advisers from Western European countries, the USA and international financial institutions hastily recommended liberal economic models although, in practice, these were unacceptable for all Central Asian countries. These models are problematic even in the West, as the 2008 financial crisis has shown. Even Kyrgyzstan, the most forthcoming country when it came to implementing the instructions of the World Bank and other Western benefactors, became hostage to attempts to cross-breed "the hedgehog with the grass snake". Eventually this combination led to a situation when the model for prosperity and stability resulted in a revolution of tulips with bitter fruits.

The first years of independence showed that the water industry was the only economic sector where the co-operation established in the Soviet time was likely to be supported, as centrifugal forces were playing a destructive role in all other sectors of economy and were affecting regional policies as a whole. Although the agreements signed by the heads of state (on February 18, 1992, March 26, 1993 and January 11, 1994) created a platform for co-operation on transboundary waters, it was impossible to keep the former status quo in respect of water allocation and water use. This was due to new geopolitical and economic forces shaping developments in the Central Asian region.

- Independence generated new political aspirations and there were many expectations that the new countries would be independent players in the world arena. Kazakhstan and Kyrgyzstan, especially at the beginning, proved to be adherents of close mutual co-operation, and worked together within the framework of different geopolitical alliances. There was a civil war fought in Tajikistan for a long time after independence, with the government struggling against various opposition groups. Tajikistan only started to introduce market principles into its economy, based on principles of close co-operation with its neighbors and the international community, after the civil war ended. Turkmenistan immediately proclaimed its independence through its policy of non-alignment to any blocks and alliances, as well as its willingness to make bilateral agreements. Uzbekistan also followed its own path, sometimes closely collaborating with all neighboring countries but simultaneously taking its own political line – an approach that in some aspects was close to the Turkmen policy. Uzbekistan also started participating in different alliances (such as CAEC, Shanghai Cooperation Organization, etc.).
- Independence immediately revealed certain advantages and disadvantages in the geographical disposition of the five countries and their natural resources. There are abundant mineral and, especially, fuel resources in Kazakhstan, Turkmenistan and Uzbekistan. These countries have considerable land resources per capita (with the exception of some densely-populated regions in Uzbekistan, such as Zeravshan Valley, Fergana Valley, Khorezm Province). Kyrgyzstan and, especially Tajikistan, lack of mineral resources and land resources, but have great hydropower potential. Moreover, all Central Asian countries with the exception of Kazakhstan and Turkmenistan are landlocked, necessitating complicated and expensive trade routes that impede ties with the world market. Agrarian (raw material) production was established during the Soviet period and remained strong, but other industrial sectors depended on Russia and this affected the progress of economic development in the region. All these factors meant that the newly independent countries were faced with a wide range of economic options. Kazakhstan was fully focused on free market approaches (with rather weak state interference in business, complete privatization including the transfer of land from public ownership, and a breakeven policy for all economic sectors including the water sector). The Uzbekistan and Turkmenistan governments retained a very strong regulative function and made gradual transitions towards market economies. Kyrgyzstan and Tajikistan took an intermediate position between these two extremes.

This considerable variation in the political and economic approaches adopted by the five states led to different transformation processes, and differing interpretation and implementation of adopted decisions and agreements depending on the actual economic and geopolitical conditions in each state. In the water sector this caused deviations from the water management principles and approaches that were established in the Soviet period.

- Due to an extreme deficit in fuel resources, Kyrgyzstan started to use the Naryn cascade infrastructure, which was constructed in the Soviet period, for the step-by-step replacement of expensive organic fuel with cheap hydropower. As a result, the operational regime of the Naryn Hydropower Scheme shifted from an irrigation regime to a hydropower production regime – that is, instead of accumulating water in winter and releasing water in summer, there was an accumulation of water in the summer and water release in the winter. In an attempt to prolong the supply of fuel resources from neighboring countries, Kyrgyzstan (in the opinion of many in Central Asia) proposed rather unfair agreements for sale of summer hydropower and barter arrangements for the supply of gas and coal from Kazakhstan and Uzbekistan at irrational prices. Although the 1998 agreement signed by Kazakhstan, Kyrgyzstan, and Uzbekistan stated the rules of play, it was difficult to implement because of the conflict of interest between power suppliers and fuel suppliers. Each tried to benefit at the expense of the other, and they could not agree on parity of conditions of delivery. As a result fuel and hydropower interests "captured" the water management schedules established for the hydropower stations on the Naryn-Syrdarya Cascade. Tajikistan adopted the same hydropower focused approach on the Vakhsh River.
- Thus, due to both internal and external pressures, irrigated agriculture – which from time immemorial was the priority for socioeconomic development and which continued to be the basis for generating livelihoods and employment (for up to 70% of rural population) – lost to a considerable degree its large and obvious economic profitability. A substantial factor here was the fall in world prices for agricultural outputs, with the price of rice falling from US$300 to US$150 a tonne, wheat down from US$200 to US$120 a tonne, and the price of cotton fiber halving (see Figure 4.5). This economic situation meant that farmers were unable to support the water

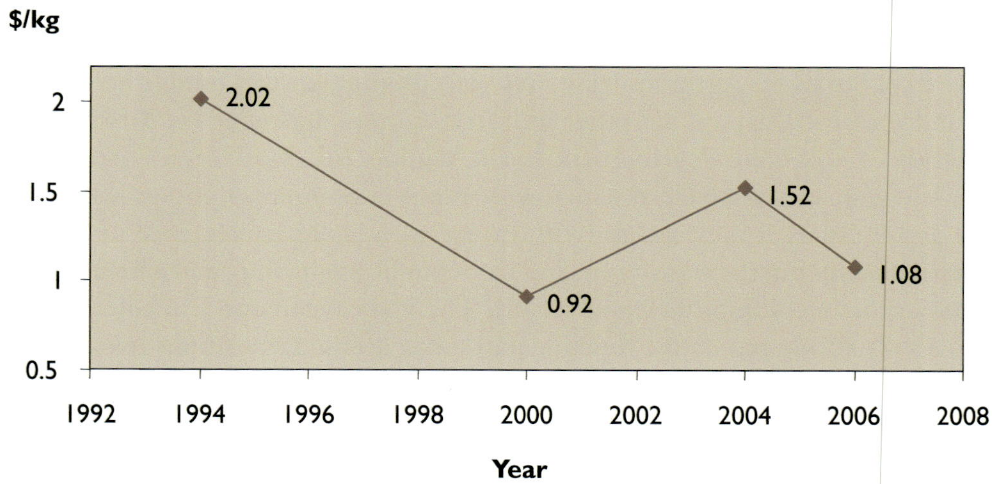

Figure 4.5 Price of cotton between 1992 and 2006 (source: *Cotton Look A*, 2007).

sector (as farm incomes were as low as US$100–200 per hectare compared with US$500–1600 per hectare in the past). However, irrigated agriculture retained its social significance, as together with related sectors it provided employment for 40% of the mainly rural population. Any irregularities in irrigation water supply caused by deviations from the co-ordinated schedules for water delivery resulted in huge social damage, which sometimes even verged on the catastrophic. The new conditions, which not only lacked guarantees for implementing the established procedure of water sharing by the riparian countries but which artificially imposed a water release regime that was unacceptable for most stakeholders along with increased prices for hydropower production (up to 8.5 dollarcent/kwh), made the existing "order" of water-and-energy exchange hardly sustainable.

- As the economies weakened in the Central Asian countries, resulting in a considerable drop in national income per capita, there was a drastic decrease in subsidies and support for the agriculture and water sector. There was much less procurement of agricultural machinery, fertilizers and other chemicals, which caused deteriorating conditions in the water infrastructure especially at the farm level. As a result water supply and land conditions drastically worsened, affecting the productivity of many crops.
- The introduction of market mechanisms into the agricultural sector (through privatization, and restructuring large state and collective farms into hundreds or even thousands of small private farms) was not accompanied by the creation of the appropriate infrastructure for the production of commodities and the water infrastructure necessary for water distribution and use. As a result many problems arose in servicing new private farmers (needing consultation, training and knowledge dissemination) and in procuring agricultural inputs.

The decrease in both overall incomes (which halved) and in profitability in the irrigated agriculture sector over the whole region resulted in impoverishment of the rural population and, at the same time, in the inability of agricultural producers themselves to protect their interests against the producers of hydropower or fuel that now act in a free market. In 1980, land profitability had reached US$2000 per hectare ha on average over the whole of Central Asia, but is now at about US$700 per hectare.

- New young staff have entered the management structures in these new economic conditions that are insufficiently trained in effective methods of managing and improving land productivity. In the past, more than half the district and provincial managers were agricultural and water specialists. At present, most local managers insufficiently realize that irrigation can only be profitable through proper water use according to actual crop requirements. It should also be noted here that local managers receive insufficient economic education and that there is interference by local administrations in the process of water allocation that interrupts equitable water distribution and impedes water supply to ecosystems, especially in dry years.
- Lack of funds has affected the hydrometric and meteorological networks, resulting in a lower accuracy of water-related weather forecasts and lowering the quality of planning and operative control over water resource use in the region. Although some international donors provide

technical assistance in this field, these actions have not been undertaken within a well-targeted program and have therefore been fragmentary and inefficient.

Unfortunately, all these problems have often not been noticed by some "uniformed" writers, who try to disseminate distrust of the actual activities of political and water organizations in the region. There has been speculation about the unwillingness of the Central Asian countries to co-operate in numerous publications and reports (see, for example, Ashok Swenn and Jerom Rivery, *Managing trans-boundary freshwater, causes for preventing conflicts*, SWP-CNP Report, August 2001) or about present and potentially future water conflicts (see, for example, Rene Kagnat, *Dark turnarounds in Central Asia*, Paris). These myths are often used for justifying the non-success of "quasi-assistance" to the region (in fact revenues go back to the sponsors) while simultaneously antagonizing other sponsors that would actually like to assist.

A significant role in the disintegration of the riparian countries is played by the ambitions of their leaders. They developed the ideology and shaped the behavior of their respective governments, especially during the initial period of independence, through their attempts to stress their identification with national interest, individuality and historical uniqueness. The books written by the heads of Central Asian states provide a combined vision of how international politics would develop – see, *We build a new state* (Nazarbaev 2000), *By Thinking about the Future with Optimism* (Akaev 1999), *Tajiks in the Mirror of History* (Rakhmonov) and *Rukhname* (Nyazov). Analyzing these books thoroughly, it is possible to find some signs of a desire for integration, although the most obvious aspiration (although often slightly shaded) is to emblaze their own country as the first among the others in a united Central Asian region. During the initial period of independence, this seemed a reaction to the pressure of the federalism of the last epoch, but it has turned out that, over time, these countries have moved further and further away from each other. So each country survives in its own way.

The figures and graphs presented below show the trends in various economic and agricultural indicators. These enable an assessment to be made of the transition towards market relations and comparisons with the previous period, and set out certain characteristics of each country.

4.2.1 Kazakhstan

Kazakhstan is potentially the richest country in the region thanks to the availability of huge reserves of oil, gas, coal, iron ore and other mineral resources along with a considerable urban population (at 56%, the highest of all countries in Central Asia). Nevertheless, the country has experienced a period of deep economic depression, both in the overall economy and in the agriculture sector in particular. The country could not avoid an abrupt drop in agricultural activities, especially cattle breeding. This was the key reason for restructuring the agricultural sector and liquidating the system of collective and state farms. However, the abundance of oil, with a total proven oil reserves ranging from 5.4 to 17.6 billion barrels, attracted many foreign investors and, in a 10-year period the country received some US$14 billion in foreign investment.

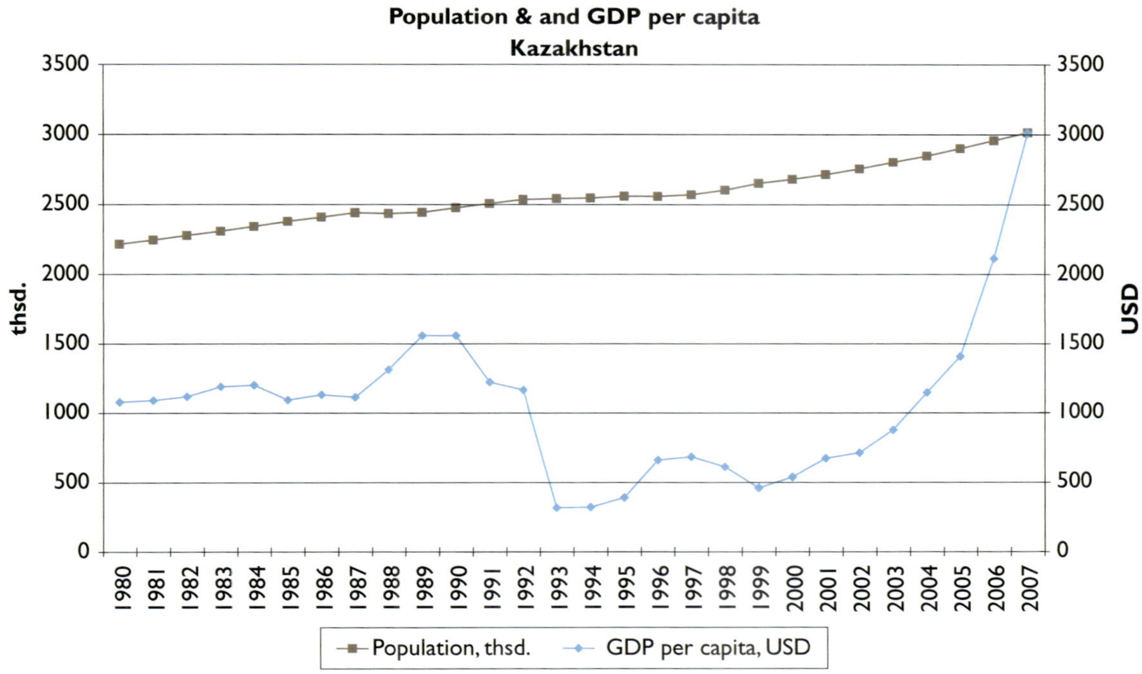

Figure 4.6a

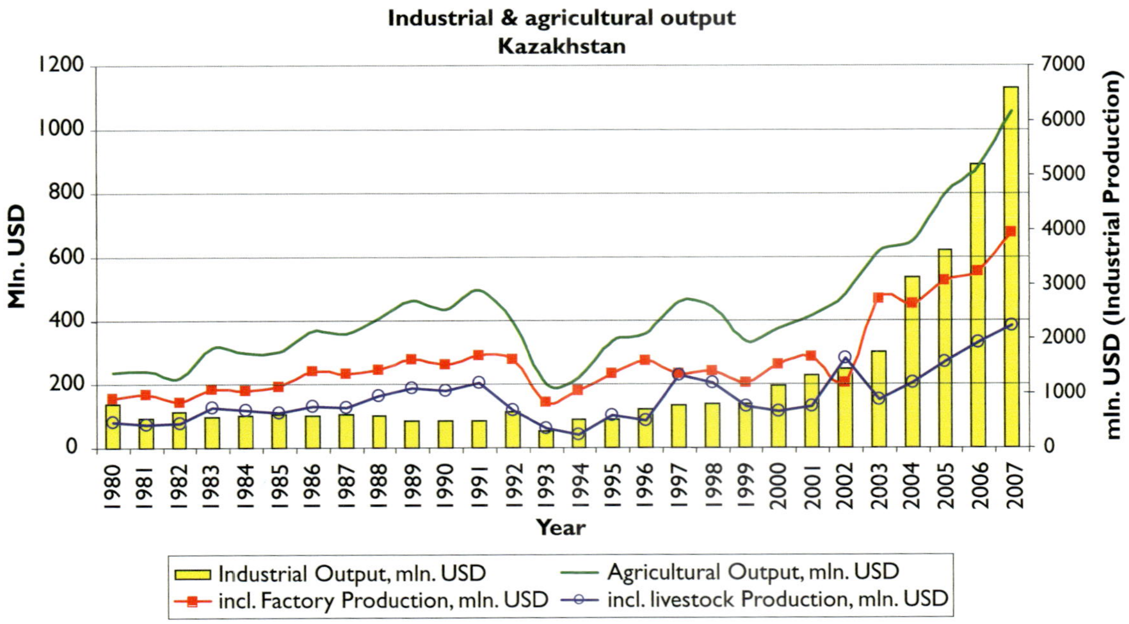

Figure 4.6b

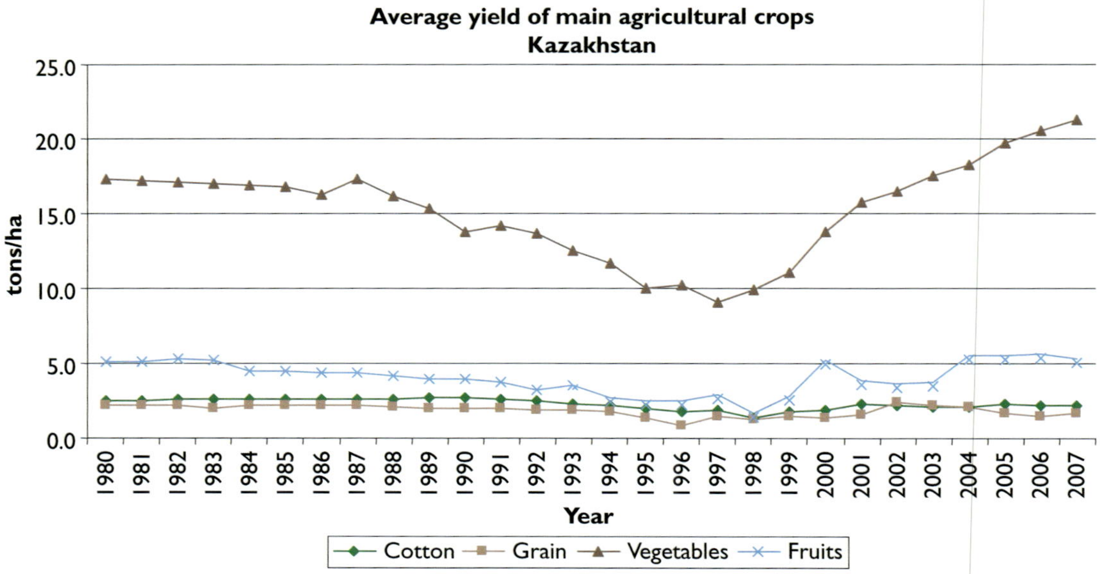

Figure 4.6c

From 1991 to 1995, the country experienced a drastic drop in GDP (Figures 4.6a and 4.6b) and a production decrease of many important goods especially livestock (Figure 4.6c)[5]. Nevertheless since 1995, Kazakhstan has seen economic growth, which has been very strong especially after 2001.

Kazakhstan has been a pioneer in the introduction of new economic mechanisms into the water and agriculture sector. The privatization of irrigated (and non-irrigated) land was implemented in a peculiar way. Initially, between 1992 and 1997, the former collective and state farms were restructured by being broken up into smaller farming units (of 20 to 600 hectares). The former administrators of farms and districts were privileged in this process. By the beginning of 1996, 93% of the land of 2,300 former collective and state farms was leased. Preferential lease terms were offered to the former administrators of farms, but rank-and-file members of collective farms exercised more caution before taking responsibility for land management because of the complicated challenges of running small farms. The Bankruptcy Act, which was adopted in 1999, immediately led to bankruptcy adjudications for many farms, and their land was transferred to members of former collective and state farms depending on the number of able-bodied people in each family. By April 1999, some 90,000 farms were legally registered, although according to the World Bank's assessment there were more than 120,000 farms (World Bank 2000).

Further reforms were implemented in a poorly developed way. Water user associations (WUAs) were initially still based on large agricultural co-operatives and farms. The 1993 Water Code governed

[5]All figures and charts are based on data from the CAREWIB system established by SIC-ICWC, and which are available at www.cawater-info.net.

the establishment of WUAs. It was supposed that WUAs would be set up as legal entities because they were responsible for managing the secondary irrigation canals. However, by liquidating co-operatives and distributing land among the farms established on former collective and state farms, farming units in South Kazakhstan Province had an average size of about 9 hectares per owner. The right of private land ownership became recognized in Kazakhstan by a law adopted on January 21, 2001. This law limited the size of any parcel of land that could be privately owned, and restricted its use to mainly construction work, horticulture and personal use. Other land resources were given out on a long-term lease basis. Prior to 2003, according to Law 404 dated April 2003, private farmers could not be WUA members. For private farmers to become WUA members, required preliminary integration into co-operative units and registration.

Then the situation became much worse, and in 1996 the ownership rights in irrigation networks were transferred to local and district authorities. It was assumed that WUAs would then buy these assets. However, due to the poor condition of the water infrastructure and the lack of experience of managing private farms, no farmers actually bought these rights. This was a direct consequence of World Bank policies and the restructuring of the agricultural sector into small farms. For example, in 1997 there were 18,000 farmers in an irrigated area of 130,000 hectares in the Makhta Aral District (former Pakhta Aral District) in South Kazakhstan Province (an average of 7.22 hectares per farm). Typically, there has been a step-by-step consolidation of farms and by 2002 the average size of the plots had increased to 13 hectares in this district, and this trend towards larger farms has continued.

According to the evidence provided by the well-researched work of Kai Wegerich (2008a), which contains a detailed description of land resource conditions in Kazakhstan, there was no order in the organization of the work of the WUAs. A quite complicated local system of water resource management was established. In each district there was a district water administration responsible for the supply of irrigation water to the WUAs. In some districts (such as the Makhta Aral District), there were associations of WUAs, which were responsible for the operation and maintenance of a large canal that delivered water to several WUAs. Farmers paid 200–750 tenge per hectare (US$1.5–6.5 per hectare) for each irrigation operation. However, there was insufficient supervision of water delivery, nor of water application on the farms. A key shortcoming was that WUAs were established mainly according to "top-down" principles and without any social mobilization. WUA directors were often appointed by the district administration rather than being elected. It is worth noting that in 2005 the World Bank and EU granted new projects to Kazakhstan that were aimed at supporting WUAs and improving the existing situation. It is difficult to say whether these projects will improve the efficiency of irrigation development, but signing these loan agreements is evidence that the government realizes the significance of introducing a proper order here and keeps alive the hope that the WUAs will be able to manage water resources in due course.

With regard to urban water supply and sanitation, more than 80% of households in Kazakhstan have been connected to the water pipeline network. However the quality of water supply services is low. Specialist organizations (vodokanals) provide water supply and sanitation services in cities in Kazakhstan. In a process of decentralization since 1993, the national government has transferred responsibility for water supply and sanitation services to urban authorities and has gradually stopped financing services in this sector. As a result, vodokanals were forced to become self-financing organizations and they have experienced many difficulties in trying to recover their operational costs.

4.2.2 Kyrgyzstan

Kyrgyzstan has been recognized by its Western partners as the most progressive nation in the region, setting the lead in terms of the rate of liberalization, privatization and transition towards a market economy, the reformation of the financial sector and the restructuring of the agricultural sector. Nevertheless, both GNP and the rate of growth of welfare and agricultural production are among the lowest in the region. The agriculture sector, which produces 45% of GNP and provides around half of total employment, is most significant. In the early 1990s, there was a dramatic drop in agricultural production followed by a small recovery which was not sustained. Cattle breeding and wool production, which were leading economic activities, have declined. Even by 2007, agricultural production had not recovered to 1990 levels (Figures 4.7a, 4.7b, 4.7c and 4.7d).

Unlike the other riparian countries, the area of irrigated land was quite insufficient for the existing population (at 0.20 hectares per capita) and it had to be restructured and leased on a long-term basis. This happened in a most democratic but rather ineffective way, as irrigated land was proportionally

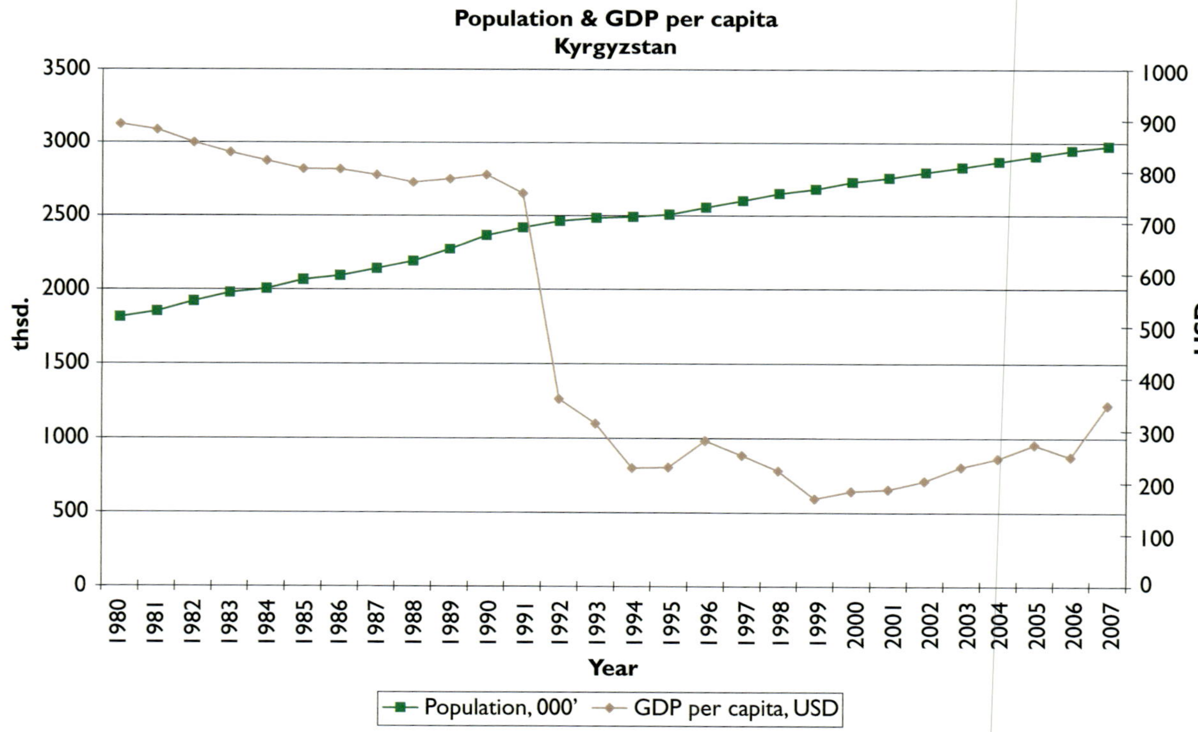

Figure 4.7a

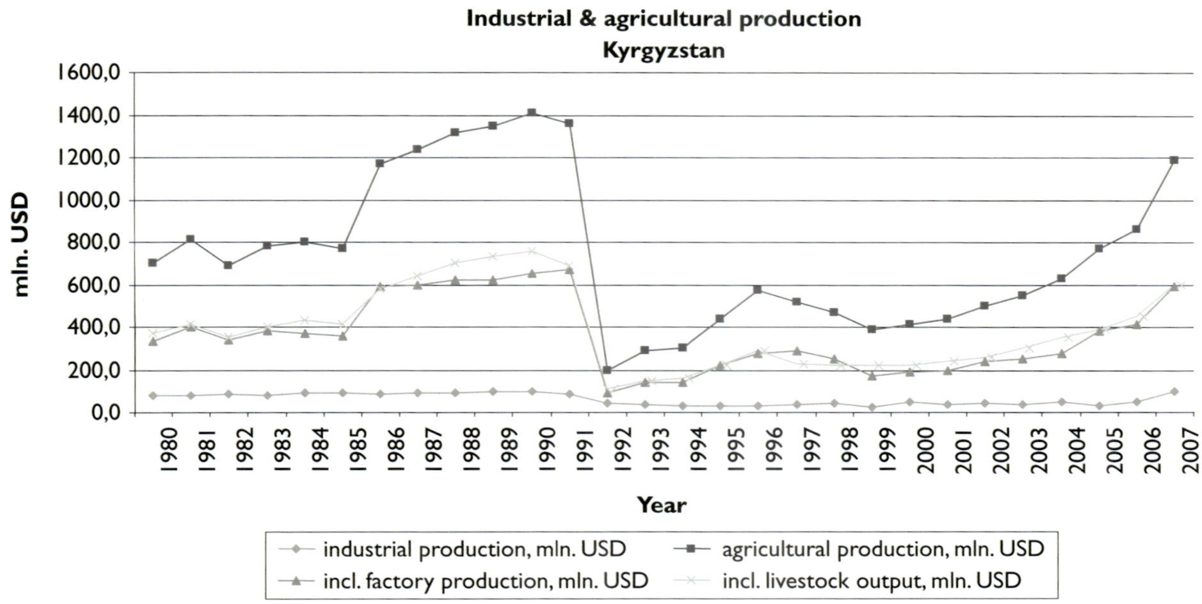

Figure 4.7b

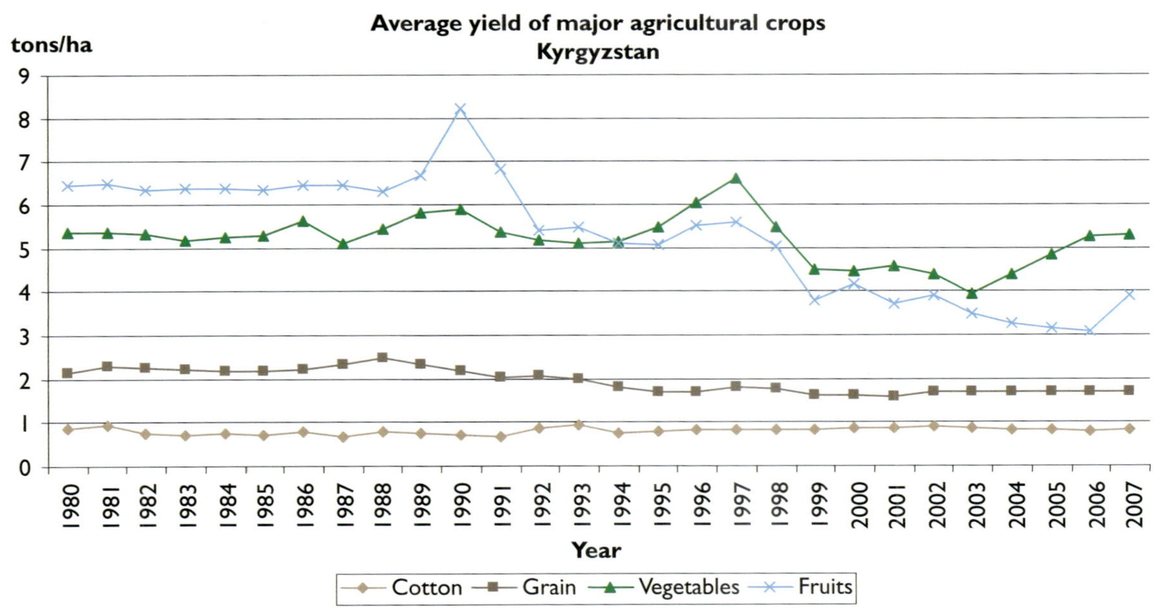

Figure 4.7c

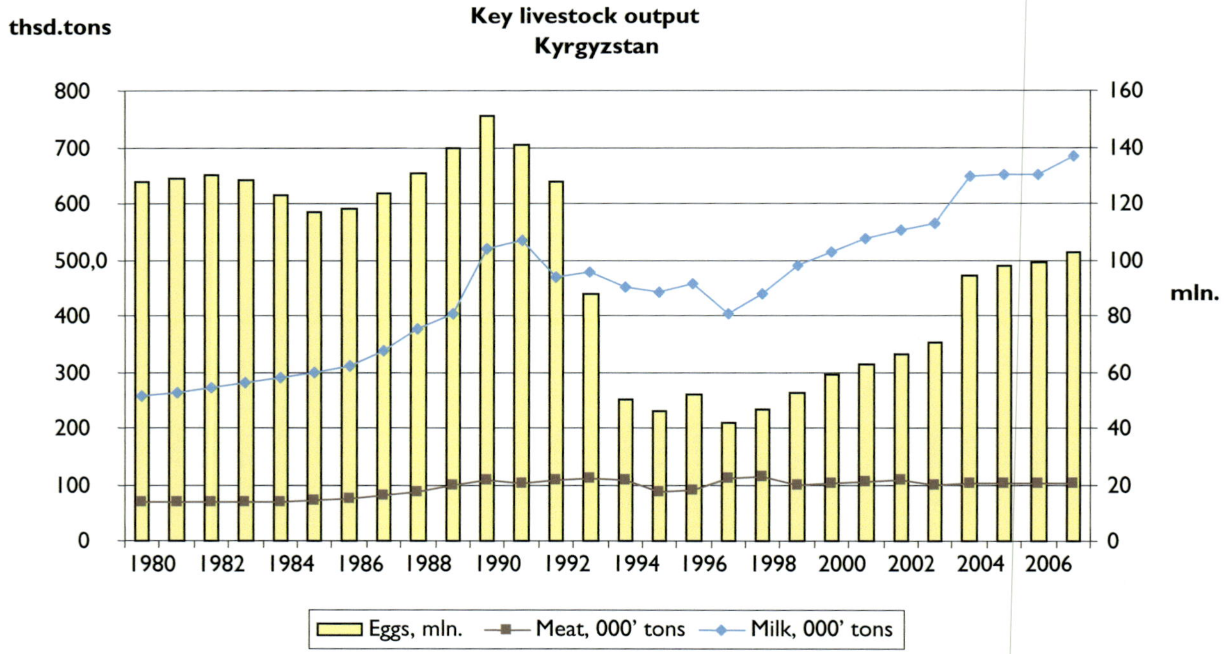

Figure 4.7d

distributed among all inhabitants working in rural areas, including teachers, medical personnel, district officials and local officers. As a result the size of a plot allocated to each user was different in each province. In the southern provinces with the greatest deficit of irrigated land (Osh, Jalalabad and Batkent provinces), an average plot measured between 0.4 and 0.5 hectares per family. Efficient agricultural production is obviously impossible under such conditions. One cannot say that this model of agricultural production (based on small plots of land) was well developed, in spite of the fact that WUAs began to be established in 1994. By 2001 a well-organized system had been created to support WUAs, with branch offices in a provincial and district network that was extensively financed by different donors. Gradually, farming units were consolidated into plots of 3–5 hectares as farmers subleased plots from leaseholders who did not engage in any agricultural activities, but even plots of this size do not provide the conditions for efficient farming, with the exception of the cultivation of vegetables, tobacco and fruit. The lack of appropriate infrastructure for servicing small farms, including appropriate marketing mechanisms, also adversely affected overall agriculture production. Kyrgyzstan did not have a system of state organized trade, as in Tajikistan, Turkmenistan and Uzbekistan, and there were no long-term contracts with farmers, which were typical for Kazakhstan, and market uncertainty created big problems for the rural population.

In Kyrgyzstan, hydropower production is another important economic activity, and one that is closely related to the water sector. Kyrgyzstan has great hydropower potential, and currently uses only about 10% of its available resources for hydropower generation. Due to a lack of oil and gas reserves and a fall in coal production, hydropower generation using the existing hydropower plants (the Naryn

Cascade) has become the main way of meeting the country's energy needs. It should be noted that power production outside the hydropower sector has almost halved since 1990 due to the weakening of the national financial capacity, an abrupt increase in fuel prices and the deterioration of the power stations. At present the hydropower sector produces some 85% of Kyrgyzstan's electricity. Given the considerable energy deficit, especially in the winter period, the Kyrgyzenergo hydropower-generating organization has changed the Toktogul Reservoir's water release regime. It has been shifted from a summer (irrigation) regime to a winter (hydropower) regime, and this has caused conflict in the Syrdarya river basin (details in the next section). It is necessary to understand that this artificially created water resource deficit impacts not only on downstream countries but also on irrigated agriculture in Kyrgyzstan itself, causing considerable losses to farms, especially in dry years.

The level of water supply and sanitation services in Kyrgyzstan is quite low. Only a third of households are connected to these systems. Another third of the population takes water from standpipes or water trailers and the remaining third of the population is not serviced at all. The greatest coverage is in Bishkek, where about 80% of the inhabitants have access to water supply and sewage systems. The proportion of the population served by water supply and sewage systems is substantially lower in smaller cities and is extremely low in rural areas. About half of the 12,750 villages do not have operational water supply systems. Only 25% of the villages in the Osh and Jalalabad provinces in the south of this country are serviced by working water supply systems.

4.2.3 Tajikistan

The economic situation in Tajikistan was complicated by the civil war. This war took the lives of about 50,000 people and forced 850,000 people to migrate. The country is currently the least developed of the Central Asian countries, socially and economically, although the world community donates the largest aid per capita to this republic. Some 80% of population lives below the poverty line. The national industrial sector is based on aluminum production, the processing of uranium and other ores, as well as on hydropower production. However 65% of the economically active population (more than 1.1 million people) is engaged in agriculture. Irrigated farming has always been highly productive in the republic, but it is both technically and economically complicated. More than 35% of the irrigation systems operate on a pumped water supply. In the period from 1990 to 1998, there was an almost a fivefold decrease in industrial and agricultural output, including in traditionally strong sectors such as vegetables and livestock (see Figures 4.8a, 4.8b, 4.8c and 4.8d).

The operational life of most of the irrigation and drainage infrastructure (pumping stations, water pipelines, waterworks, irrigation canals, etc.) is at least 30 years, and this infrastructure is at risk. The operating conditions of the water systems are rapidly deteriorating, resulting in pumping stations with a decreased operating capacity, in increasing water losses from irrigation canals, and in low efficiency of water use at field level. In Tajikistan, the power sector is subsidized to a considerable degree, and power and pumped water were traditionally supplied to farms free of charge. The pumped water supply systems consume much energy, since they lift water from 50 to 250 m in height, as well as supply water across a very uneven terrain (World Bank 2003a).

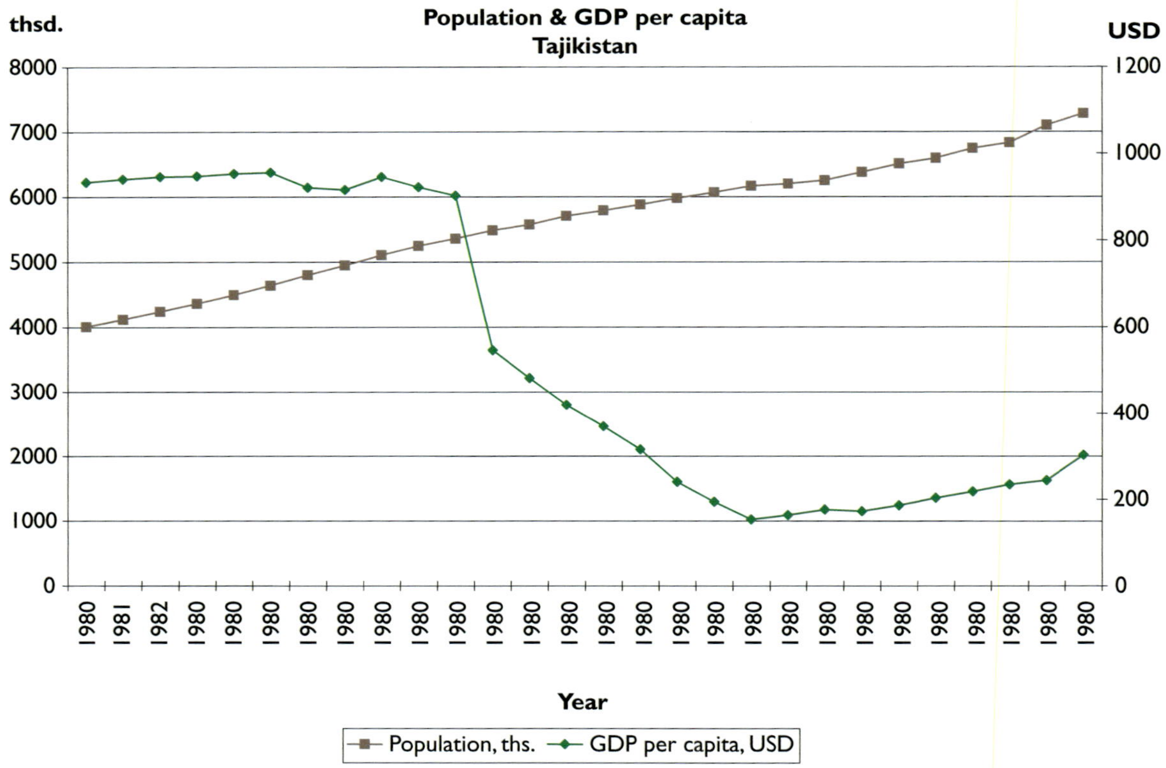

Figure 4.8a

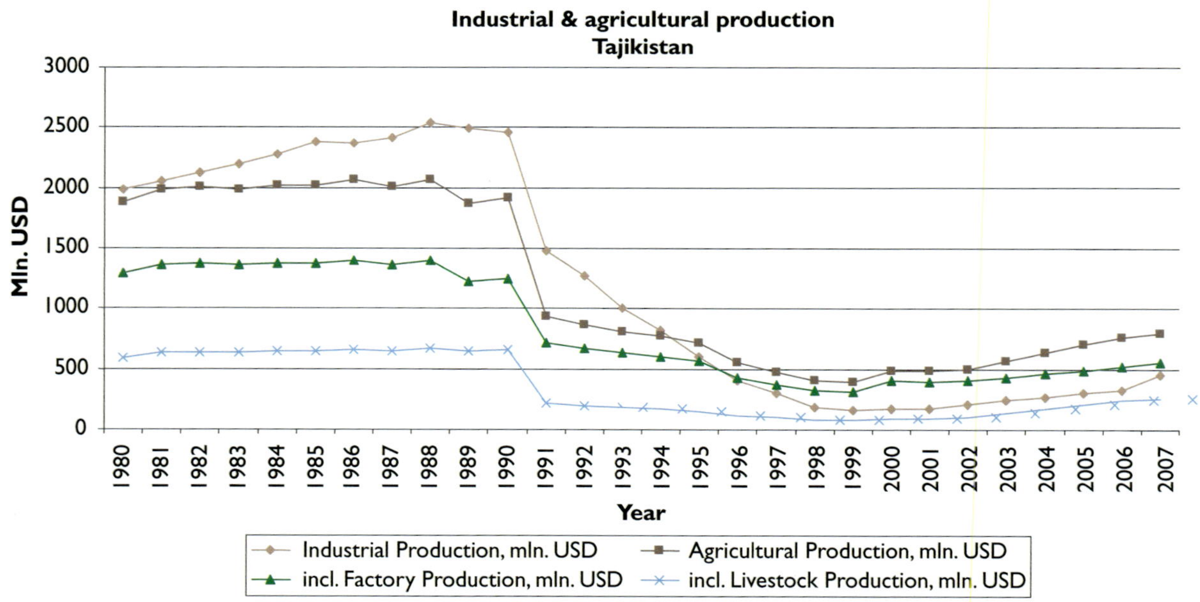

Figure 4.8b

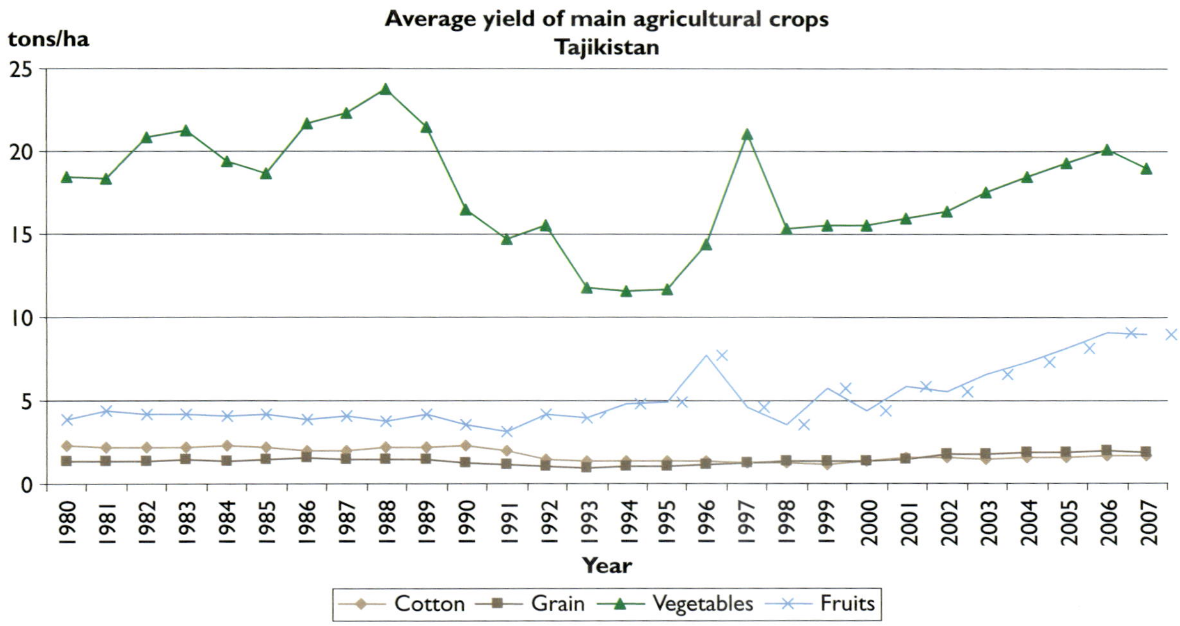

Figure 4.8c

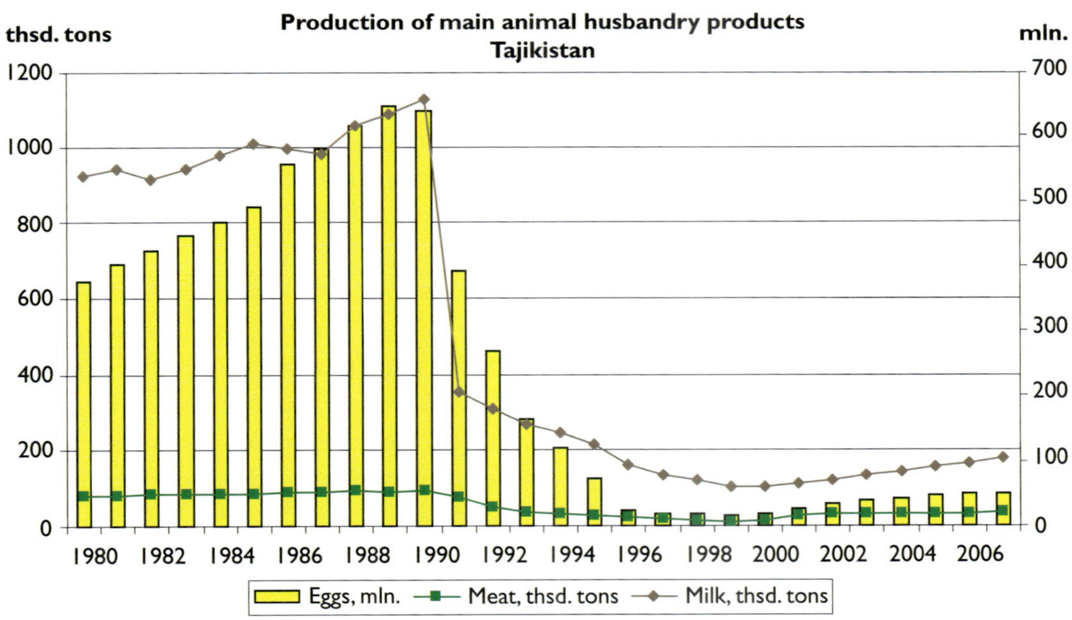

Figure 4.8d

In Tajikistan the agriculture sector has been of key economic importance, and was actually the leading sector exports. Nevertheless, the republic could not prevent substantial falls in agricultural output in the 1900s, due to the complexity of the water supply systems and owing to a protracted and insufficient restructuring of agricultural production. Unlike Kazakhstan and Kyrgyzstan, the Tajik government did not dare to adopt a privatization policy and leasing system, and it made some half-hearted decisions. It provided insufficient financial and material support to the agriculture sector, and overall there was not enough progress in agricultural development. At the end of the first decade in the new millennium, the agriculture sector barely survives under a state-run trade system for cotton and wheat.

There remain some highly productive and economically sustainable collective farms from the Soviet time period, such as the Samatov and Urinkhojaev in the Soghd Province. These are really cooperatives, that include as members a few villages with thousands of inhabitants. The population of these collective farms are very firmly opposed to the wish of local and republican authorities to restructure the farms and change their integral structure. In the framework of the IWRM-Fergana Project, an attempt was made to preserve the Samatov collective farm as a model farm, turning it into a large modern federation of WUAs with multiple production activities.

4.2.4 Turkmenistan

Turkmenistan is the potentially richest country in this region. With enormous mineral reserves and land resources and a relatively small population, Turkmenistan suffered little from the disintegration of the USSR. Its oil potential is evaluated at 2.5 billion barrels and there is 1.6 trillion m^3 of gas. Nevertheless, the country could not avoid a downturn in GDP in 1991 for much the same reasons that saw GDP fall in the other riparian countries of the Aral Sea basin. While industrial production fell by some 20% after independence, by 2000 the country had recovered GDP per capita to the levels of 1990 and by 2006 GDP per capita was double the 1990 levels (Figures 4.9a, 4.9b, 4.9c and 4.9d).

Early on, the government pursued a policy of limited privatization in the agriculture sector, implementing it mainly in the agricultural processing industries. Most of these agricultural enterprises were associated into the Turkmen Cooperative Alliance (TCA), which united more than 700,000 members at republican, provincial and local levels. The TCA encompasses all enterprises involved in processing tomatoes, cucumbers, honey, vegetable oil and fruit and also some cotton-processing factories. The TCA is also responsible for supplying the rural population with consumer goods, and so the rural consumer co-operatives have become an integral part of the TCA. Immediately after independence, Turkmenistan started to expand its local cotton processing capability, using foreign investment to construct large textile factories in the form of joint ventures. As a result, in the period from 1991 to 2001, the amount of cotton processed within the country as a proportion of the cotton cultivated Turkmenistan rose from 6% to 42%.

The state remains the owner of land resources, and has transferred the right to grant land use to provincial authorities. All land users, both private and co-operative, rent their plots from the provincial

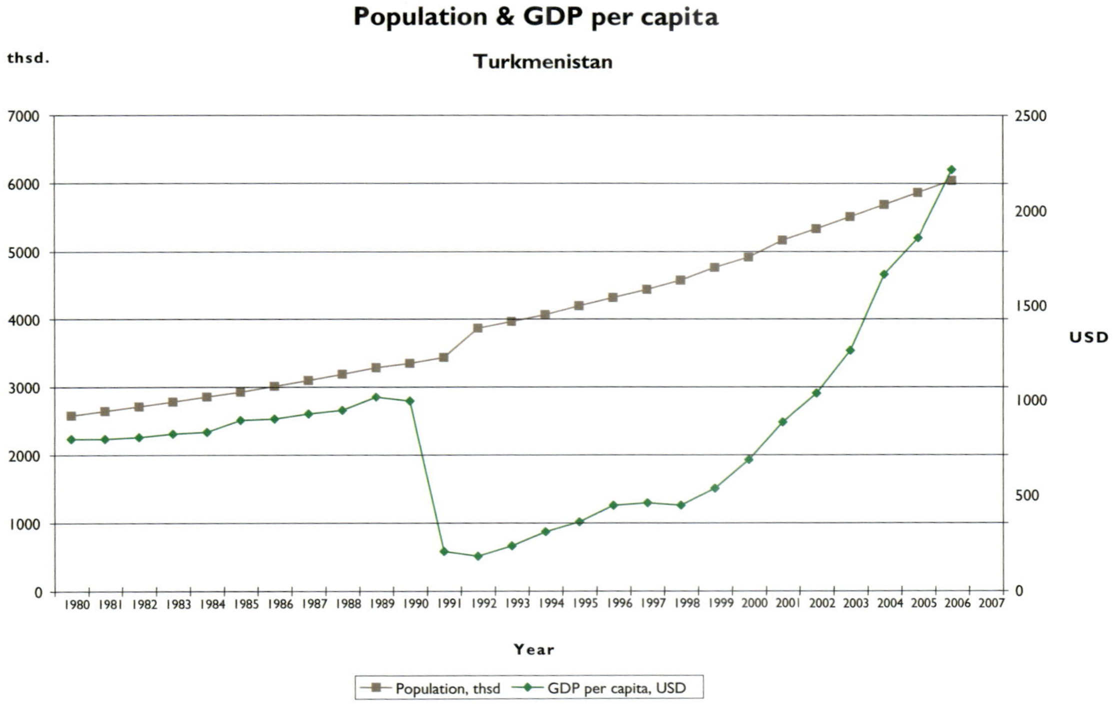

Figure 4.9a

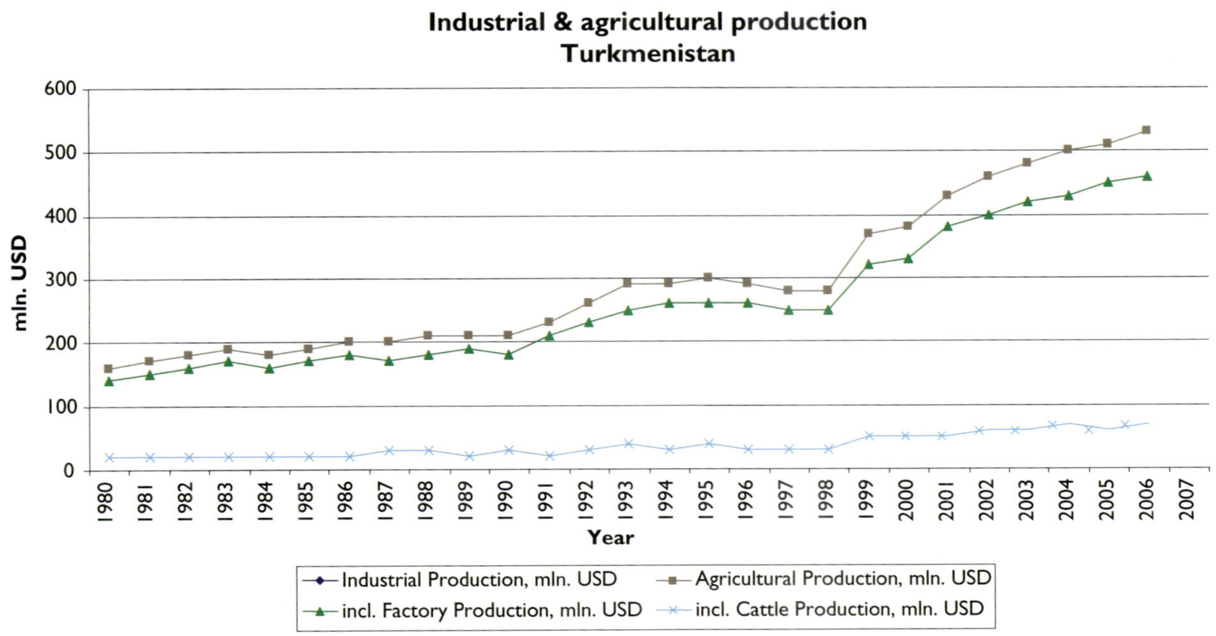

Figure 4.9b

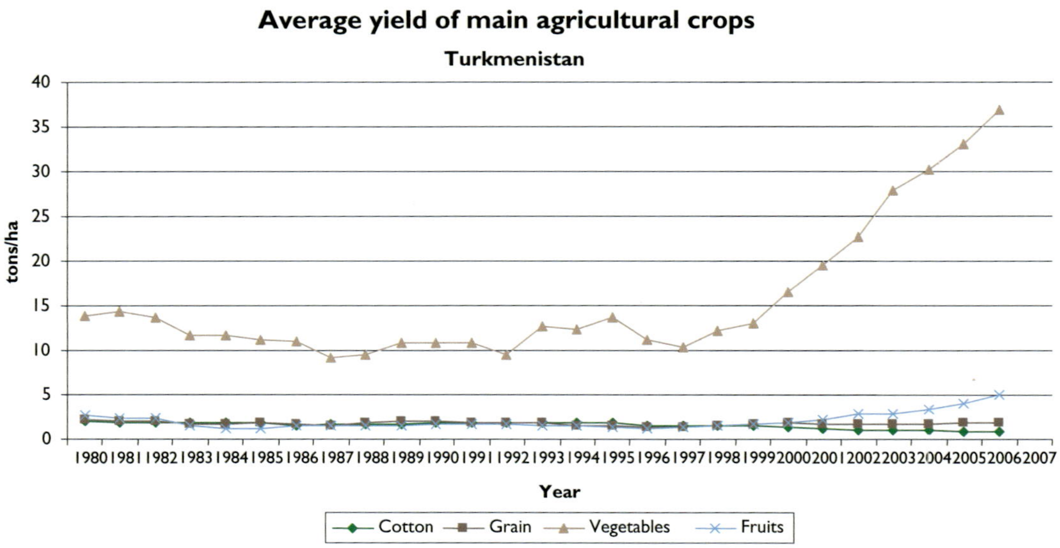

Figure 4.9c

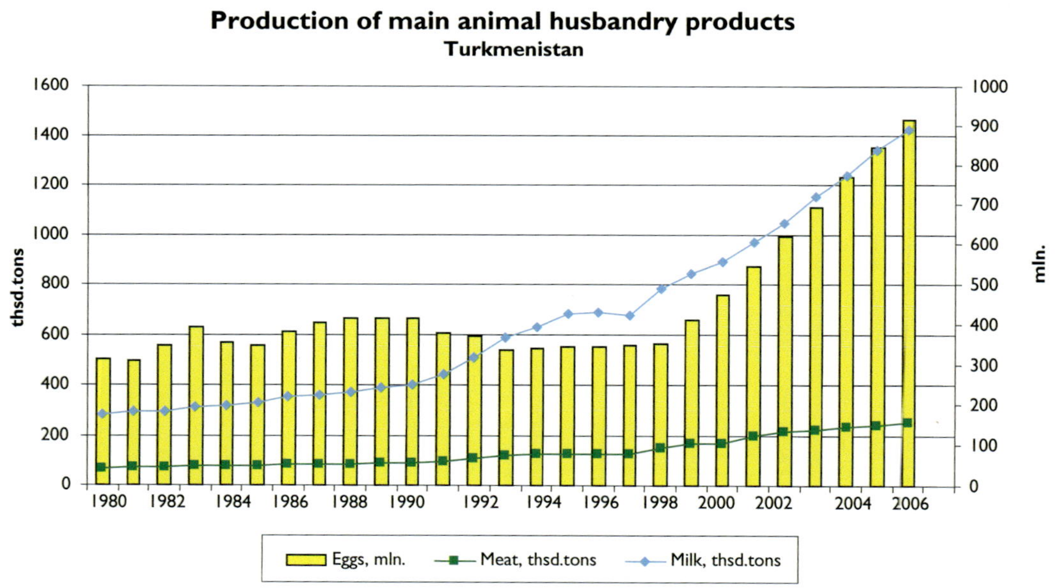

Figure 4.9d

authorities. Since 1995, most of the collective and state farms have been reformed into large farmer associations, which lease land to their members. Farmers who wish to do so can buy machinery for agricultural activities from the state. Up to 50 hectares of "dry" land can be leased by a farmer on the condition that development will be implemented by the farmer himself. The irrigated land on former

Table 4.6 Provision of primary foodstuffs in Turkmenistan.

Production-consumption ratio

Foodstuffs	1991	1992	1993	1994	1995	1996	1997
Bread	0.15	n.a.	n.a.	n.a.	0.85	0.71	0.95
Vegetables	1.54	1.23	1.00	1.30	1.30	1.36	0.95
Water Melon	2.30	1.56	1.38	1.78	1.46	1.80	n.a.
Oil	2.54	1.91	1.82	n.a.	1.71	1.68	1.70
Meat and meat products	0.75	0.72	0.67	0.70	0.69	0.83	0.80
Milk and dairy products	0.61	0.63	0.84	0.88	0.90	1.40	1.54
Eggs	1.70	0.84	0.83	0.95	0.95	1.28	1.37

Note: n.a. means “not available”.

collective and state farms was distributed among farmers through a joint stock share organization, and the size of the allocation which was determined by a farmer’s length of service, professional skill and other factors. The most successful farmers – leaseholders who have got their own machinery and equipment and who achieve a high farming productivity over a period of 10 years – can receive a plot of up to 3 hectares for private ownership. According to the available data (Stulina 2006, page 121), some 249,000 hectares had been transferred into private ownership by the mid 2000s. If a farmer leaves the joint stock company, the farmer’s land is redistributed to other members and the farmer has the right to receive a price for his joint stock share from these members. This should take place with the consensus of all other members of the company.

The key reforms in the agrarian sector have been aimed at increasing the prosperity of the rural population, which accounts for 56% of the total population of the country. Some 40% of the rural population are directly involved in agricultural production. Government policy has aimed for self sufficiency in foodstuffs, especially in grain, milk and meat, through the “10 years of sustainability” program (which ran from 1995 to 2005). The Dekhkan Bank – and its branches, the Cotton Bank and the Grain Bank – was established to support this program. The bank purchased agricultural outputs and issued loans at low interest rates. In 1999 the agriculture tax was reduced and, as a result of all these initiatives, food security was achieved by 2000 (Mkrtychan et al. 2000).

According to data from the WUFMAS project, the net profitability of Turkmen farms was 1.5 times higher than those in Uzbekistan due to significant subsidies to cotton and grain producers and despite the state dominated trade and fixed prices (WUFMAS 1998). The government particularly ensured that farmers received preferential credits and 50% compensation for the costs for all agricultural inputs (seeds, fertilizers, machinery, etc.). The farmers paid low prices for petroleum products and technical services, and were exempted from all taxes if they fulfilled their contract obligations for production. Further subsidies encouraged the private farms to produce other necessary foodstuffs. In 1991, the private sector produced 98% of the potato output, 72% of vegetables, 90% of meat, 95% of milk, and 93% of eggs.

In a pattern untypical of the other riparian countries, the average household income in rural areas was generated from the agricultural activities of the Dekhkan-related cooperatives and private farms (36.5%), from personal garden plots and livestock (42.1%), from pensions (14.9%) and from secondary employment (6.3%).

Statistical data of 2003 for the Central Asian countries as a whole show that 86% of urban households and 14% of rural households had access to safe water (World Bank 2003). However, a field survey of rural households in Turkmenistan's Akhal and Mary provinces conducted as part of the Gender Aspects of Integrated Water Resources Management project has shown that 47% of rural households had a water tap in their houses or a water post in their yards and 52% of the rural inhabitants took potable water from nearby wells. Only 1% of the rural population actually did not have access to safe water resources. Moreover, 100% of the respondents answered affirmative to a question about easy accessibility to a water supply (Stulina 2006). It is perhaps obvious from this result that rural people have different attitudes on water standards to urban dwellers.

4.2.5 Uzbekistan

Uzbekistan is the most populated republic in Central Asia with about 30 million inhabitants. In the past (as recently as 1980), the country had high rates of population growth (up to 3.6% per year), which subsequently decreased to 2.2% per year and at present stands at some 1.6% per year. Nevertheless, even a population growth of 300,000 people a year requires considerable additional financial, land and water resources and generates pressure to increase land productivity. The rural population (which amounts to 61% of the national population) is a key target group for the government because about 30% of GDP is produced in rural areas (down from 41% in the Soviet period). In addition, the rural economy generates almost half of Uzbekistan's foreign currency earnings, up to 26% of its export volume and some 42% of all jobs. Some 30% of industrial enterprises are also related to the agriculture sector. Field surveys by experts of the SIC ICWC in Tashkent Province have shown that each US dollar of net profit earned in irrigated farming is accompanied by US$1.5 of net earnings in related sectors and by US$0.7 of net earnings in the service sector [25]. Obviously, a sustainable water supply is most critical for the country. Along with other riparian countries, the republic was affected by the collapse of the USSR and it experienced lower agricultural production and overall economic decline (Figures 4.10a, 4.10b, 4.10c and 4.10d) in the first years of independence.

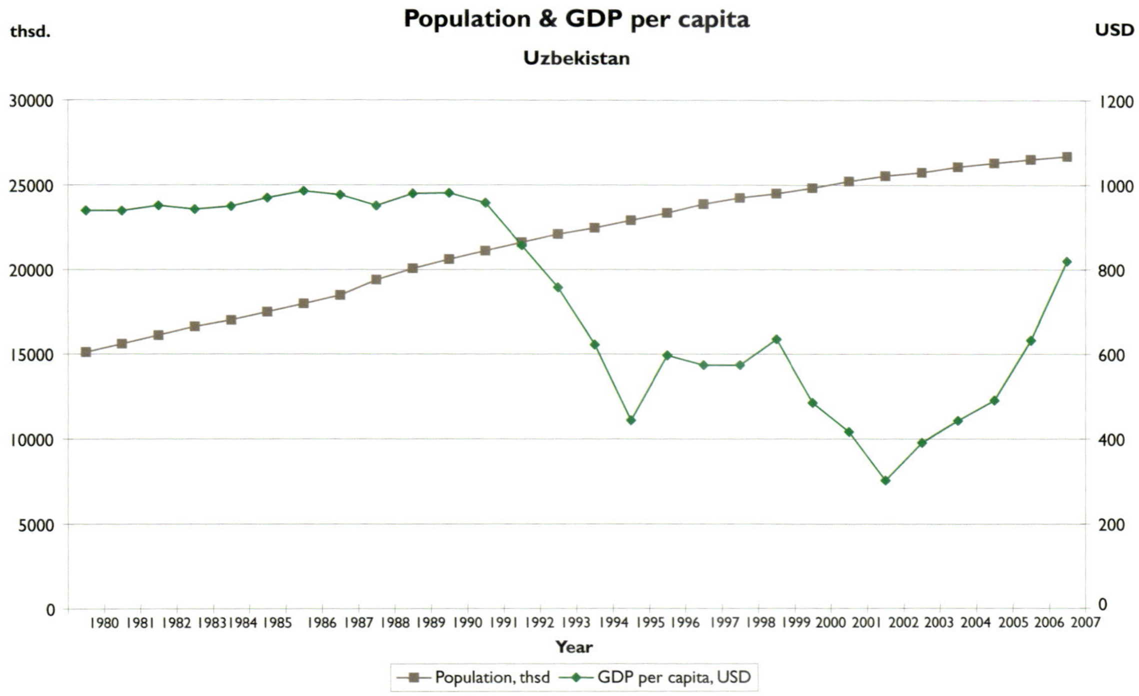

Figure 4.10a

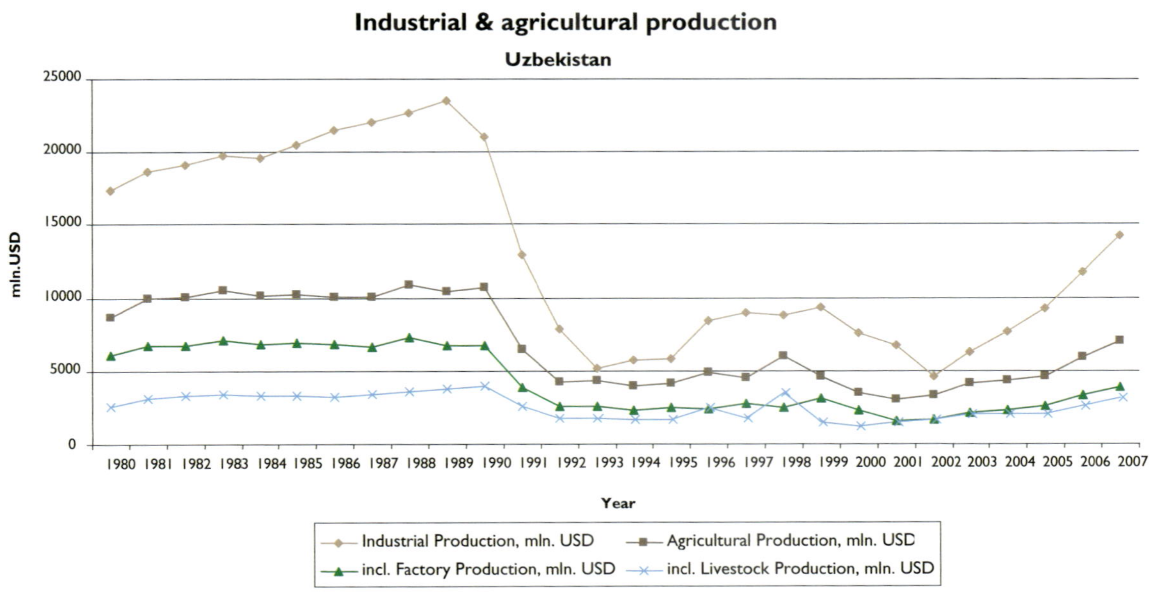

Figure 4.10b

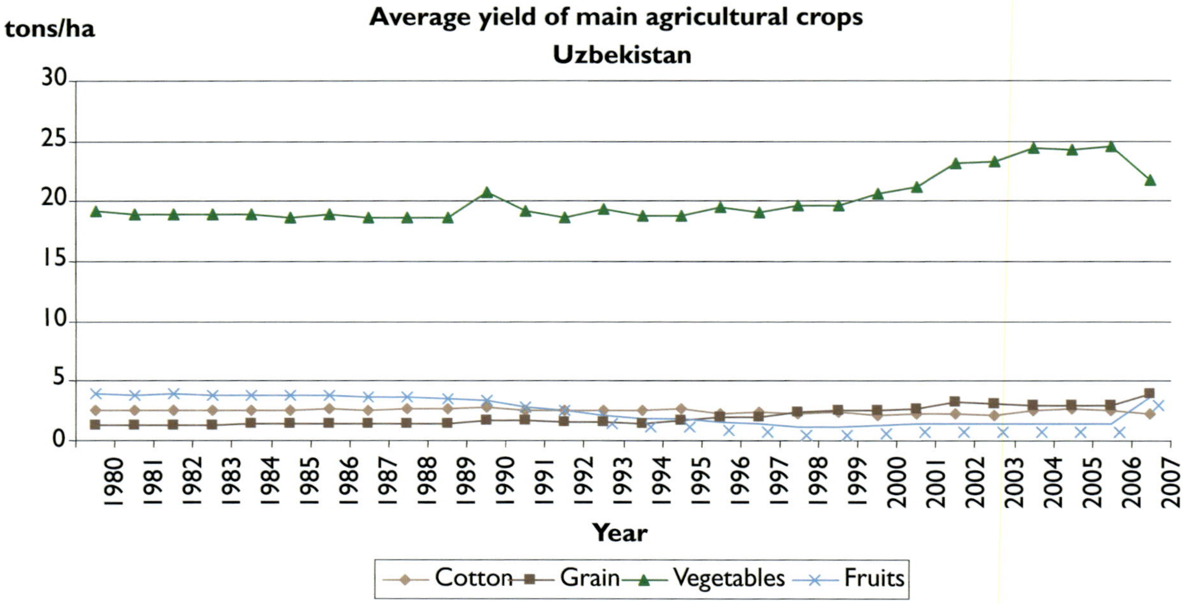

Figure 4.10c

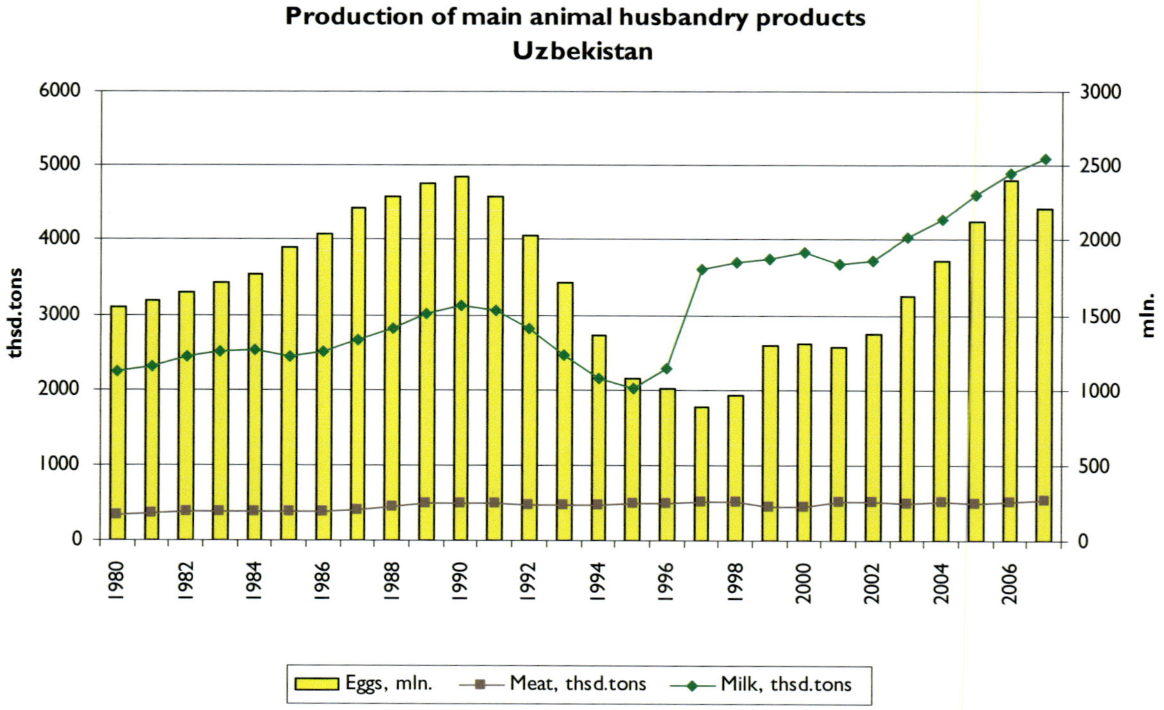

Figure 4.10d

The Uzbekistan government adopted a policy of gradual reform in the agriculture sector. Land remains under state ownership. It was first rented to cooperatives (shirkats) and then to private farmers after strengthening their capacity and skills. It is necessary to note that for many years after independence the agriculture sector was subject to a process of continuous reform that created a certain instability and did not give farmers the confidence to make large investments into agriculture. In addition, the state monopoly on grain and cotton acted as constraint on agricultural development. Although, the government purchases farm output and makes it look like it provides a guaranteed market for agricultural products, the prices paid by the state are quite low. Since 2001, the state monopoly for agricultural products has been reduced to 50% of the market, and the remaining output could be sold at free market prices. However, up to now, the profitability of grain and cotton production for farms remains rather low, and net earnings are only US$160–200 per hectare on average.

Since the Soviet time, the size of so-called homestead land (a plot of cultivated land adjoining one's house) remains in the range of 0.12 to 0.20 hectares of irrigated land or around 0.5 hectares of rain-fed land. Although the productivity of this land is quite high (in Tashkent Province it generates up to US$1,187 per hectare), and this would argue for a more radical transition towards private land use, privatization has not been implemented. In the first years of independence, private cotton or grain farms had to be no less than 10 hectares in size, but following recent government decisions, local authorities are now encouraged to promote a process of enlarging average farm size areas to 40–60 hectares for cultivated crops and up to 5 hectares for orchards, vineyards and vegetable production. According to farmers, these areas are still too small for good agricultural practice [41].

It is important to note that according to the findings of the gender survey (Stulina 2006), the rural household budget consists of income from farm employment (28%), personal garden plots (36%), state funds in the form of pensions, benefits, grants, etc. (15%), secondary employment (18%) and financial aid from relatives (13%). Thus, land in Uzbekistan generates less income than in Turkmenistan (by 15%). In the period after independence Uzbekistan has diversified its agricultural production to a certain extent, reducing the area under cotton (by more than 25%) and increasing the area under grain (be more than three times).

Uzbekistan has not completed the reform of the agriculture sector and this undoubtedly still impacts on the sustainability of the operations of the newly established WUAs and of the water sector as a whole.

In spite of specific development issues in each of the riparian countries and the different methods that have been used to restructure the agricultural sector and increase productivity, all countries share the approach of developing water users associations (WUA). In essence, this idea is the reverse of the water and land associations that existed at the start of the Soviet rule. It addresses the same problems, but it requires a structure for servicing many small-scale water users. The liquidation of collective and state farms, land privatization and land leasing had destroyed the existing system of on-farm water use. Small-scale water users found themselves in a situation where nobody was concerned about planning and managing of water delivery, where nobody was concerned about cleaning irrigation and drainage canals and nobody was concerned about water distribution to and among users. Water management organizations supplied water to the boundaries of former collective farms, but from that point, water users themselves had to organize the whole process of water distribution and use. It was therefore necessary to establish new managing structures, which, on the one hand, would be subordinated to water users but, on the other hand, would be able to provide equal and fair access to water for each user. As mentioned above, the first efforts in this direction were undertaken in Kazakhstan, but only in 2004 were WUAs legally registered in their present form (and still far from perfect) in this country.

The government of Kyrgyzstan was also an evolutionary pioneer, issuing a decree in 1997 and after long experiments and trial and error adopting a law "On Water Users Associations" on March 15, 2002. A "top-down" strategy was adopted here, establishing WUAs through a special department under the Ministry of Agriculture and Water Resources. The activity of this department was financed by the World Bank. The practice of establishing and reorganizing WUAs was researched in the IWRM-Fergana project, which was jointly carried out by SIC ICWC and IWMI. This project has shown that WUAs tend to be poorly managed and inefficient bodies if they are created by this "top-down" approach without any direct (of "bottom-up") participation of water users and if they ignore hydrological principles and other organizational issues. The researchers found that WUAs had to be reorganized on hydrological principles and by matching existing settlements and water infrastructure with water accounting, management and allocation systems. As a result, three WUAs established within the command area of the Aravan Akbura Canal in Osh Province were reorganized into six new WUAs.

The need to establish and strengthen WUAs is increasing rapidly, both to accommodate the process of farm restructuring (with many farms being subdivided into smaller units) and to tackle a growing water deficit. This has been particularly reflected in the practical initiatives and legislation enacted in Tajikistan and Uzbekistan. At the same time, water users often act autonomously and ahead of any enabling (legal and material) environment.

WUAs based on hydrological principles and supported by a well-equipped and financially sustainable infrastructure can be considered as cooperative organizations, and they are quite effective at managing water distribution and improving water use practice. They can be gradually transformed into multi-sectoral units that can efficiently increase output in the rural areas of these arid regions. Already some WUAs have created an extension service to train farmers in advanced agricultural practice, marketing and diversification into different fields, such as seed farming, aquaculture, etc.

Japanese-style multi-sectoral cooperatives, where farmers play a role as key agricultural producers and the associations carry out many other related activities such as relations with local authorities, marketing, supplier of seeds and agricultural chemicals, leasing of agricultural machinery (rather than simply focusing on water distribution), are an ideal organizational structure for a network small farms. At present, such associations can organize the processing of agricultural outputs, distributing the profit from different activities among all participants in a multi-stage process according to personal labor inputs.

4.3 The Complicated Issue of Water Sharing in the Aral Sea basin

Although the Aral Sea Basin is often characteristic as one of the basins in the world with the most scarce water resources, this does not really correspond to the facts because the available water resources are, in principle, quite sufficient. The available water resources per capita considerably exceed the UN indicators for "water famine" (1600 m^3 per person per year) or an "extreme water deficit" (1000 m^3 per person per year), even taking into account current population growth rates. The SIC ICWC, together with European experts, has prepared an assessment of available water resources as part of the WARMAP project, which was financed by the European Union. This assessment can be found in the *analysis report for the regional strategy on rational and effective use of water resources in Central Asia* (UN SPECA, 2001).

In Central Asia, water resources originate from renewable surface water, groundwater and return water of anthropogenic origin. The water resources mainly belong to the basins of two rivers – the Amudarya and the Syrdarya. Some rivers, such as the Kashkadarya, Zeravshan, Murgab and Tejen, form separate basins with closed catchment areas bordering on to the Amudarya, but have lost their historic hydrological links with the main watercourse. In Kazakhstan and Kyrgyzstan, water resources are also formed within other hydrographic systems. There are seven separate river basins in Kazakhstan and four separate river basins in Kyrgyzstan.

Several other big rivers of interstate significance are located outside the Aral Sea basin. These rivers are of particular significance.

- The Chu River is 1067 km long, with a 62,500 km^2 catchment area, and has its origin in the Tien Shan in Kyrgyzstan. It flows northward into southern Kazakhstan where disappears within the Ashikol Depression.
- The Talas River is 661 km long, with a 52,700 km^2 catchment area, and originates in Kyrgyzstan. It ends within the Muyun-Kum sands in Kazakhstan.
- The Tarim River is 2030 km long, with a catchment area of around 1,000,000 km^2, and has its origins in Kyrgyzstan and Tajikistan. For most of its course, the river is within the People's Republic of China.
- The Irtysh River is 4240 km long, with a catchment area of 1,643,000 km^2, and crosses the eastern part of Kazakhstan before flowing into the Ob River in Russia.

The integrated use of water resources in inland basins, such as the Ili River basin in Kazakhstan, the Issyk-Kul Lake basin in Kyrgyzstan and others, is of great significance for the socioeconomic development of the independent countries of Central Asia.

Keeping in mind the need for regional water management co-operation with countries such as Iran, Afghanistan, Russia and China, it seems reasonable to extend the scope of this section of the book to take into account the interests of all riparian countries of the Aral Sea basin in Central Asia.

Table 4.7 Average natural river runoff in the Amudarya basin in the period 1934 to 1992 (three hydrological cycles), km^3/year. (SIC ICWC)

River basin	River runoff formed within each riparian state					Total in the Amudarya basin
	Kyrgyzstan	Tajikistan	Uzbekistan	Turkmenistan	Afghanistan and Iran	
Panj	–	21.089	–	–	13.200	34.289
Vakhsh	1.604	18.400	–	–	–	20.004
Kafirnigan	–	5.452	–	–	–	3.452
Surkhandarya	–	0.320	3.004	–	–	3.324
Kashkadarya	–	–	3.232	–	–	1.232
Zeravshan	–	4.637	0.500	–	–	5.137
Murgab	–	–	–	0.868	0.868	1.736
Tejen	–	–	–	0.560	0.561	1.121
Atrek	–	–	–	0.121	0.121	0.242
Rivers of Afghanistan	–	–	–	–	6.743	6.743
Total (km^3)	1.604	49.898	4.736	1.549	21.593	79.280
Total (%)	2.0	62.9	6.0	1.9	27.2	100

According to origin, renewable groundwater reserves can be divided into two categories – groundwater that is naturally formed over a catchment area and groundwater formed due to deep percolation under irrigated areas. Some 339 aquifers have been explored and approved for use within the Aral Sea Basin. The total regional renewable reserves of ground water have been estimated at 43.49 km^3, including 25.09 km^3 in the Amudarya basin and 18.4 km^3 in the Syrdarya basin. Aquifers have an important hydraulic interconnection with surface runoff. This interconnection becomes apparent when the surface runoff decreases with increasing abstraction of groundwater. Taking this into consideration, the amounts of groundwater reserves that can abstracted are usually approved by the national water commissions. The total volume of proven reserves available for use is some 16.94 km^3 (see Table 4.9). The annual volume of groundwater abstraction in 1999 amounted to 11.04 km^3, although at the beginning of the 1990s it exceeded 14.0 km^3. It should be noted that some aquifers are formed within the territories of

Table 4.8 Average natural river runoff in the Syrdarya basin in the period 1951 to 1974 (two hydrological cycles), km^3/year. (SIC ICWC)

River basin		River runoff formed within each state				Total in the Syrdarya basin
		Kyrgyzstan	Kazakhstan	Tajikistan	Uzbekistan	
Naryn		14.544	–	–	–	14.544
Qaradarya		3.921	–	–	–	3.921
Small rivers in an area between Naryn and Qaradarya		1.760	–	–	0.312	2.072
Right bank of the Fergana Valley		0.730	–	–	0.408	1.188
Left bank of the Fergana Valley		3.500	–	0.855	0.190	4.545
Rivers in the middle reach		–	–	0.150	0.145	0.295
Chirchik		3.100	0.749	–	4.100	7.949
Akhangaran		–	–	–	0.659	0.659
Keles		–	0.247	–	–	0.247
Arys and Bugun		–	1.183	–	–	1.183
Rivers in the lower reach		–	0.600	–	–	0.600
Total	(km^3)	27.605	2.426	1.005	6.167	37.203
	(%)	74.2	6.5	2.7	16.6	100

neighboring countries (this occurs in the Golodnaya Steppe, Kazalinsk, Kafirnigan, Fergana and other basins), therefore there is a need for interstate collaboration for regulating groundwater use as well as for groundwater pollution control.

Thus, assuming population growth rate and of the replenishment of renewable water resources at present levels, water availability in Central Asia (modeled under different scenarios) should be quite sufficient for the next 25 years. This means that the region should be able to cover its needs for water for its economic development (see Figure 4.11) (Dukhovny 2007a). Even by 2030, under a high water use scenario, the available water resources per capita in the region will amount to 1430 m^3 per

Table 4.9 Cumulative natural river runoff in the Aral Sea basin, km³/year (SIC-ICWC).

State	River basin		Aral Sea basin	
	Syrdarya	Amudarya	km³	%
Kazakhstan	2.426	–	2.426	2.1
Kyrgyzstan	27.605	1.604	29.209	25.1
Tajikistan	1.005	49.578	50.583	43.4
Turkmenistan	–	1.549	1.549	1.2
Uzbekistan	6.167	5.056	11.223	9.6
Afghanistan and Iran	–	21.593	21.593	18.6
Total	37.203	79.280	116.483	100.0

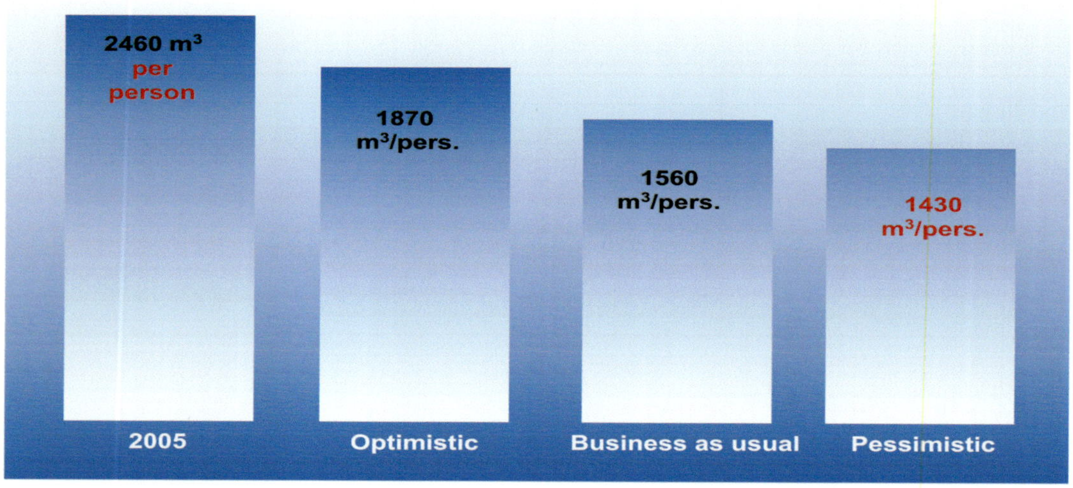

Figure 4.11 Water scenarios for 2030 (after SIC-ICWC).

person per year, which is much more than in countries with a real high water deficit such as Israel and Jordan.

But why then is the fear of conflict and confrontation a permanent feature of reports from both inside and outside the region? The instability of water resources and environmental degradation are indeed prominent issues. Concerns about security and sustainability in the region are found in many publications (see, for example, Petrov 2009). Three groups of threats are to be highlighted: i) internal threats (incompleteness of political reforms and the unsatisfactory socioeconomic situation), ii) regional threats (contradictions and challenges in political, socioeconomic and environmental policies) and iii) external threats (geopolitical forces that play a role in Central Asia and the situation in the neighboring countries). All suggest that the five riparian countries (plus Afghanistan) are "doomed" to co-operate as the only way to withstand regional threats and as a tool to facilitate the settling of internal problems in each country. It is therefore important to specify the real threats and the ways that these can be overcome. Let us consider regional challenges and the water problems that may cause the loss of this natural resource under the present conditions of use.

First consider the Amudarya river basin (Figure 4.12a). The origin of the runoff is mainly located along the Panj, Vakhsh and Kafirnigan tributaries. However, significant decreases in the available water supply[6] do not take place here. Some water losses occur in the Vakhsh cascade of reservoirs (Nurek, Baypaza and Sanguda); this loss is estimated at 0.42 km^3 a year (Sredazgiprovodkhlopok 1984). Downstream from the confluence of the major tributaries, there are points of water abstraction which are used for supplying water to a number of irrigation schemes along the middle reach of this river. These include the Syrkhandarya irrigation system through the Amu-Zang Canal, the Karshi Steppe through the Karshi Pumping Stations Cascade, the irrigation areas in the Zeravshan river basin through the Amu-Bukhara Canal, and the irrigation areas in South Turkmenistan through the Karakum Canal. As is typical for the middle reaches of rivers where there are few possibilities to construct in-channel reservoirs, seasonal runoff is regulated through a separate system of reservoirs. The Khauzkhan, Zeyd, Kopetdag, Kuyamazar, Turdakul, Shorkul and Talimarjan reservoirs were constructed on the canals taking water from the middle reaches of the Amudarya. These have a combined storage capacity of 6.76 km^3. The middle reach is characterized by both water losses from the riverbed, which vary from 3.5 to 6.8 km^3/year, and water losses from the system of reservoirs as well as from the Karakum Canal, which amount to 2.5 to 3 km^3/year). Owing to the continuous operational regime of the Karakum Canal over a long period, its bottom layer sediment has become almost impervious to water, and the total water losses are now estimated to be only 15% to 18% of the total flow.

The flow of the Amudarya in its lower reaches is regulated by the Tuyamuyun Reservoir. Depending on the management regime applied, water losses here vary considerably, ranging from an estimated 2.6 to 10 km^3. The total in-channel water losses along the Amudarya, without accounting for a disputed difference in river flow balance, are estimated to amount to 6 to 8 km^3/year on average (see Table 4.11) (Sorokin, 2002).

The unstable differences in the river flow balance cannot only be linked to water losses or to unrecorded water withdrawals, which create considerable difficulties for operational water management practice. As a rule, the dynamics of filling and discharging water in the riverbed itself (along a

[6]The term "available water supply" means the average annual volume of water that is used, or can be potentially used, for different economic purposes. It is formed by surface river runoff and groundwater minus natural and dynamic operational water losses and in-stream environmental flows.

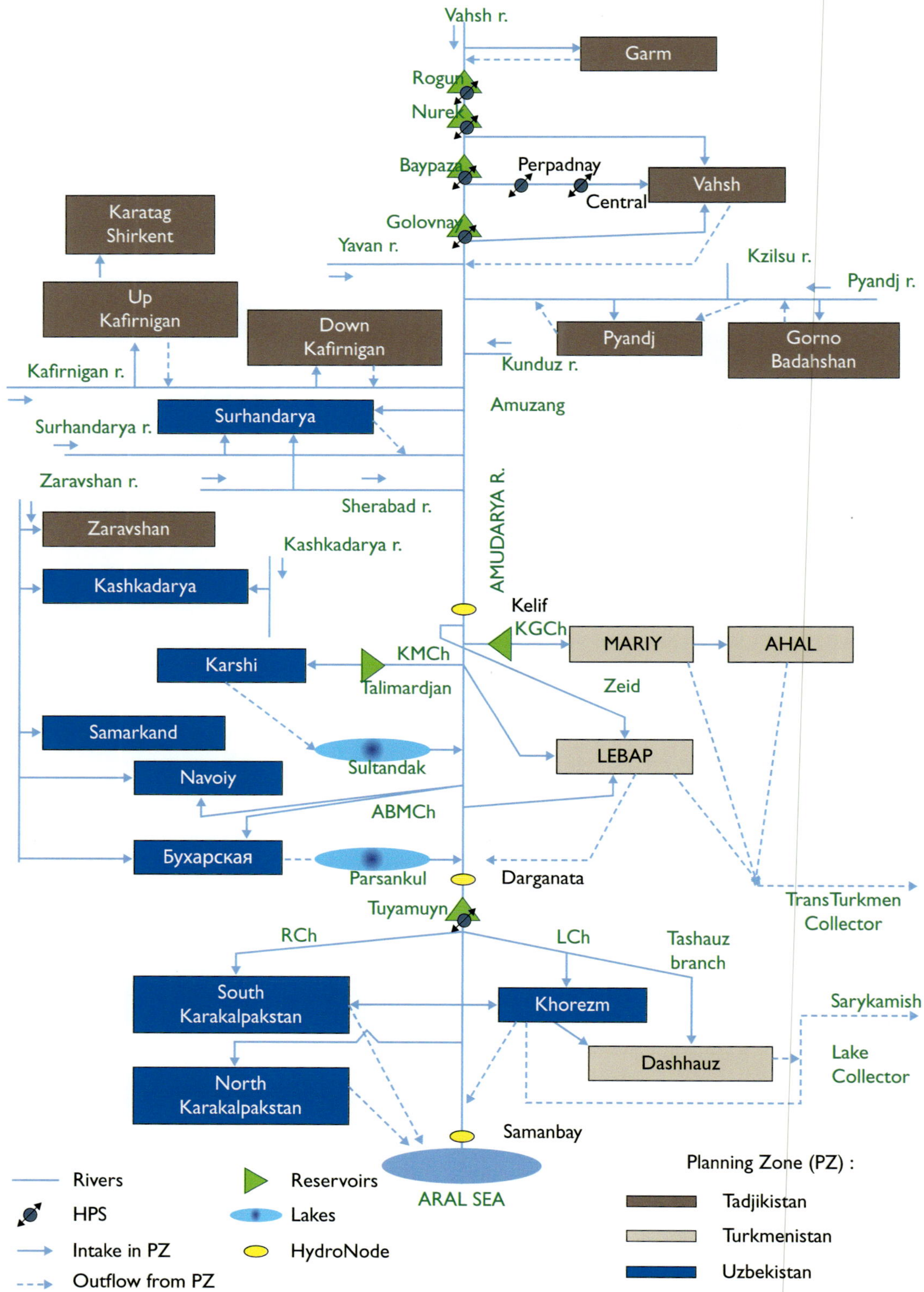

Figure 4.12a Schematic view of the Amudarya river basin (after SIC-ICWC).

Table 4.10 Groundwater reserves and use in the Aral Sea basin, million m^3/year.

State	Estimated reserves	Reserves approved for use	Actual abstraction in 1999	Groundwater use by sectors					
				Water supply	Industry	Irrigation	Tubewell drainage	Pilot pumping out	Other
Kazakhstan*	1,846	1,27	0,293	0,200	0,081	0	0	0	0,012
Kyrgyzstan*	1,595	0,632	0,244	0,043	0,056	0,145	0	0	0
Tajikistan	18,7	6,02	2,294	0,485	0,200	0,428	0,018	0	0,060
Turkmenistan	3,36	1,22	0, 457	0,021	0,036	0,150	0,060	0,001	0,015
Uzbekistan	18,455	7,796	7,749	3,369	0,715	2,156	1,349	0,12	0,400
Total	43,956	16,938	11,037	4, 118	1,088	2,879	1,427	0,121	0,487

Table 4.11 Water losses from the Amudarya riverbed, km^3/year.

Period (hydrological year, season)	Estimated water losses		
	Kelif to Darganata	Darganata to Samanbay	Kelif to Samanbay
1970 to 1979	3.8	2.5	6.3
1980 to 1989	4.3	3.8	8.1
1990 to 1999	4.6	4.3	8.9
1999 to 2000	3.5	3.0	6.5
2000 to 2001 including:	2.9	2.5	5.4
– over the non-growing season	0.8	0.5	1.3
– over the growing season	2.1	2.0	4.1

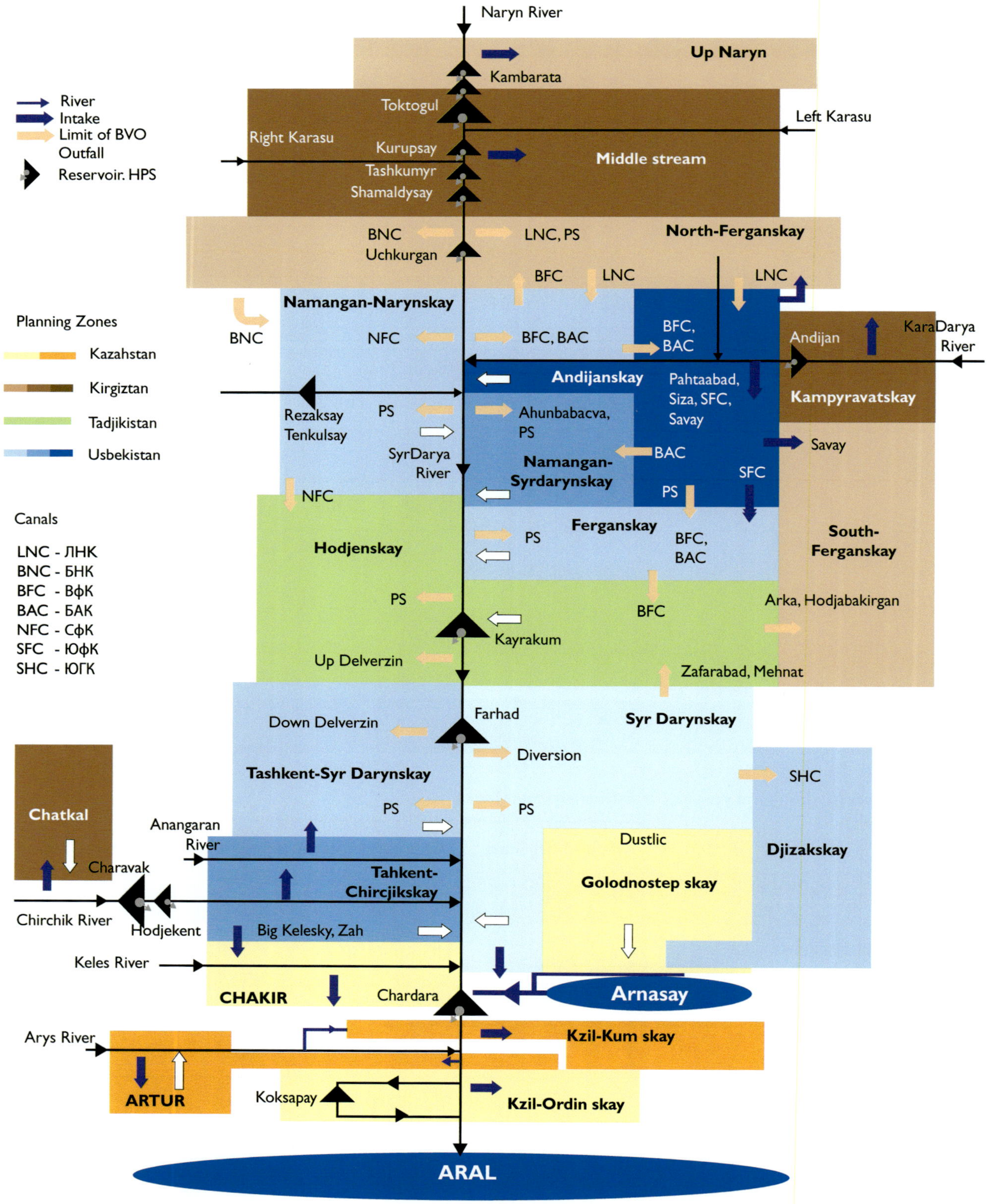

Figure 4.12b Schematic view of the Syrdarya river basin (after SIC-ICWC).

distance of this water of more than 800 km) are not evaluated under the present system of operational water management, although taking this factor into account could considerably improve the accuracy of the actual river flow balance calculation (the difference now is around 2 to 3 km^3/year). It would be necessary to establish an online monitoring system with automated gauging stations on the river trunk and head intake structures for a more precise definition of water losses from the riverbed. This would enable more accurate balances to be drawn up for the river runoff.

Using best estimates, and adding in a necessary environmental base flow (8 km^3 in an average year) and water losses from reservoirs (3 km^3) to the water losses detailed above, then the available water supply in the Amudarya amounts to 42 to 45 km^3/year, and the available water supply to the whole Amudarya basin (including the runoff in all closed river basins) is 60 to 63 km^3/year.

Let us now consider the Syrdarya river basin (Figure 4.12b). It is a little easier to specify the natural water losses from the riverbed here (in-channel water losses do not exceed 5 km^3/year and the water requirements of the delta and related ecosystems amount to 2 km^3/year). However, it is much more complicated to compile a record of the dynamic losses of the river runoff due to the actual water release regimes.

The Toktogul Reservoir has a total design storage capacity of 19.5 km^3, with a dead storage capacity of 5.6 km^3 and an active storage capacity of 14.0 km^3. Construction work was started in 1962 and completed in 1982. The first part of the hydroscheme came into operation on a temporary basis in 1974. The scheme for the integrated use and protection of the water resources in the Syrdarya river basin envisaged that the Toktogul Reservoir should serve as a reservoir for the purpose of irrigation and energy production. It also has the added function of regulating the flow of the Naryn River throughout the year for irrigation purposes in the Syrdarya river basin as a whole and for the seasonal regulation of the flow into the Fergana Valley. In its report on putting the Toktogul Hydropower Station into commercial operation, the State Commission wrote: "*The primary function of the Toktogul Hydroscheme is the compensatory regulation of the Naryn River flow throughout the year for the purpose of improving the availability of water for irrigation in the Syrdarya river basin, which is a key water consumer, and for meeting the needs of the communal and industrial sectors for water. Using this hydroscheme for the purpose of energy production is only an incidental use.*"

In the approved project framework, the Toktogul Reservoir, along with the Andijan, Charvak and Chardara reservoirs, was to be used for compensatory regulation of the river flow in co-ordination with the operational regime for the whole cascade. The project provided for base water releases at a rate of 100 m^3/sec from the Toktogul Reservoir and not less than 50 m^3/sec from the Chardara Reservoir. This provision was fixed in paragraph 4 of the operational rules for the Naryn-Syrdarya Reservoirs Cascade, which were issued in 1985. The use of water releases from the Toktogul Reservoir for energy production (at a rate of 80 m^3/sec in excess of minimum base flows) was also envisaged.

The reservoir came into full commercial operation in 1986 at the end of a cycle of dry years (1974–1985). The reservoir started to operate according to the operational regime agreed under the project by providing for accumulation of water in the winter and appropriate increases in water releases

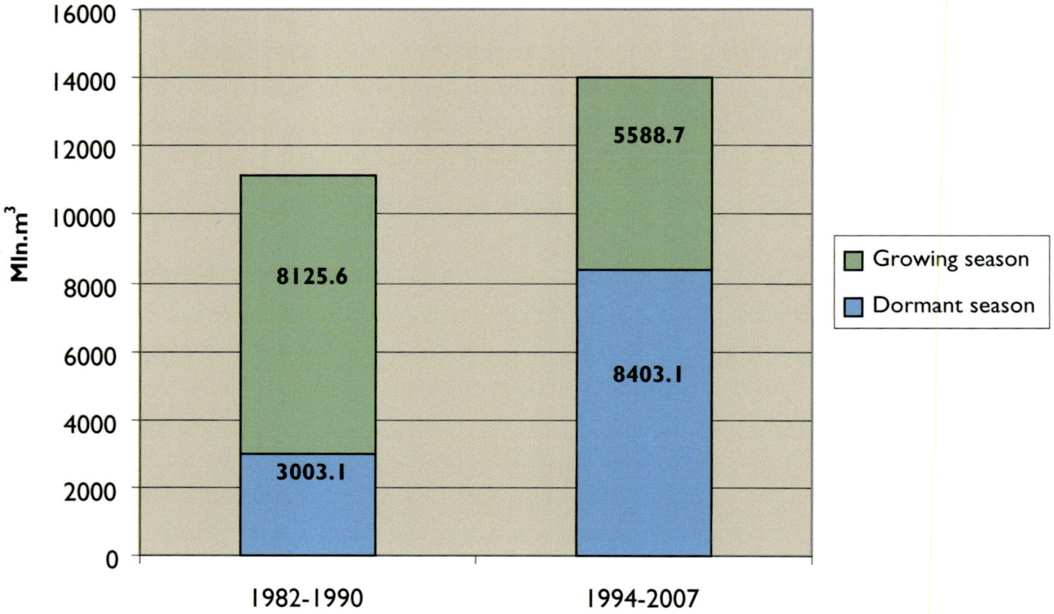

Figure 4.13 Average annual volume of water released from the Toktogul Reservoir over the periods 1982 to 1990 and 1994 to 2007 (SIC-ICWC).

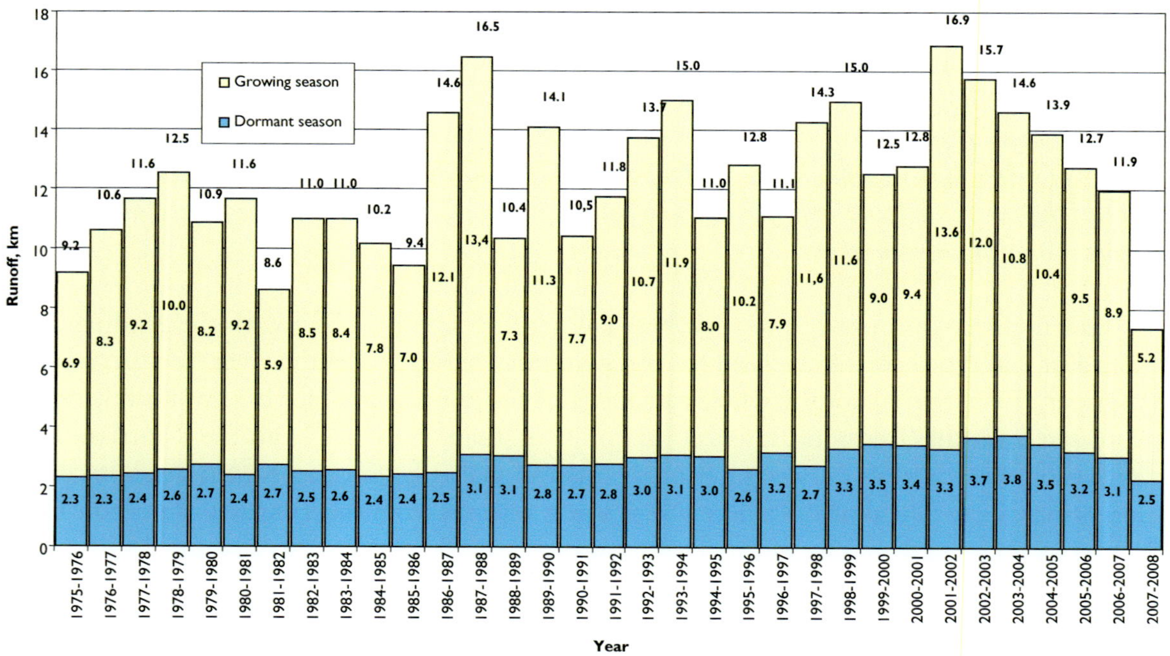

Figure 4.14 Inflow into the Toktogul Reservoir over the period 1975 to 2008 (SIC-ICWC).

in the summer. In the period from 1982 to 1990, the average volume of water releases during the winter (October to March) amounted to 3 billion m^3 compared with 8.1 billion m^3 of summer water releases (Figure 4.13). Over the same period, inflows into the reservoir were 2.7 billion m^3 in the non-growing season and 9.3 billion m^3 in the growing season respectively. Under this operational regime, the Toktogul Reservoir was filled to the design volume by 1988, allowing year-round regulation of the river flow and accumulating 13–17 billion m^3 of water by the beginning of the growing season.

This situation started to change in 1992, when there was a step-by-step shift from the established operational regime of the Naryn-Syrdarya Reservoirs Cascade, which saw a gradual increase in the water accumulated in the summer period with a considerable increase in winter water releases. Since 1990, there has been a steady increase in winter water releases, infringing upon the interests of irrigated farming and other water consumers needs including the ecosystems of the delta.

In 1994, this trend had turned into a management principle, and as a result the reservoir was operated according to a regime for energy production and there was an abrupt reduction of the volumes of water released for irrigation. Volumes of winter water releases have increased, to a peak of 9.7 billion m^3 during the autumn and winter of 2007–2008, and volumes of summer water releases have been as low as 3.6 billion m^3 during the dry years of 2001 and 2002 (Figure 4.15).

Figure 4.15 clearly shows that by operating the Toktogul Reservoir from 1981 to 1991 according to a regime designed to accumulate water for year-round regulation (especially the accumulation of water in 1986 to 1988 when there was abundant water) this regime should have enabled any consequences of the later drought periods to be overcome smoothly.

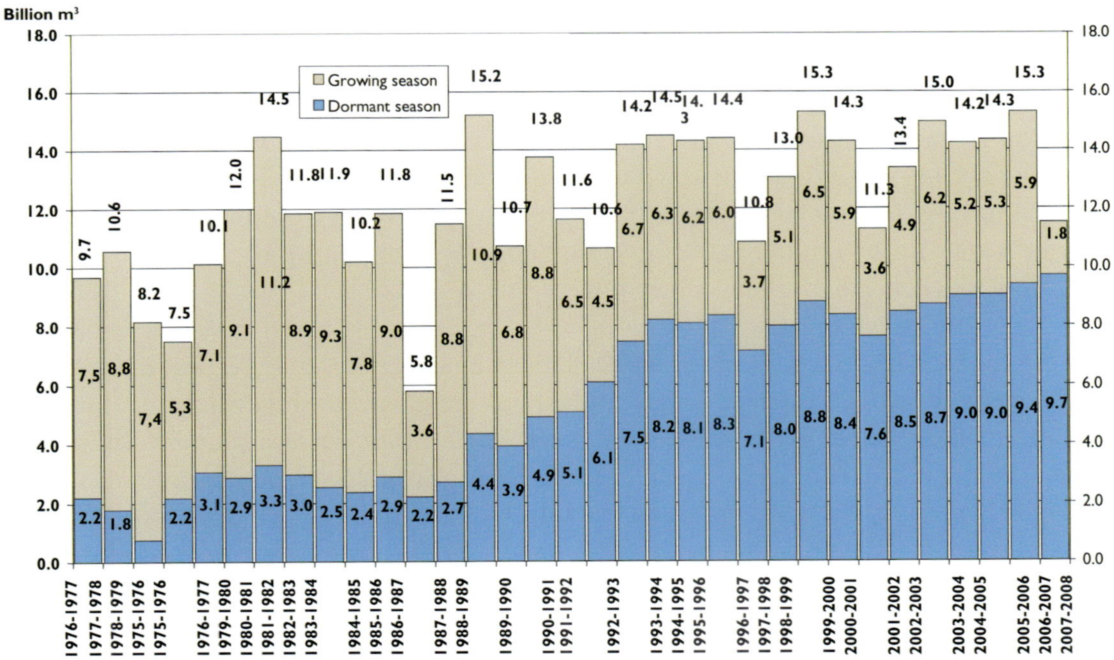

Figure 4.15 Water releases from the Toktogul Reservoir (SIC-ICWC).

It is necessary to note that despite of the relatively low water availability in the period from 1988 to 2002, it was possible to avoid a reduction of water supplies to the irrigated farming sector and infringing upon the interests of all water users. This was achieved thanks to the well co-ordinated work of the water organizations in the region and the Syrdarya BWO. This was even achieved in the dry years (2000 and 2001), when water availability in the lower reaches of the basin was down by 50%. However, the following wet years (2002 to 2006) were not used for water accumulation as envisaged by the operational rules for the Naryn-Syrdarya Reservoirs Cascade. As a result, the droughts in 2007 and 2008 created a catastrophic situation with regard to irrigation water availability in the Syrdarya river basin because prior to the start of the growing season, the drawdown of water in the Toktogul Reservoir had reduced the water volume to almost the dead storage level.

This situation arose because of the changing of the operational rules developed for the Toktogul Reservoir, that shifted from year-round regulation towards seasonal regulation for the benefit of energy production. Under this regime, the reservoir is filled at the beginning of the autumn-winter season and is empty at the beginning of the growing season. The release regime was based on actual annual inflow (inflow was balanced with water releases, taking into account water losses), and the reservoir was operated to meet direct commercial objectives only.

This regime of seasonal regulation for the benefit of energy production only created a situation where even in wet years (for example, 2003 and 2005) there was insufficient water available for irrigation in June and July (see Figures 4.16 a, b, c and d). This happened because the Kyrgyzstan hydropower producers exerted pressure on Uzbek and Kazakh energy resource suppliers to sign profitable multilateral agreements under which Kyrgyzstan would procure oil and natural gas in return for hydropower. For example in 2004, in spite of sufficient water to supply irrigation water supplies at the agreed levels for planned water use during the growing season, water availability was severely constrained in the Uchkurgan to Kayrakum irrigation area. In June, only 84.4%, 78.4% and 81.5% of the planned water supply was received in the respective ten-day periods of the month. In the Kayrakum to Chardara irrigation area, only 83.9% of planned water supply was received in the first ten-day period, 73.7% in the second ten-day period, and 87% in the third ten-day period. Irrigation water supplies only returned to planned levels in July. This outcome, the result of pressure from energy producers and "hydro-egoism", led to a long search for a commercial compromise and the drawing up of a new agreed operational regime for 1994 was delayed until the end of June of that year.

During the catastrophic high water levels of 1969, 24 km^3 of river water was discharged into the Arnasay Depression through the emergency spillway of the Chardara Hydroscheme. However when necessary water releases were not implemented, there was a drastic decrease in the volume and level of water in the Arnasay-Aydarkul Depression. Water salinity reached 9 g/l, because this lacustrine system was mainly fed by drainage water discharged from the Golodnaya Steppe irrigation system plus limited water releases from the Kly River. When the Toktogul Reservoir shifted towards a winter water release regime, the unproductive discharges of river water into the Arnasay Depression immediately increased because the carrying capacity of the riverbed downstream from the Chardara Hydroscheme is very limited during the winter (at 400 to 600 m^3/sec). Table 4.12 shows resulting volumes of water entering into the Arnasay Depression.

Thus, river flow losses from the Syrdarya due to water discharging into the Arnasay Depression during these winter high waters was some 3.4 km^3/year on average, which considerably exceeded natural water losses. As a result, the available water supply of the Syrdarya River was reduced to only 27 km^3/year. During following years the Committee of Water Resources of Kazakhstan implemented work to increase the carrying capacity of the riverbed downstream from Chardara to 800 m^3/sec and, as a result, water

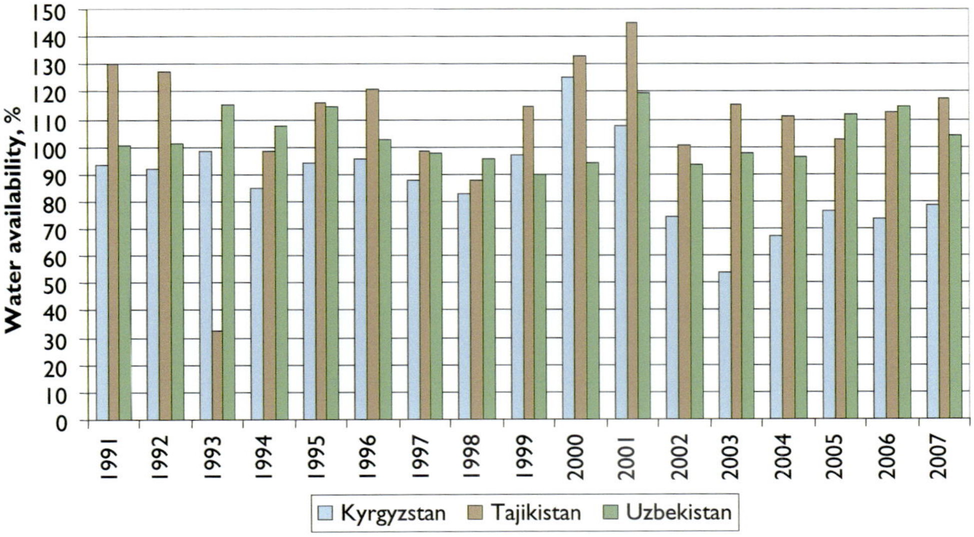

Figure 4.16a Average water availability (as a percentage of planned supply) over the April to September period in the area between the Toktogul Reservoir and the Kayrakum Reservoir (SIC-ICWC).

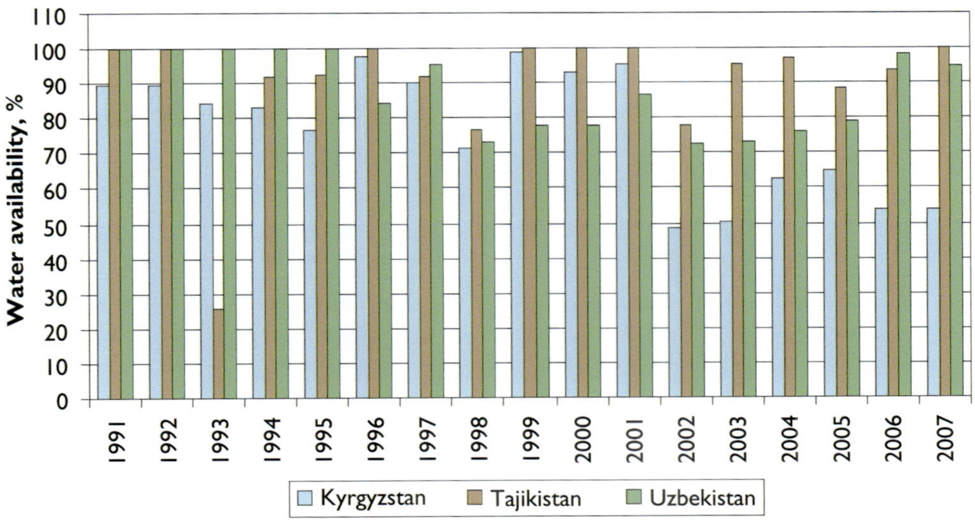

Figure 4.16b The lowest level of water availability (as a percentage of planned supply) received in the area between the Toktogul Reservoir and the Kayrakum Reservoir during the critical April to September months (SIC-ICWC).

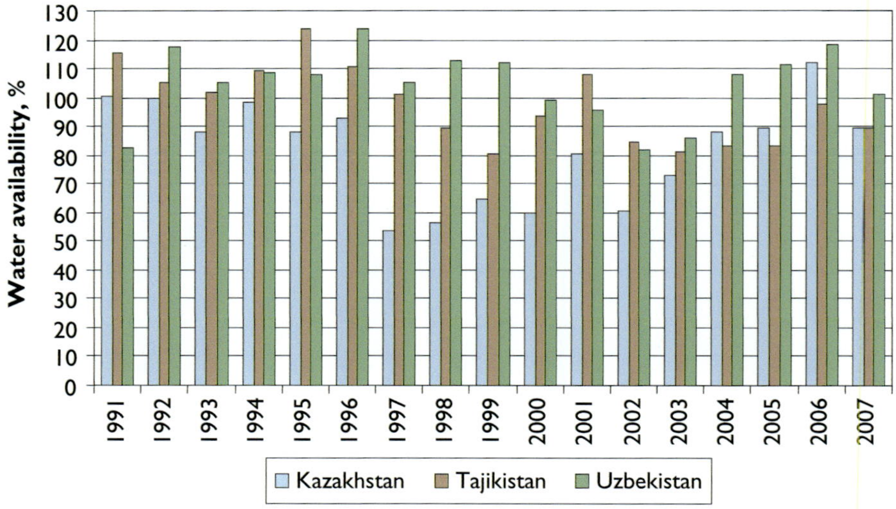

Figure 4.16c Average water availability (as a percentage of planned supply) over the April to September period in the area between the Kayrakum Reservoir and the Chardara Reservoir (SIC-ICWC).

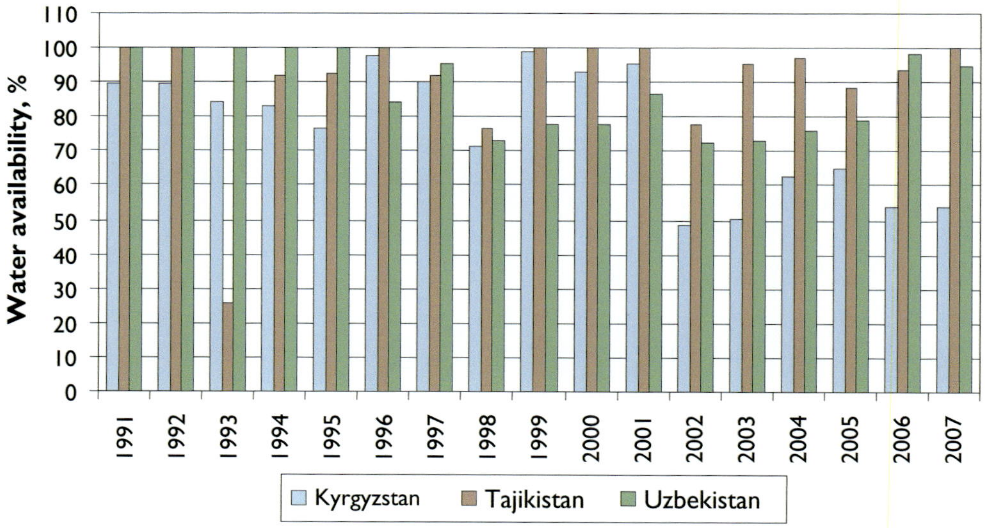

Figure 4.16d The lowest level of water availability (as a percentage of planned supply) received in the area between the Kayrakum Reservoir and the Chardara Reservoir during the critical April to September months (SIC-ICWC).

releases into the Arnasay Depression were reduced over the winter. At the same time, water discharge into the Syrdarya delta and into the Northern Small Aral Sea has considerably increased (Table 4.13).

Thus, combining the results for the Amudarya and the Syrdarya, the total available water supply in the Aral Sea basin consists of 90–93 km^3/year of surface water and 17 km^3 of groundwater (in total, 107–110 km^3/year). However, this does not take account of the volume of return water (Table 4.14).

Table 4.12 Water releases into the Arnasay depression after the Toktogul Reservoir shifted towards a winter water release regime, in km^3 **(Carewib Database).**

1993	1994	1995	1996	1997	1998	Total
2.65	9.27	4.00	1.21	1.24	3.13	21.44

Table 4.13 Consequences of water releases from the Toktogul Reservoir and consequent occurrence of high water in the winter, km^3/year (SIC-ICWC).

Year	Water releases from the Toktogul Reservoir	Water releases from the Kayrakum Reservoir	Water releases from the Chardara Reservoir
2002–2003	8.5	12.4/9.5*	6.4
2003–2004	8.7	15.0/11.5	10.0
2004–2005	0.0	15.3/13.0	11.8
2005–2006	9.0	14.9/10.0	11.0
2006–2007	9.4	12.2/10.4	8.5
2007–2008	9.7	12.4/9.5	8.2

***taking into account water pumped from drainage tubewells.**

Return water is an additional reserve available for use. However, return water is a primary polluter of water bodies and the environment due to its high salinity. Drainage water disposed from irrigated areas forms about 95% of the total volume of return water, and only some 5% of return water originates from industrial and community waste water.

With the development of the irrigation and drainage systems in the region, there has been a steady increase in the volume of return water, especially during the intensive growth of irrigation between 1960 and 1990. In the 1990s, the volume of return water entering the river systems stabilized, and it even decreased to some extent due to a temporary reduction in the area of land under irrigation and the degradation of drainage systems. In the 1990s, the total volume of return water ranged from 28.8 km^3/year to 33.5 km^3/year. About 13.5–15.5 km^3 of return water annually entered the Syrdarya river basin and about 16–19 km^3 entered the Amudarya river basin. More than 51% of the total volume of return water enters the rivers through the drainage canals and about 33% flows into desert depressions. Only 16% of return water is reused for irrigation because most return water is unsuitable give the high pollution and salinity levels.

It should be kept in mind that with an increase in the efficiency of irrigation systems and improvements in the practice of water use, the volumes of return water will decrease and salinity will rise. Therefore, no more than 50% of the volume of return water currently used can be included in an assessment of the future available water supply. Thus, the total annual volume of water resources available for use under present conditions, taking into account return water and all unproductive losses of river runoff, amounts to 114–117 km^3 in the basin as a whole or, in terms of the present population, 2,380 m^3 per person per year, which is 10% less than shown in Figure 4.11.

In concluding this analysis of water resources in the Aral Sea basin, it is useful to specify the measures actually necessary for increasing the volume of available water supply:

- obtaining a better estimate of the water losses from the Amudarya riverbed from the integrated hydrological studies being implemented with the participation of the hydrometeorological services and water organizations of all riparian countries and through automated monitoring at all gauging stations and water intake structures;
- restoring the year-round regulation regime of the Syrdarya runoff for irrigation use, or constructing additional storage reservoirs for seasonal runoff regulation in order to compensate for runoff losses that now amount to 3.5 km^3/year on average;
- completing the system of supervisory control and data acquisition (SCADA) on the Syrdarya;
- improving return water management by both BVOs (river basin management organizations);
- improving the co-ordination between all primary water users that utilize the water resources of both rivers (for hydropower generation, water supply, irrigation, etc.).

Another critical aspect for water use in the basin is strict compliance with the 1992 interstate agreement between the riparian countries and the ICWC executive bodies. Tables 4.14 and 4.15 provide an analysis of some of the implementation aspects of this agreement to the year 2000.

Table 4.15 shows that only the share of Kazakhstan was lower than the established water limits, and the use of water resources by all other countries exceeded their allocated shares over the period under consideration (excluding the dry years). However, the actual annual deviation from mean annual values was within a range of 5%, and this within the allowed deviations established by a decision of the ICWC for both BWOs.

Tables 4.17 and 4.18 show the situation in the most recent decade (the 2000s) using data collected by the Central Asian Regional Environment and Water Information System (CAREWIB). It should be noted that apparently problem-free supply when looking at average supply over a year or growing season can be quite misleading, and a different picture when analyzing monthly water availability (or even supply over a ten-day period) at particular sites or in individual countries. Figures 4.16c and 4.16d show the dynamics of average and minimum water availability during the most critical irrigation months in an area between the Kayrakum Reservoir and the Chardara Reservoir, which provide the most challenges for water resource management. In this period any deviation below the required water supply affects crop yield and the economic indicators for agricultural production.

Table 4.14 Return water (source and disposal) in the Aral Sea Basin, mean values in the period 1990 to 1999 according to estimations of SIC ICWC), km^3/year.

Country	Drainage water in irrigated areas*	Community and industrial waste waters	Total volume of return water	Return water disposal		
				Into rivers	In natural depressions	Reuse for irrigation
Kazakhstan**	1.60	0.19	1.79	0.84	0.70	0.25
Kyrgyzstan**	1.70	0.22	1.92	1.85	–	0.07
Tajikistan** (total)	4.05	0.55	4.60	4.25	–	0.35
In the Syrdarya basin	*1.05*	*0.14*	*1.19*	*0.92*	–	*0.27*
In the Amudarya basin	*3.00*	*0.41*	*3.41*	*3.33*	–	*0.08*
Turkmenistan	3.80	0.25	4.05	0.91	3.10	0.04
Uzbekistan (total)	18.40	1.69	20.09	8.92	7.07	4.10
In the Syrdarya basin	*7.60*	*0.89*	*8.49*	*5.55*	*0.84*	*2.10*
In the Amudarya basin	*10.80*	*0.80*	*11.60*	*3.37*	*6.23*	*2.00*
Total in the Aral Sea basin	**29.55**	**2.90**	**32.45**	**16.77**	**10.87**	**4.81**
Syrdarya basin	**11.95**	**1.44**	**13.39**	**9.16**	**1.54**	**2.69**
Amudarya basin	**17.60**	**1.46**	**19.06**	**7.61**	**9.33**	**2.12**

***taking into account water pumped from drainage tubewells.**
****data presented in national reports of the SPECA project.**

During the first years after independence until 1996, the average water availability in all three riparian countries in this area was never lower than 85% of planned supply and at minimum was still 65% (minimum value observed in Kazakhstan in the command area of the Dustyk Canal). However, from 1997 to 2003, the situation drastically changed, with average water availability dropping down to 60% and with minimum values as low as 30%. There is a clash of interests at two levels here. In the area between the Toktogul and Kayrakum reservoirs, water availability never dropped below 55% of the planned supply, even during critical periods of water shortage. This situation can be explained because water availability in this site depends only on water releases from the Toktogul Reservoir, and reflects a certain level of co-ordination between the interests of Kyrgyzstan, Kazakhstan and Uzbekistan. At

Table 4.15 Actual distribution of water resources in the Syrdarya river basin, 1992 to 1999 (Carewib Database).

Country	1992–1993		1993–1994		1994–1995		1995–1996		1996–1997		1997–1998		1998–1999		Average		Limit, %
	km^3	%	km^3	%	km^3	%	km^3	%	km^3	%	km^3	%	km^3	%	km^3	%	
Uzbekistan	11	50.7	10.36	49.1	9.82	48.1	11.54	51.9	11.95	54.1	11.98	53.99	12.46	54.5	11.3	51.76	50.5
Kazakhstan	8.46	39	8.42	39.9	8.42	41.2	8.48	38.1	8.1	36.7	8.2	36.95	8.32	36.4	8.34	38.32	42
Tajikistan	2.05	9.45	2.15	10.2	1.99	9.75	2.04	9.17	1.87	8.47	1.83	8.25	1.88	8.22	1.97	9.07	7
Kyrgyzstan	0.18	0.83	0.19	0.9	0.19	0.93	0.18	0.81	0.17	0.77	0.18	0.81	0.21	0.92	0.19	0.85	0.5
Sub-total	21.69	100	21.12	100	20.42	100	22.24	100	22.09	100	22.19	100	22.87	100	21.8	100	100
To the Aral Sea	**7.1**	–	**9.25**	–	**6.5**	–	**3.9**	–	**4.9**	–	**5.88**	–	**7.3**	–	**6.38**	–	–
TOTAL	28.79	–	30.37	–	26.92	–	26.14	–	26.99	–	28.07	–	30	–	28.18	–	–
To the Arnasay Depression	**1.30**	–	**9.32**	–	**4.92**	–	**1.00**	–	**1.29**	–	**2.19**	–	**4.12**	–	**3.45**	–	–

Table 4.16 Actual use of planned limits of water withdrawals from the Amudarya, 1993 to 1999 (Carewib Database).

States-Users	1993–1994		1994–1995		1995–1996		1996–1997		1997–1998		1998–1999		Average		Limit, %
	Actual data (records of the ICWC meetings)														
	km^3	%	km^3	%	km^3	%	km^3	%	km^3	%	km^3	%	km^3	%	
Kyrgyzstan	0.15	0.29	0.13	0.26	0.16	0.31	0.17	0.34	0.14	0.10	0.14	0.27	0.15	0.29	0.29
Tajikistan	7.32	14.20	7.01	14.22	7.41	14.32	7.51	15.00	7.23	14.67	7.46	14.44	7.32	14.47	15.17
Turkmenistan	22.76	44.15	21.15	42.90	21.46	41.46	21.02	41.98	20.91	42.43	21.82	42.25	21.52	42.53	42.27
Uzbekistan	21.32	41.36	21.01	42.62	22.73	43.91	21.37	42.68	21.00	42.61	22.23	43.04	21.361	42.71	42.27
Sub-total	51.55	100	49.30	100	51.76	100	50.07	100	49.28	100	51.65	100	50.60	100	100
To the Aral Sea	**11.2**		**8.9**		**3.1**		**4.9**		**0.52**		**8.1**		**6.12**		
TOTAL	**62.75**		**58.20**		**54.86**		**54.97**		**49.80**		**59.75**		**56.72**		

Table 4.17 Actual water allocation in the Syrdarya basin, 1999 to 2008.

Country	*1999–2000*		*2000–2001*		*2001–2002*		*2002–2003*		*2003–2004*	
	km^3	%	km^3	%	km^3	%	km^3	%	km^3	%
Uzbekistan	12.9	48.3	12.9	51.9	11.8	40.5	12.2	47.4	12.6	50.5
Kazakhstan	11.5	43.1	9.7	38.8	15.1	52.2	11.5	44.5	10.3	41.1
Tajikistan	2.02	7.6	2.1	8.3	1.9	6.4	1.9	7.4	1.9	7.7
Kyrgyzstan	0.29	1.1	0.3	1.0	0.3	0.9	0.2	0.7	0.2	0.7
Sub-total	26.7	100	25.0	100	29.0	100	25.8	100	25.0	100
The Aral Sea	2.63		3.7		4.8		8.2		10.5	
Total	29.3		28.7		33.8		34		35.5	
Arnasay Depression	3.04		0.36		1.2		4.7		2.9	

Table 4.18 Actual use of water quotas in the Amudarya basin, 1999 to 2008.

Country	***1999–2000***		***2000–2001***		***2001–2002***		***2002–2003***		***2003–2004***	
	Actual data from the Amudarya BVO									
	km^3	%	**km^3**	%	**km^3**	%	**km^3**	%	**km^3**	%
Kyrgyzstan	0.05	0.10	0.04	0.11	0.03	0.06	0.00	0.01	0.01	0.02
Tajikistan	7.86	18.09	7.51	20.61	7.19	15.21	6.74	13.20	7.62	14.37
Turkmenistan	17.23	39.66	13.74	37.72	19.30	40.83	21.47	42.04	22.35	42.15
Uzbekistan	18.31	42.15	15.14	41.56	20.75	43.90	22.86	44.76	23.05	43.47
Sub-total	43.45	100	36.43	100	47.27	100	51.07	100	53.03	100
The Aral Sea	4.81		0.60		4.55		12.59		6.41	
TOTAL	**48.26**		**37.03**		**51.82**		**63.66**		**59.44**	

	2004–2005		*2005–2006*		*2006–2007*		*2007–2008*		Average	
	km^3	%	km^3	%	km^3	%	km^3	%	km^3	%
	13.7	46.7	14.2	55.0	13.9	45.5	10.5	54.3	12.7	48.5
	13.6	46.4	9.6	37.2	14.7	48.2	7.3	37.8	11.5	43.7
	1.9	6.3	1.9	7.2	1.7	5.7	1.4	7.1	1.8	7.0
	0.2	0.6	0.2	0.7	0.2	0.6	0.2	0.8	0.2	0.8
	29.3	100	25.8	100	30.5	100	19.3	100	26.3	100
	10.1		9.2		6.0		5.9		5.5	
	39.5		34.9		36.5		25.21			
	2.2		0.3		0.8		1.0			

	2004–2005		***2005–2006***		***2006–2007***		***2007–2008***		**Average**	
	km^3	%	**km^3**	%	**km^3**	%	**km^3**	%	**km^3**	%
	0.01	0.02	0.01	0.01	0.02	0.03	0.01	0.02	0.02	0.04
	6.94	13.59	7.53	13.91	7.66	15.65	7.66	19.64	7.41	15.72
	21.56	42.22	22.35	41.28	19.71	40.27	15.43	39.55	19.24	40.79
	22.56	44.18	24.26	44.80	21.56	44.05	15.91	40.79	20.49	43.45
	51.07	100	54.15	100	48.95	100	39.01	100	47.16	100
	15.63		6.05		2.22		1.62	4.9		
	66.70		**60.20**		51.17		40.63			

the downstream site (between the Kayrakum and Chardara reservoirs), two additional factors come into play. These are the mutual obligations of Tajikistan and Uzbekistan regarding maintenance of the Karakum Reservoir and the supply electric energy to North Tajikistan. The countries also have to pay regard to the water supply through the interstate canals (BFC, NFC, Dalversin and SFC) and carry out mutual obligations to Kazakhstan and Uzbekistan. In addition, an intermediate organization of local administrative bodies (the Syrdarya Khakimiat) interfered in the water sharing process on the Dustyk Canal for its own benefit.

In recent years, one more disruptive factor has been the management regime at the Naryn Hydropower Stations Cascade. This is run by the Kyrgyzenergo organization. In spite of the need for daily regulation of energy production, during the night (after 6 p.m.) this organization sharply decreases the volume of water released from the last unit of the cascade (the Uchkurgan Hydropower Station) into the river near the entrance to the Fergana Valley, although this type of daily regulation could also be implemented at the upstream hydropower stations in this cascade.

In 2005 international experts noted unplanned fluctuations in the water levels in the upstream reservoir of the Uchkurgan Hydroscheme. This was caused by daily fluctuations in water releases by as much as ±200 m^3/sec according to the energy consumption schedule at the Uchkurgan Hydropower Station. In 2008, this phenomenon led to a catastrophe, since the river flow was completely blocked during the night while the Kyrgyz managers were trying to deal with power cuts. This operational regime of the Naryn Hydropower Stations Cascade considerably disrupts water diversion into the irrigation systems of the Big Fergana Canal and the Northern Fergana Canal, which supply water to the agricultural lands in the Fergana Valley. The water level in the river can fluctuate from 0.5 to 2.5 m over a 1 to 3 hour period (Figure 4.17), lowering the volume of water diverted through the irrigation canals and infringing the design regulations developed for operating these hydraulic structures. Everyday, in an area of 350,000 hectares in the Fergana Valley, farmers and other water users are faced with the consequences of poorly-regulated water supply, which drastically reduces their ability to use water in an efficient manner !

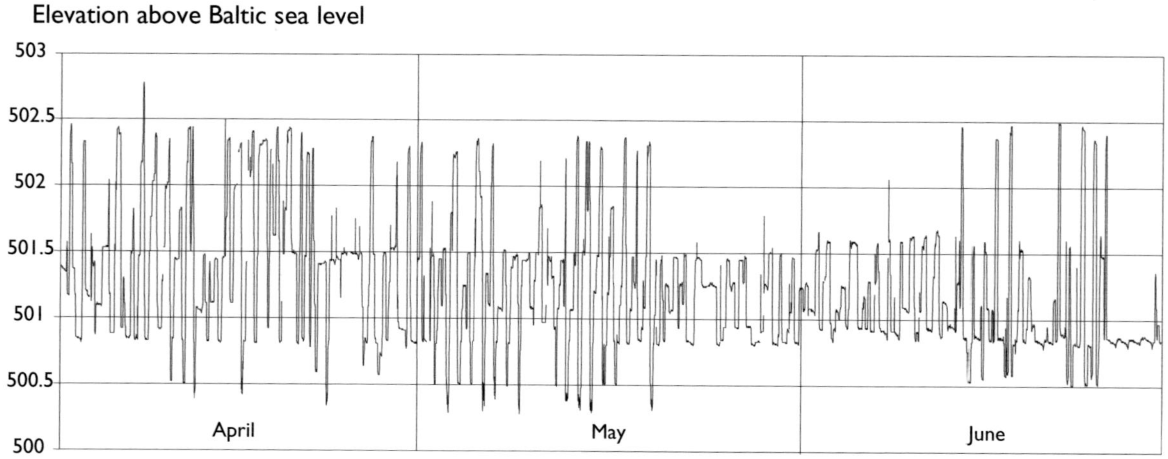

Figure 4.17 Water level fluctuations in the Naryn River downstream of the HPS cascade at the beginning of the 2008 growing season (SIC-ICWC).

The situation in the Amudarya river basin is more stable, although it is less monitored. In addition to the difficulties related to specifying runoff losses and accounting for the dynamics of water masses under changing water flow regimes, the interrelation of river flow with the surrounding terrain plays a specific role. For example, groundwater is replenished when the water level is high in the riverbed, and there is an inflow of groundwater from adjacent territories into the riverbed when the river's water level is low. Another complicating factor is the design of the multi-head water intake structures of the Karakum Canal and Amu-Bukhara Canal and the special regime of water diversion into the Karshi Canal, which creates backwater effects at this location of the river. This hampers the specification of the river flow characteristics and, as a result, lowers the accuracy of water distribution.

In dry years, national policies (rather than an interstate strategy) are predominantly aimed at maximizing crop yields within zones with a higher potential for income generation. These are located

Table 4.19 Intra-national distribution of water shortage in 2000. **(Report FY2003 OESI, by Dukhovniy, Khorst, 2003)**

Republic, river reach, province	**Shortage (km^3)**	**Shortage (% of limit)**
Tajikistan		
Upper reach	0.7	11
Republic as a whole within the Amudarya basin	**0.7**	**11**
Turkmenistan		
Middle reach	1.8	17
Dashhowuz Province	2.8	55
Republic as a whole within the Amudarya basin	**4.6**	**30**
Uzbekistan		
Middle reach	0.8	15
Khorezm Province	1.2	36
Karakalpakstan	3.7	59
Republic as a whole within the Amudarya basin	**5.7**	**37**

along the upper and middle stretches of the river. This restricts the water supply to the lower reaches (Khorezm Dashhowuz, and Karakalpakstan) where water consumption per hectare under crops is considerably higher. The economic results from two dry years (2000 and 2001) as presented in Table 4.19 and Figure 4.18 show the typical consequences of this policy (Report FY2003 OESI, by Dukhovniy, Khorst, 2003).

The problems of water allocation and management can be attributed to these clusters of reasons, and they need addressing by appropriate measures in order to improve water availability in a sustainable and uniform way.

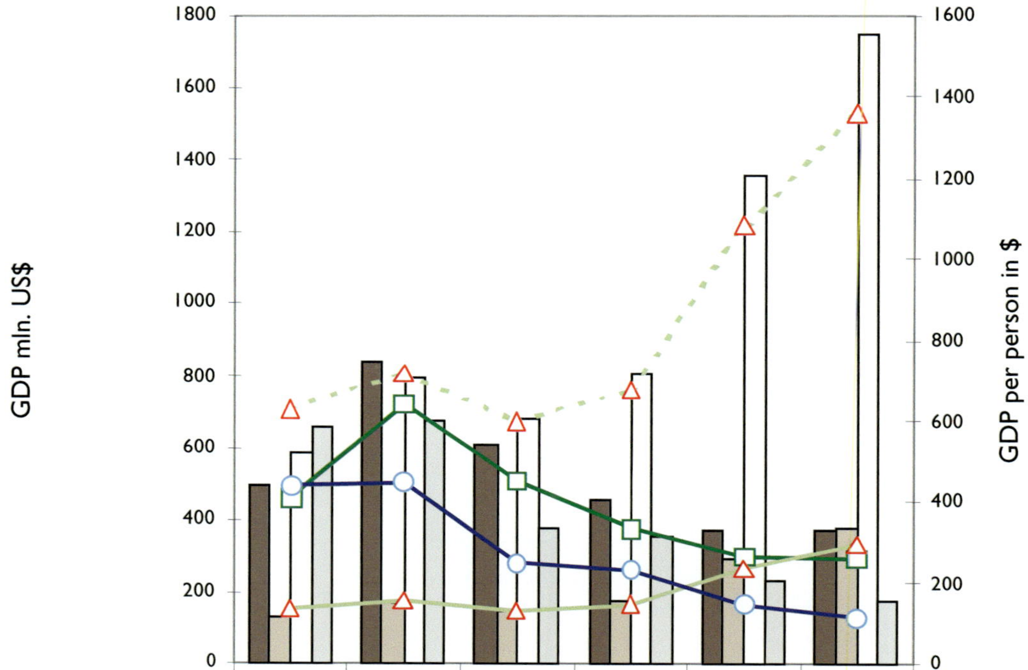

	1995	1999	2000	2001	2002	2003
Khorezm province(mln.$)	495.0	841.6	606.1	454.3	369.9	369.2
Dashoguz province**(mln.$)	126.5	171.5	148.4	174.9	294.4	378.2
Dashoguz province*(mln.$)	584.5	792.4	685.4	807.9	1360.0	1747.4
Karakalpakstan(mln.$)	658.1	672.3	378.8	356.7	232.1	175.8
Khorezm province($/men)	408.3	640.6	452.9	333.8	267.5	263.2
Dashoguz province**($/men)	136.0	155.6	129.9	146.1	235.1	294.4
Dashoguz province*($/men)	628.4	718.8	600.2	675.1	1086.1	1360.2
Karakalpakstan($/men)	439.0	447.3	248.1	231.1	149.3	113.0

Years

Figure 4.18 Dynamics of GDP and GDP per capita (SIC-ICWC)
*estimated using the black market dollar rate; ** estimated using the official exchange rate)[7]

[7]A combination of the black market dollar rate and official exchange rate.

- **Legal aspects**: weakness and insufficient clarity in the provisions of the relevant signed agreements; lack of mechanisms for practical realization of these agreements; lack of mechanisms for specifying appropriate sanctions and ensuring the commitments of the all parties; and inadequate quotas dating from the Soviet period.
- **Institutional aspects**: weak mandates and limited functions for the regional organizations; overlapping activities between regional and national agencies, which needs to be addressed by efficient co-operation; lack of free access to information on water management and water monitoring; insufficient and disproportionate funding; and interference by national authorities in the activities of the regional organizations.
- **Hydrological aspects**: few studies of in-stream water losses, which requires upgrades of the monitoring system; poor control of waste water discharges into rivers and water diversion from rivers; better forecasts required of river runoff and monitoring of flow rates; a need to collect information on changing climatic parameters that affect river runoff; improved monitoring of water quality in streams; and measures needed to ensure sanitary and ecological flows.
- **Hydropower aspects**: a need for better regimes of summer and winter water releases from storage reservoirs; lack of clear and fixed commitments regarding the winter needs of upstream countries for electrical energy; unco-ordinated pricing policies for electricity and natural gas (coal, crude oil); a need to co-ordinate the activities of hydropower producers with other water users (through integrated river basin management principles).

4.4 Irrigation or Hydropower, or Irrigation and Hydropower?

Comparing past and present use of the water resources in Central Asia shows that much firmer guidelines for water resource management were developed in the unified country with proper economic balancing of supply and demand in the region as a whole. These policies allowed floods and droughts to be overcome without any conflict, although this was achieved at the expense of the environment – nature became "the victim". As a consequence, the Aral Sea, the rivers in the region, and vast areas in the vicinity of the Aral Sea were turned into "hot spots" for salinization. What has changed since the Soviet period? The reserves of water resources have not changed considerably and industrial and agricultural production has fallen significantly, but both within and outside the region fears of future water conflicts have become stronger.

The resolution of the conference held in 2006 to mark the 15-year anniversary of the ICWC noted that, in spite of the periods of droughts and disastrous floods, the ICWC has managed to avoid any serious conflict. However, one can say that clear signs of change could already be observed. There were serious sounds of disagreement in the Syrdarya basin in 2007 and 2008, when water resources management practically became "hostage" to the erroneous policy for water use adopted by the Hydropower Department of the Ministry of Industry and Power of Kyrgyzstan. This would not have happened in the Soviet time. For example, during the catastrophic drought in 1975, the decision was taken to blow up a closure on a temporary diversion tunnel (which provided dead storage for the Toktogul Reservoir) in order to increase the irrigation water supply in spite of the priority that had been given to hydropower production.

However, there is a great difference between national laws and international agreements. As Wouters and Vinogradov [21] have noted, regional bodies do not have the powers to make international water agreements compulsory. Of course, they are not able to do this at present, but they can (and must) introduce international and regional water laws based on universal values and principles of voluntary contribution, equality and co-operation. The means not only preparing legal instruments (agreements, treaties and conventions) that specify the procedures, rights and duties for all riparian water users and water consumers, but also establishing the appropriate institutional framework for transboundary water resources management. Special powers should be delegated to the regional institutions along with responsibilities for executing the provisions of interstate agreements and treaties, as well as for elaborating, co-ordinating and approving appropriate mechanisms that will create a strict framework for transboundary water use and management.

Key factors, therefore, are a sincere and persistent will from top management and trust between nations. There should be a means of meeting the co-ordinated aspirations of national governments and parliaments, and these and other powers should demonstrate their good will by concrete actions. Eventually the water users themselves should also be involved in the system of interstate water governance by means of transforming the state-managed system into a mixed public-participatory version. This is key to achieving integrated water resource management in a river basin.

The International Crisis Group (ICG 2002) states that the causes of the contradictions and tensions in the present system of transboundary water management in Central Asia are:

- lack of coherent water management;
- failure to abide by or adapt to water quotas;
- tensions over barter agreements and payments;
- uncertainty over future infrastructure plans.

Undoubtedly the mutual tensions in the riparian countries are not only attributable to these causes. There also are other reasons for disagreement, such as lack of equitable representation of all riparian countries in the regional organizations, the deployment of key water management bodies in Uzbekistan (SIC of ICWC, Amudarya BVO in Urgench and Syrdarya BVO in Tashkent) and incomplete coverage of all aspects of water resource management by the ICWC itself. The ICWC for example, is not responsible for groundwater and return water management and operation of certain parts of the water infrastructure. So each water management organization also suffers from its own shortcomings, leading to further tensions. This is especially true for the two BVOs. In this case, the following issues are of interest:

- insufficient financial resources due to the fact that only Uzbekistan and, to a lesser degree, Turkmenistan and Kazakhstan finance the activities of interstate bodies, though with insufficient amounts;

- lack of permanent access to water infrastructure on the rivers due to country border restrictions;
- insufficient information on water diversions from the rivers and return water into the rivers (a considerable number of water intake structures are monitored by the BVOs or are selectively monitored only);
- lack of technical capability and powers to restrict water withdrawal or to close a water intake in case of unplanned and excessive water diversion;
- dependence of BVOs on personnel from local authorities and the administrative pressure exerted by these authorities on the organizations.

Actually some provisions in the first interstate agreements did have the mechanisms for settling these issues. In particular the agreement of January 11, 1994 (articles 10 and 11) provided for free access to all water infrastructure for personnel and workers of the ICWC and BVOs, as well as obligatory implementation of ICWC decisions by all categories of water users. However, these provisions were not followed-up by legal bilateral documents and, as a result, the water resources management organizations face extreme difficulties.

Nevertheless, the deeper causes of tensions and conflicts need to be kept in mind. Kyrgyzstan and Tajikistan, with their weaker economic development, have specified hydropower generation as a feasible long-term source of income, using their location in the upstream zone of river runoff to meet growing energy needs through the low-cost production option of hydropower. There is a tendency to create a certain mechanism of "hydrological pressure" on downstream countries, by manipulating water releases from storage reservoirs as a means of dictating prices for water or for acquiring preferential terms for goods in other economic spheres hardly related to the water sector such as fuel, and this is turning water resources into commercial goods. Moreover, some rather vague provisions in the international water agreements create an opportunity for different interpretations of some regulations. The resolutions of the Dublin Conference are even being used by those that want to transform water into a commodity like oil, gas, etc. The initiator of this policy in the region is Turdakun Usubaliev, the former First Secretary of the Central Committee of the Communist Party of Kyrgyzstan. He has published a book on this matter, suggesting that this policy would rehabilitate the economy of this republic.

The recommendations of the 1992 European Convention and the 1997 UN Convention concerning the boundaries of national sovereignty and regional governance, and setting out an interpretation of concepts such as "equitable and reasonable water use", "responsibility for damage," and "co-ordinating actions on transboundary waters" have a non-obligatory nature, and each party to these conventions can therefore "listen attentively" to these recommendations but act in their own way. Clearer definitions and interpretations of agreements and provisions concerning specific river basins are very important, and appropriate examples that might act as templates for the region include the USA-Canada Boundary Waters Treaty (1909), the Convention on the Protection of the Rhine (1998), and the Indus Waters Treaty signed by India and Pakistan in 1960 .

As noted in Section 4.1, in 1994 and 1995 the members of ICWC and IFAS proposed a number of measures that could considerably strengthen the legal and institutional basis for co-operation and lower the risks of losing manageability of the water resources in the Aral Sea basin. Section 4.3 clearly shows where the main seeds of discord are arising, and these conclusions have indeed been clearly confirmed by the International Crisis Group.

While the first water agreements signed by riparian countries (1992, 1993 and 1994) remain in effect, they require support by relevant mechanisms for executing the agreed rights, duties and financial relations, and they have not been co-ordinated for several different reasons and they have had some unintended consequences. For example, the 1998 agreement on the Syrdarya river provided for barter relations to trade "water resources for hydropower" for "fossil fuel" and merely created specific problems for sustainable water use in the basin.

Since the start of the ICWC, plans for supplying the water quotas according to the 1992 agreement were followed quite exactly, but already by 1993 and 1994 there were certain tensions with hydropower organizations. These problems mainly related to implementing the scheduled water releases from the Toktogul Reservoir. At the twenty fourth ICWC session (October 23, 1999; protocol 24, question 1, clause 3), this became apparent when a change was made to the agreement. The agreement was changed from "*To approve the specified limits for water diversion from the Amudarya and Syrdarya rivers, and the operation regime for the reservoir cascades on these rivers*" to "*To approve the operation regime for the Naryn/Syrdarya Cascade as a proposal for joint discussion at the meeting of experts representing the water management and hydropower departments to prepare a draft inter-governmental agreement*". Only later were the limits for water diversions for riparian countries and some zones established in ICWC resolutions. The operational regimes for water releases from the reservoirs were presented as recommendations that were to be approved at a session of the Hydropower Council and through co-ordination with the owners of the hydroschemes. Thus a "bomb" was laid under articles 2 and 11 of the 1992 agreement, which predominantly provided for preservation of the agreed order and established procedures for water resource conservation and use, as well as pointed out that the ICWC resolutions were compulsory for all water users and water consumers.

After the disintegration of the USSR, the supply of fossil fuel resources (coal, natural gas and crude oil) to Kyrgyzstan, which were made in the Soviet time, ceased. The government of Kyrgyzstan started to insist on certain compensatory supplies of fossil fuel and energy resources as a precondition to support the appropriate water release regimes from the reservoirs (which balanced hydropower and agriculture needs). In 1995, an annual system was introduced to co-ordinate the conditions for supporting the irrigation water release regime from the reservoirs. This was based on compensation granted by Kazakhstan and Uzbekistan to Kyrgyzstan in the form of supplying fossil fuel and electricity. In accordance with this approach, the electricity produced at the Naryn/Syrdarya Hydropower Plants in excess of Kyrgyzstan's own needs and through water releases made during the growing season for downstream countries should be supplied to those countries. This should also be in accordance with the requirement for year-round regulation of the Syrdarya runoff by the Toktogul Reservoir and should be supplied in equal quantities to Kazakhstan and Uzbekistan. Compensation for this supply of electric energy should in the form of the supply of equivalent amounts of energy resources (coal, natural gas, crude oil and electricity) to Kyrgyzstan. Compensation could also be paid with other products or with money according to the agreement between the parties and compensation could be paid for the accumulation of multi-year water reserves in the storage reservoirs.

By 1996 and 1997, serious deviations were observed from the compensatory measures as stipulated in the agreements. In particular, this concerns the agreements between the governments of Kazakhstan, Kyrgyzstan, Tajikistan and Uzbekistan (December 21, 1995) on the operation regime of the Naryn/Syrdarya Cascade. It is significant that in order to generate additional economic benefits, Kyrgyzstan began to decrease the volume of water released for its own needs from 1997. In 1996–1997, it released 4.5 km^3 but by 2000 this had fallen to 2.5 km^3, and this allowed additional water releases to be made for irrigation purposes in downstream countries which generated increases in the compensatory payments of energy resources to Kyrgyzstan. From 1995 to 1998 Uzbekistan met its commitments relating to natural gas supplies, but this changed in 1998 when Kyrgyzstan sharply cut back on its obligations for water releases from the Toktogul Reservoir during the growing season. In the same period, Kazakhstan met its commitments relating to the supply of coal and electricity (Table 4.20), excluding delivery of electricity in 1998. At the same time, the Bishkek Thermal Power Plant did not completely meet obligations for generating electricity in the winter period, creating a complicated economic situation, and, as a result, the irrigation water supply ceased to be stable and secure.

Changes to, or deviation from, the barter agreements can take place for several reasons: i) delays in signing the agreements on supplying fossil fuel and electricity owing to prolonged price disputes; ii) non-execution of commitments related to the compensatory payments of coal, oil, and gas; and iii) specifying prices for "released water" which depend on the price of gas.

In the 10-year period 1995–2005, the average price of electricity in the Central Asian market ranged from 1 to 3 cents per kilowatt-hour as a result of the low cost of electricity generation at the Ekibaztuz and Mary Thermal Power Plant and in the hydropower sector. Therefore, the attempt by Kyrgyzstan to charge Kazakhstan 2 cents per kilowatt-hour for electricity in the summer and to charge Uzbekistan 4 cents per kilowatt-hour caused disagreement, and both countries refused to pay these prices. The International Crisis Group report (ICG 2002, page 15) confirms that in 2003 Kyrgyzstan raised electricity prices to 3.36 cents per kilowatt-hour as part of its demands for compensation. In 2008 and 2009, in the first offer to Uzbekistan, the price of electricity was set at 8.8 cents per kilowatt-hour, allegedly to reflect the price of natural gas, although a Japan International Cooperation Agency report (JICA 2009) found that the Bishkek Thermal Power Plant did not consume more than 5% of the natural gas used by Kyrgyzstan (Figure 4.19) and the vast majority of natural gas was used by domestic consumers. As noted in Section 4.3, these price demands caused instability and disrupted the activities of water management organizations, especially in dry years. The 1988 agreement that closely linked water releases to a barter arrangement involving electricity and fuel resources had the unintended consequence of creating complete uncertainty in the operation of the Naryn/Syrdarya Cascade and prioritizing the hydropower sector. This was especially obvious in the increasing commercial demands made by Kyrgyzstan, which were opposed by the other parties to the agreement and which caused the signing of protocols to be delayed by 5 to 6 months each year. In 2000, the protocols for that year were not signed until July. A similar delay was observed in 2009, and often the water and energy barter was not implemented in full by the contributing parties.

The tendency to overprice electricity distorted water prices. (The current net income of farmers does not exceed 6 cent/m^3 for irrigated farming, and it is clear that farmers cannot spend all their income

Table 4.20 Supply of water and energy resources according to inter-governmental agreements, 1995 to 2005.

	Period	1995		1996		1997		1998	
Capacity of the Toktogul Reservoir, billion m^3	1 Jan	17.7		13.9		13		10.2	
	1 Apr	14.2		10.4		9.8		7.3	
	1 Oct	15.5		15.2		11.8		15.1	
Discharges from the Toktogul Reservoir, billion m^3	Plan	6.5		6.5		6.5		6.5	
	Actual	6.3		6.2		6.1		3.7	
Export of energy to Kazakhstan and Uzbekistan		Kaz	Uzb	Kaz	Uzb	Kaz	Uzb	Kaz	Uzb
	Plan	1100	1100	1100	1100	1100	1100	1100	1100
	Actual	782	928	995	1077	710	1615	468.8	489
Supplies to Kyrgyzstan									
Natural gas, million m^3	Plan	200		500		630		772	
	Actual	200		476		632		748	
Karaganda coal, thousand tons	Plan	985		600				566.7	
	Actual	450		202				150.4	
Black oil, thousand tons	Plan							20	
	Actual							23.8	
Electricity, million kW hours	Plan			635		0	400	250	200
	Actual	415		635		11.4	433.5	150	74.9
Diesel fuel, thousand tons	Plan								
	Actual								
Transformer oil, tons	Plan								
	Actual								
Turbine oil, tons	Plan								
	Actual								

	1999		2000		2001		2002		2003		2004		2005	
	13.5		14.5		11.9		10.4		15.6		17.2		19.8	
	10.4		11.0		8.7		7.5		12.4		14.8		13.7	
	16.3		13.7		12.1		17.4		19.5		19.2		18.8	
	6.5		6.5		5.9		5.3		7.2		5.5		6.2	
	5.1		6.5		6.2		3.7		4.9		6.2		5.2	
	Kaz	Uzb	Kaz	Uzb	Kaz	Uzb	Kaz	Uzb	Kaz	Uzb	Kaz	Uzb	Kaz	Uzb
	1100	1100	1100	1100	1100	1100	1100	1100	1100	1100	1100	1100	1100	1100
	585.3	970	673.2	1926	912.4	1038	334		499		1111	1153	1237	589

	500		422		700									
	331		430.5		594.9									
	566.7		362.5		618		500		512.5		500		500	
	572		331.1		465.5		152.4		539.8		500			
			55		20	20	15		15		15		15	
			27.9		9.8	16.5	6.3		0		15			
			0											
			185											
					23									
					15.4									
					540									
					425									
					240									
					222									

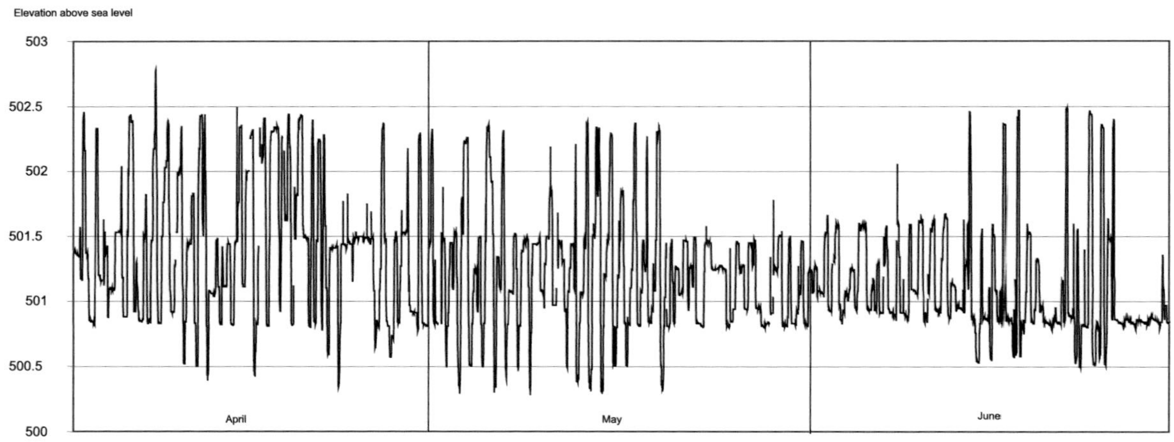

Figure 4.19 Example of an electricity energy balance by country, 2007 (JICA 2009).

to pay for water.) The 1998 agreement also completely ignored ecological aspects, and it did not account for the specific requirements of dry and wet years. The agreement was generally only aimed at compensating Kyrgyzstan for the export of electricity that had been generated as a result of the water release regime during the growing season, providing guarantees of fossil fuels or services on the basis of barter deals. At best the 1998 agreement was a compromise, attempting to balance the needs of the hydropower sector with the requirements of the environment and irrigated farming within a framework of regional collaboration.

Kyrgyzstan also had reservations about the agreement. In numerous publications and in speeches by Kyrgyzstan officials, it was argued that the downstream countries did not want to take into account the operating costs of the Toktogul Reservoir which had to be covered by Kyrgyzstan. It is necessary, therefore, to understand clearly the actual operating costs of the Toktogul Reservoir and the Naryn Hydropower Stations – both the complete costs and for whom these costs are incurred.

Under the present regime, in which the winter water releases greatly exceed the summer releases, the Toktogul Hydroscheme operates to the absolute detriment of irrigated farming. This is important, because in the past the natural river runoff in the middle and lower reaches of the river was higher than the river runoff as regulated by the Toktogul Reservoir over the last 15 years. Under these circumstances, why does Kyrgyzstan require compensation? On the contrary, Kyrgyzstan should compensate the countries located in the middle and lower reaches of the river for the losses they have incurred not only due to the cancellation of the original design regime of the Toktogul Reservoir but also because of the loss of natural river runoff that has become disrupted by the Naryn Cascade. A key problem is that Kyrgyzstan's current policy is based on the doctrine of "absolute territorial sovereignty" and ignores the transboundary nature of this river. This policy should be changed by legal and political means, and by organizing public awareness campaigns to inform the general public in Central Asia about the real

situation. It is necessary to come to the understanding that to refuse collaboration in establishing equal rights for water use leads to the need to compensate neighbors to cover losses resulting from illegitimate (or non-agreed) actions.

A key factor in the conflict between upstream and downstream countries – as noted by many of those that predict future water wars – consists of the attempts to link water releases with energy or rather with the procurement of fossil fuel and energy. This approach is basically harmful if we assume that water is not a commercial commodity, while energy and fossil fuels are goods directly subjected to market forces. Prices for energy and fossil fuel depend on supply and demand, and also on the hidden forces generated by the monetarist policies that rule the modern capitalist world. Demand for fossil fuel varies throughout the year in a rather unpredictable way, while demand for water, which basically is a source of life and a necessary input for food production, is rather stable over the seasons. In principle, water is nature, while fossil fuel and energy are classical commodities that generate surplus value. One type of fossil fuel can be replaced with another, and the same can be said about energy sources. Water is a unique and irreplaceable good that predetermines not only economic welfare but also social and environmental well-being. Therefore, if we want our world to be sustainable under conditions of a growing population and climate change, we should regard water as an irreplaceable natural good that provides life and prosperity rather than a mere commodity within a mix of water and energy resources. Water resource planners should take into account opportunities for balanced production of hydropower and co-ordinate the other needs of riparian countries in transboundary river basins in accordance with the regulations of Article 5 of the 1997 UN Convention. Using water as a commodity to generate income, including through energy production, should not deprive people of their rights to use water for drinking, domestic needs, food production and recreation.

During 2005 to 2008, the ADB RETA 6163 Project provided a platform for regular dialogue between the riparian countries for the purpose of preparing and co-ordinating a new agreement for the Syrdarya basin, including the necessary procedures to implement any agreement. These dialogues made the position of Kyrgyzstan clear, regarding the present management of the Syrdarya, and of Tajikistan, in its plans to construct the Rogun and Dasht /Djun hydropower plants and the Yavan hydropower plant on the Zeravshan River. The consolidated position of Kyrgyzstan and Tajikistan was first presented at the preparatory meeting of experts, and then finally formulated by the presidents of these countries at the summit of the heads of Central Asian states during the IFAS meeting of April 28, 2009. Kurmanbek Bakiev, then President of Kyrgyzstan declared that his primary concerns were the costs incurred for preserving water resources used by all riparian countries and the funds needed for protecting the glaciers (a main source of water) that are rapidly reducing, rather than the Aral Sea. Although the President of Tajikistan, Emomalii Rakhmon did not ask for funds for protecting glaciers, he said that the interests of Tajikistan were focused on the future construction of hydropower plants. These views rather confirm the opinion expressed by the President of Uzbekistan, Islam Karimov, that Kyrgyzstan and Tajikistan were working towards a future policy of a hydropower "decree" and they refuse to recognize the entitlement of downstream countries to a fair and well-argued share of water resources in accordance with international water laws.

Those regional and international voices that argue that the present operation of the Toktogul Reservoir and overall water resource management, with a focus on meeting "short-term energy needs" without any long-term planning, should be stopped [24], are gradually getting more support. The JICA study (JICA 2009) has shown that it is impossible to meet current water needs for winter energy production and summer irrigation under the present approach because of the imbalance of demand and resources. To improve this overall situation, a package of measures needs to be implemented to:

- save water as a resource;
- increase the efficiency of water use;
- reduce electricity consumption;
- develop alternative sources of power supply (regional energy supply exchange, new hydropower plants).

Similar ideas are presented in a USAID project (PA Consulting Group 2002), which suggests focusing first on reducing electricity power losses. The project estimates that these amount to some 42% of the total supply and they could be reduced to 10%.

Until recently, the Kyrgyzstan and Tajikistan governments paid little attention to these opportunities. However, Mr Davydov, who was appointed Minister of Energy in Kyrgyzstan, in November 2008, has shown that energy consumption can be considerably decreased. As a result of his policies, winter water releases have halved from the 2007 volumes. Thanks to these measures, it became possible to keep some 6.3 km^3 of water in the Toktogul Reservoir (almost 1 km^3 more than the dead storage capacity).

The interest of foreign researchers in the problems of water resource management in the Syrdarya basin has produced some serious work, including specific theoretical studies aimed at searching for new compromises. Wegerich (2008b) has noted that trade-offs made by downstream countries in accordance with the requirements of Kyrgyzstan did not guarantee adherence to the agreements on water release regimes needed by the downstream countries during the growing season. Wegerich studied the barter agreement on electric energy procurement in exchange for supply of black oil, coal and natural gas in the Syrdarya basin and the agreement to reimburse part of the operating costs of the Kirov Reservoir in the Chu-Talas basin. In both cases, over the period 2000 to 2005, Kyrgyzstan did not answer to its commitments for water release according to the agreed schedule and rates. More effective mechanisms are required, because commercial and short-term interests carry much more weight than contractual commitments if there are no mechanisms for control and for penalties in the case of non-compliance.

Tobias Siegfried and Thomas Bernauer (2008) have also found that even in the wet years of 2002 to 2005, the average coefficient of application for the agreed water releases was 0.51 because the decisive factor and driving force governing water release from the Toktogul Reservoir was demand for electricity rather than the agreed water releases under the signed agreements. Given that winter demand for electricity has increased by 25% since 1991, the focus of the operational regime of the Toktogul Reservoir on hydropower interests becomes even more obvious; a fact also noted by the Japanese researchers.

In his survey, Keely Lange (2001) has noted the following reasons for the failure to adhere to the agreed schedule of water releases from the Naryn/Syrdarya Cascade to downstream countries:

- the use of water resources as a part of "the vicious circle related to the procurement of natural gas, coal and other resources";
- disagreements about electricity prices and the prices of natural gas and coal;
- a failure to pay for the actual delivery of natural gas and coal or a failure to deliver the agreed supply of natural gas and coal;
- the occasional linking of the delivery of water to an absolutely inappropriate return supply, such as flour in 1999; and
- endeavors by Kyrgyzstan to gain indemnification for loss of profit arising from generating some electricity in summer (from irrigation water discharges) rather than concentrating generation – although whether these one-sided actions are really feasible is very doubtful.

In view of these discussions, it is necessary set out some general principles and provide some more precise definitions. Demands for water releases should be considered in association with settling fuel and energy problems, but they should not depend on them. The interests of Kyrgyzstan and Tajikistan should be recognized, but water supply and irrigation in downstream countries should not become "hostage" to the energy well-being of the upstream countries. The provision that "hydropower is one among others but not a privileged water user" is clearly stated in all international water agreements. The annual planning and provision of water supply should not therefore depend on the uncertain situation in the fuel and energy markets, because this jeopardizes agricultural production and food security, makes the struggle against poverty and famine more difficult, and reduces the overall well-being of the rural population.

It should be noted that this uncertainty plays against water supply for food production as guaranteed in accordance with Article 11 of the International Covenant on Economic, Social and Cultural Rights and follow-up comments (see comment 15) of the Committee on Economic, Social and Cultural Rights (Hodgson 2004). In principle, the approach that barters energy supply in the winter in exchange for additional water releases in the summer can be applied, but it is necessary to exclude the price of electricity generated and supplied in the summer and avoid any link to fuel prices from these arrangements. Electricity generated in the summer in excess of the domestic demands in Kyrgyzstan – and produced for the benefit of irrigation – should be compensated for by supplying Kyrgyzstan with electricity in winter (rather than natural gas or coal) at identical prices, even although in principle summer electricity undoubtedly has a lower price on the Central Asian market as it is a less scarce resource. Under such a scheme, speculation could be avoided.

Each country would see the benefits of water use based on its own strategy. The upstream countries would concentrate on generating income from hydropower production, and only see irrigated farming and other sectors as secondary income. The downstream countries would plan their business strategies around income generated from irrigated farming and the associated sectors. However, keeping in mind the difference in the peak water season of each strategy (in winter for energy or in summer for agriculture) it is necessary to specify mutually beneficial options to compensate for the losses due to competing demands for water use. This behavior finds its roots not only in international water law but

also in common laws that require economic actors "to prevent possible damage" to associated sectors. Sorokin and Averina (2002) have presented this approach in their work on evaluating the activities of the consortium under different scenarios for managing water, fuel and energy resources. By using the Pareto principle, they show that it is possible to achieve a compromise solution, which minimizes the total losses of the parties involved. Moreover, it is necessary to keep in mind that energy is often presented as "a lost effect" – that is, it is argued that losses are incurred by not keeping back water to generate electricity in winter. But these are not actual losses and are often farfetched, and they certainly cannot be justified by previous economic history. They also do not take into account the losses suffered

Table 4.21 Flood damage on the Syrdarya, 2004–2008 (JICA 2009).

	2004	**2005**	**2006**	**2007**	**2008**
Damage (million tenge)	280	853	–	927.2	–
Damage (million US$)*	2.38	7.25	–	7.88	–
Flooded area (hectares)	55,733	30,460	–	93	–
Flooded:					
residential areas (units)	2	6	–	2	–
country houses (units)	3	5	–	5	–
schools (units)	–	1	–	–	–
residence buildings (units)	805	74	–	269	–
private land (units)	–	–	–	577	–
Evacuation (people)	2,085	31,824	–	1,500	150
Resettlement (people)	289	–	–	420	–
Disturbed living conditions (people)	–	–	–	700	–
Allocated resources and equipment by the Ministry of Emergency (people/technical equipment)	134/20	222/32	92/19	108/26	289/94
Explosion works (explosion/expenditure per ton)	– /4	122/17.73	127/19	93/28	56/14.35

* 1 Kazakhstan tenge = 0.008501 US dollar (2008).
Source: Ministry of Emergency Situations, Kazakhstan.

by Kazakhstan following flooding as a result of winter water releases for generating hydropower by Kyrgyzstan in wet years (2004 to 2007) (see Table 4.21).

An attempt to search for a compromise can be undertaken by comparing different ratios of net profit and loss under various regimes of water release, ranging from a maximum winter regime (10 km^3 in winter and 2 km^3 in summer; a ratio of 5:1) to a maximum summer regime (8 km^3 in summer and 4 km^3 in winter; a ratio of 0.5:1).

Figure 4.20 shows that both the consolidated net revenue in the irrigation and hydropower sectors and gross incomes in the irrigation sector and hydropower sector reach a maximum in the case of the

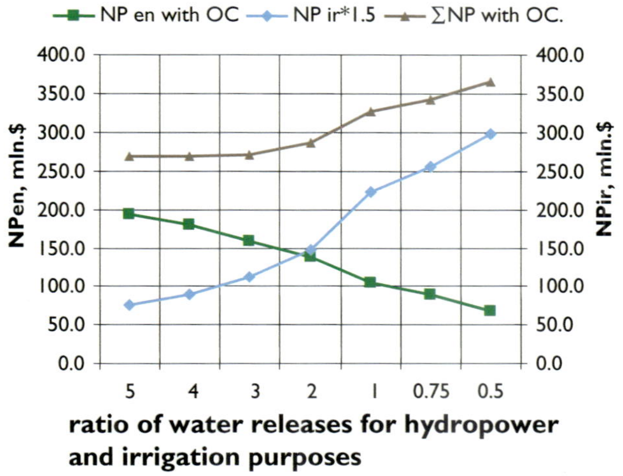

Figure 4.20a Evaluating net profit and losses curves in a hydropower and irrigation trade-off.

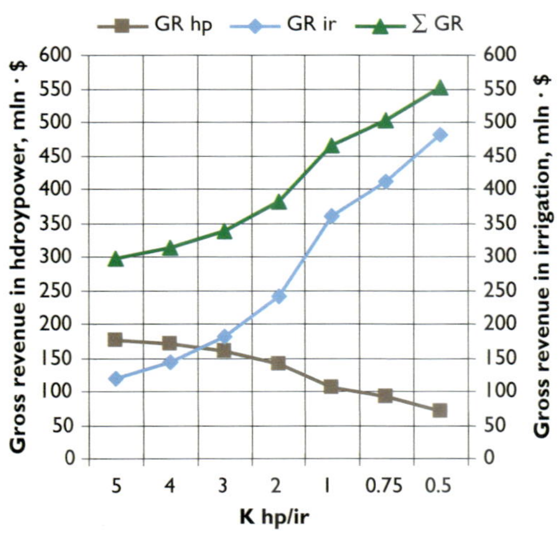

Figure 4.20b Comparison of gross revenues in hydropower and irrigation.

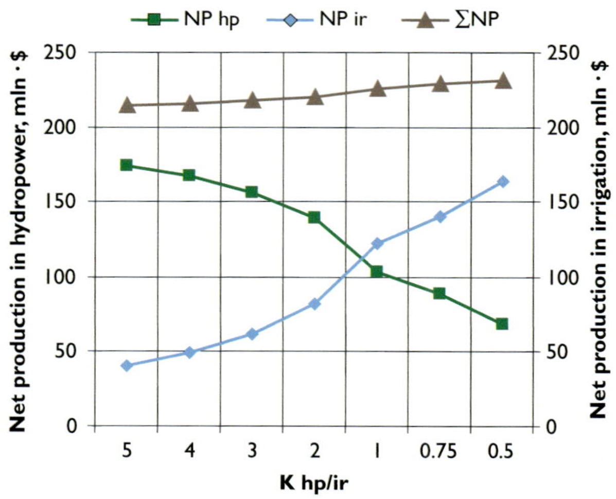

Figure 4.20c Comparison of net production from hydropower and irrigation.

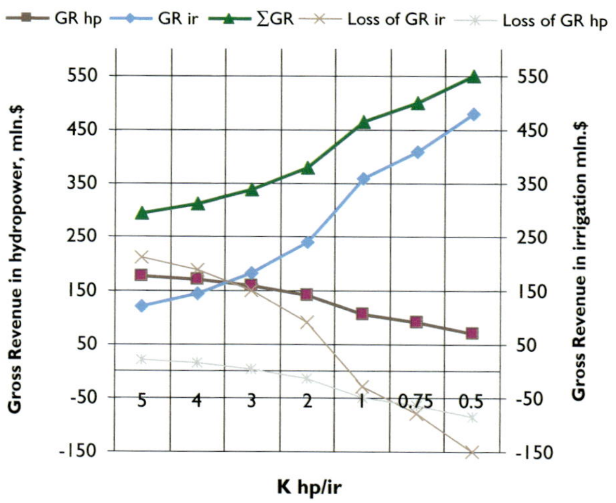

Figure 4.20d Evaluation of gross revenues taking into account opportunity losses.

irrigation option, where there is a ratio of 0.5:1 for water releases for hydropower and irrigation. In the case of the hydropower option (hp/irr = 5) we find a maximum gross income for the hydropower sector (US$194.4 million) and a minimum for the irrigation sector (US$120 million), with respective values for net revenues of US$174 million in the hydropower sector andUS$40.8 million in the irrigation sector. However, under this option, losses in the irrigation sector amount to US$210 million. If the total gross income is adjusted to take this into account, the remaining income adds up to US$ 60 million only. But under a gradual transition towards prioritizing irrigated farming, the consolidated gross income reaches a maximum (more than US$500 million). By considering the overall losses of all participants of a water and energy economy under different water release regimes from the reservoirs, optimal solutions can be found. Under this approach, the optimal regime is actually located at the point where the net production curves plotted for both economic sectors cross. An optimal solution is provided with a ratio of 1:1 for water releases for hydropower and irrigation purposes (hp/irr = 1, when 6 km^3 of water is released in winter and 6 km^3 is released in summer.

Petrov (2009) reached a similar conclusion about the inefficiencies of the current operational regime of the Naryn/Syrdarya Cascade and the ability to get much more benefit from an operation regime that meets the needs of irrigated farming. As a representative of the power sector of his national economy, Petrov is aware of all the details of power production costs, and for a long time, he held the post of Deputy Minister of Energy of the Republic of Tajikistan.

The method elaborated by Sorokin and Averina (2002) envisages payment for runoff regulation. It should be noted that similar approaches have also been suggested by Kyrgyzstan (Mamatkamov and Asanbekov 2000) and the World Bank's "water power nexus". However, both these documents contain a proposal to include losses incurred by Kyrgyzstan to the sum of US$123.5 million due to construction of the Toktogul Hydroscheme under price adjustments, but it is not clear how these figures have been derived. The World Bank suggests an amount of US$20 million as an annual compensation.

According to the SIC ICWC assessment, the cost of seasonal regulation of river runoff by the Toktogul Reservoir should be calculated by compiling "a complete profit and loss account", and it would amount to 1.4 cent/m^3. The potential payment by Uzbekistan and Kazakhstan for runoff regulation (when water availability is less than in an average year) is estimated at US$35 million per year. The cost of runoff regulation without taking a "a profit and loss account" approach amounts to 3 cent/m^3, and payment for runoff regulation would not exceed US$ 8 million per year. These figures should be compared with the cost of electricity as supplied by Kyrgyzstan – it is claimed that it supplies 2.2 billion kWh at a price of 4 cent/kWh, which corresponds to 6 km^3 of water releases, and an income of US$90 million.

It is understandable that the cost of year-round regulation of river runoff is increasing considerably, and it depends on the duration of the period with high water availability and the ratio of years with water accumulation to the years with too little water. In this case, the cost of year-round regulation of river runoff might increase to 5–7 cent/m^3. However, such a solution should aim at stabilizing the multi-year regimes.

Improved co-ordination of irrigation and hydropower options in the region can be achieved through the approach proposed by Japanese specialists (JICA 2009). They suggested a suite of measures aimed at energy saving and rational water resource use in the form of two parallel clusters of improvements (Figure 4.21):

- enhancing mutual trust through the continuation of the dialogue between riparian countries in the region;
- focusing on the long-term perspectives;
- respecting the individual interests of each country and proving support in accordance with their priorities;
- eliminating sources of confrontation, and improving understanding through transparency of information, frankness of positions, etc.

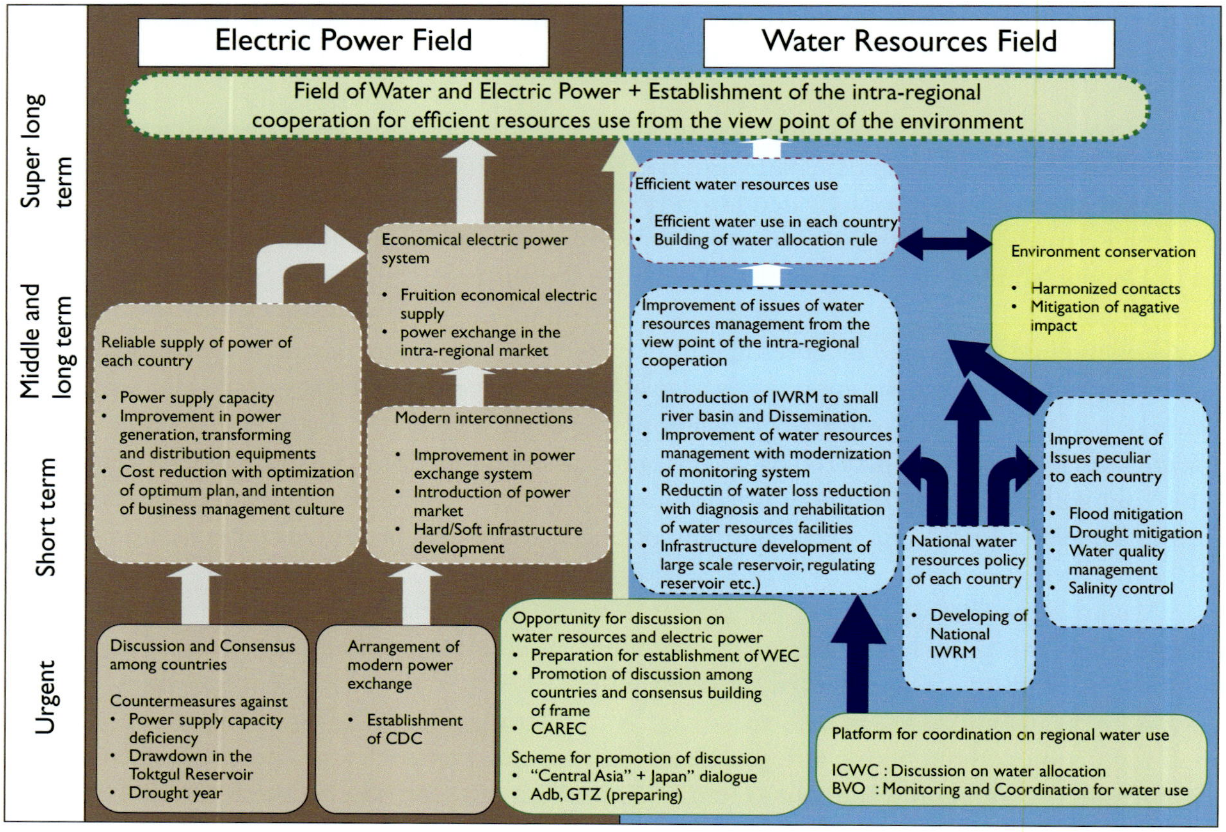

Figure 4.21 Long-term vision for development of the fields of water resources and electrical power.

The establishment of a regional water and energy consortium plays a special role in the attempts to achieve this long-term consensus. It is supposed that the consortium will perform the following functions:

- organize and implement a financial mechanism that can solve problems related to the shortage of funds to buy electricity and fuel resources to compensate for water delivery and that can guarantee timely payments;
- act as an insurance agency that is able to cover possible losses due to objective causes not related to human activities – losses due to subjective (human-induced) causes should be covered by the party at fault in the shape of penalties.

SIC ICWC has prepared draft statutes for an interstate water and energy consortium that provide key directions for the activities, functions, organizational arrangements and operational procedures of the consortium. According to these statutes, the consortium will be established as a body capable of providing advanced mechanisms for financing and supporting mutually profitable water-energy barter policies between riparian countries. These are some of the provisions of these draft statutes.

- National ministries and departments of riparian countries, their national corporations and companies, enterprises and organizations in the fuel and energy sectors and in the water sector are to be contributing founders of the consortium.
- Relevant international organizations and other legal entities of any form of business organization can be members of the consortium. Admission to membership is open for organizations of other countries that share the targets and tasks of the consortium.
- The council of managers (hereinafter "council") is headed by a chairman who is appointed on behalf of the host country of the council's seat for a period until the next session and who is in charge of the consortium's activities.
- The executive committee acting with the authority of the consortium charter and approved by the council is the executive body of the consortium. A technical director appointed by a council resolution will be in charge of the executive committee.
- Annual demands for water resources and electricity (by volume and terms of delivery) are approved by resolutions of the Interstate Co-ordination Water Commission and the Energy Council as established by the Central Asian countries. On the basis of these demands, the consortium schedules the optimal regimes for the operations of the hydropower plants and water storage reservoirs, and the supply of fuel resources, in order to achieve the lowest operating costs and the maximum water supply and to meet specific environmental needs. After co-ordinating with the river basin organizations (BVOs) and the United Dispatch Center for Energy (UDC Energy), the prepared schedules are handed over for implementation. The BVOs and the UDC transform these schedules into plans for water distribution and for water release from reservoirs.
- Mutual settlements between consortium members are based on agreed prices for fossil fuel and electricity that are approved by a resolution of the council. The business plan of the consortium has to observe the interests of all members in an equitable manner.
- The organizational structure, number of officers and service personnel and the salary administration system of the executive committee are approved by the Council.
- The size of the contributions to the authorized fund of the consortium is specified in the consortium charter on the basis of agreement with all parties.

The Syrdarya and Amudarya BVOs and the UDC Energy have to provide the technical assistance for implementation of the planned measures and, as executive bodies, facilitate proper management of resource flows. At the same time, the consortium will be the financial mechanism for solving problems related to any member having a shortage of funds to buying electricity and fuel resources and for compensating for water services, which should guarantee timely payments and regulate the procurement of electricity and coal in the summer as an indemnity for the reduced production of electricity in the winter. The consortium should become a collaborative organization, which will be able to ensure a well-managed system of payments and cash flow between Central Asian countries, as well as take care of the sustainable functioning of the water and energy system. The consortium should be organized as a union of participants for the exchange of water and energy resources, the members of which have access to funds to help them to organize the supply of fuel and energy and to cover possible losses.

The revolving fund of the consortium will depend on the cost of electricity generated through the spring and summer water releases from the Toktogul and Kayrakum reservoirs that are implemented to meet the needs of irrigated farming in Uzbekistan and South Kazakhstan in excess of the needs for electricity in Kyrgyzstan. These funds should correspond to the cost of fuel and energy resources procured to cover the possible winter electricity deficit in the Republic of Kyrgyzstan

The aim is procure sufficient volumes of electricity or fuel (coal and natural gas) for the thermal power plants efficiently. It might be decided to buy Turkmen natural gas for the Bishkek Thermal Power Plant because natural gas transportation is less costly than the transportation of coal. Another possibility is that electricity might be produced at the sites of coal or gas production, and transported via the integrated power transmission grid that already exists in the region. The interests of all transit countries should be taken into account in such a case.

During the growing season, it is necessary to find buyers for electricity generated by the Toktogul and Kayrakum hydropower plants in excess of the needs of the host countries. By searching for the most profitable regimes and transactions, and while correctly managing this business this should improve the compensation mechanism, which has not functioned in the last five years.

The consortium will agree its actions with the riparian countries if disputes arise related to disregard of the interstate agreements. It will also facilitate the settling of inter-state disputes in the sphere of water resources management and use. In all its activities, the consortium should be guided by the following key principles:

- the activities of the consortium are limited only by its financial capability to support the efficient implementation of all agreements reached to ensure sustainability of the water and energy sector;
- it will adopt a policy of non-interference in relations within sectors in the riparian states;
- it will take a compensatory approach to decision-making at national level if inter-state agreements are affected;
- the assessment of the economic situation should be co-ordinated with all executive bodies of the ICWC (including the BWOs and SIC).

In July 2004, the drafts of the conception and meeting protocol were signed by all parties and it was decided to pass this draft of the IWEC conception as coordinated with the World Bank to the Committee of National Co-ordinators of the member states of the CAC organization for review. However, different interpretations of the role of the consortium has hindered the final decision-making process. Kyrgyzstan and Tajikistan want to see the consortium as a body that funds the construction of hydropower generation facilities and helps develop the hydropower sector. Kazakhstan and Uzbekistan want to see it as the mechanism for regulating financial and economic relations in the process of planning and implementing the annual and multi-year operational regimes of water storage facilities. This divergence in expectations has practically stopped the realization of the idea, although it is considered realistic and meaningful. It is supposed that the consortium can integrate both functions under joint operation and maintenance of the Naryn/Syrdarya and Vakhsh/Amudarya cascades and manage the water release regimes in the interest of all the riparian countries. The water release regimes will be clearly regulated and should, at the same time, serve the various specific commercial interests.

Exactly such an opportunity was explained in *Evaluating of the Rogun Reservoir's impacts on the Amudarya runoff regime* (Dukhovny and Sorokin 2009). Table 4.22 convincingly shows that the irrigation regime in operation at the Vakhsh Cascade is the most profitable regime. The additional profit, compared with the present operational regime of the Nurek Reservoir, amounts to US$245 million a year, including US$188 million for the hydropower sector and US$56.83 million for irrigated farming.

Table 4.22 Supply of water and energy resources according to inter-governmental agreements, 1995 to 2005 (Carewib).

Options	Annual losses in irrigated farming and associated sectors	Reduction (–) or increase (+) in annual losses under the present operational regime of the Nurek Reservoir taking into account income generated from energy production	Energy production by the Rogun HEP in monetary terms	Additional profit versus the present operational regime of the Nurek Reservoir
Preserving the present operational regime of the Nurek Reservoir	94.71	–	–	–
Hydropower regime, El. 1240	211.3	116.59	162.35	45.76
Hydropower regime, El. 1290	174.6	79.89	194.71	114.82
Irrigation regime, El. 1240	59.2	–35.5	159.39	194.89
Irrigation regime, El. 1290	37.85	–56.86	188.41	245.27
Irrigation and hydropower regime, El. 1240	76.18	18.53	194.84	176.31

It is obvious that, within the framework of this consortium, a financial input of the agricultural sector to the sum of US$ 6 million a year would compensate for lost profits in the hydropower sector under maintenance of a regime necessary for irrigated farming. Moreover, regulation of multi-year runoff will be ensured, which cannot be provided under any regime that prioritizes hydropower. Experiences around the world also confirm the advantages of joint and balanced use of water and energy resources (Risbekov 2009).

- It would separate the functions of managing the reservoirs from the functions of hydropower production and transform the hydropower sector into a regular water consumer.
- It would allow the joint use of hydropower facilities based on redistribution of costs and incomes (the experience learnt in South Africa and South America).
- It would deliver institutional improvement through the equitable participation of riparian countries in water resource management (according to representation, financing and rights).

This approach will allow water resource management to be self-sufficient and will ensure effective co-ordination with the hydropower sector.

4.5 Agreements are the Legal Basis for Future Sustainable Management

International water law is often too much of a simplification and leaves options for all participating countries to develop their own policies, although the need for these laws cannot be denied. Undoubtedly, most countries located in the upper watersheds, making reference to the Harmon doctrine of absolute territorial sovereignty and citing the USA-Mexico Treaty as a precedent, try to fix their right to be "dictators" of river flows. However, as shown in the previous sections of this chapter, gradually and with great difficulty, all over the world an understanding of the benefits of collaboration replaces the strategy of absolute territorial sovereignty, which does neither provide for sustainable development nor for secure national existence. This is due to many reasons, including international prestige and the dependence on downstream riparian countries in other spheres of development, such as transport, finance, fuel resources, etc.

Apart from the unclearly formulated provisions in international water laws, there are several provisions in international laws in general that clearly state the legitimate rights of neighboring countries. The principle "don't cause harm, but if you do then pay" should be highlighted here. In spite of their sovereignty, all countries should act in accordance with this principle and avoid actions that can result in direct or indirect mutual damage.

The legal regulations on which we want to focus attention concern the right to water. The general comment 15 to the International Covenant on Economic, Social and Cultural Rights Article 11(1) specifies "a number of rights, emanating from and indispensable for, the realization of the right to an adequate standard of living including adequate food, clothing and housing" (ECOSOC 2002). The right to water clearly falls within the category of guarantees essential for securing an adequate standard of living, particularly since it is one of the most fundamental conditions for survival. This should also be seen in conjunction with other rights enshrined in the International Bill of Human Rights, including its most important principles of the right to life and human dignity.

Private correspondence with Dinara Ziganshina, a doctoral candidate of the International Water Law Institute in Dundee who specializes in the water issues in Central Asia, states that principles of international law are increasingly significant and can be used to counter attempts to prove the viability of "hydro-selfish" actions within a country's own territory that are described in international water law publications (Wouters and Vinogradov 2002). This requires mutual commitments regarding co-operation, consultation and information exchange and efforts to achieve peace, international security, sustainable development, nature conservation while meeting national needs and requirements based on the principles of international law. In this connection, it is important to follow the idea of balancing the different often competing interests, because these require mutual concessions that should be stated in regional agreements as well as in the international water laws.

Given the uneven distribution of natural water resources throughout the planet, and the aspiration and necessity to provide equal access to water resources for all purposes (regional, national and communal), international water laws require strengthening and there is a case to establish special monitoring instruments under the UN or the UN Security Council, because water is so essential for life. A precise definition is needed of "fair and reasonable use of transboundary waters", as each country currently understands in its own way and by virtue of its own interests and because each national government considers its own activities as equitable and reasonable. Another matter is to define the scope and limits of national sovereignty and transboundary (regional or basin) management. A further aspect is to find ways to observe and assess the agreements on environmental flows. In addition a commitment should be embodied in international law to avoid any action that can adversely affect neighboring countries (rather than relying on the manifestation of goodwill).

Time is now ripe for implementation of all these regulations so that countries can no longer hide behind the Harmon doctrine and use it as an instrument of political or economic leverage against downstream countries. In Central Asia, the 1992 agreement, as was noted earlier, has created a good basis for further improvement of regional co-operation:

- parties are actively participating in ICWC activities in many different ways (regular sessions, working groups for improving the management framework, joint implementation of the regional projects, putting IWRM into practice, etc.);
- parties, with the technical assistance of the BVOs, maintain a system of water-sharing between riparian countries at a rather high level in accordance with the principles of water-sharing established during the Soviet period;
- a unified information system for water resources management in the two main river basins has been established and is being further developed;
- the Amudarya and Syrdarya BVOs maintain a system of joint planning and use of transboundary water resources in co-ordination with national organizations based on the principle of seasonal planning and monitoring of the process of water-sharing;

- the heads of state participate in discussions at regular meetings held to review water problems in the Aral Sea basin and issues related to the activities of IFAS (last meeting, at the time of writing held April 28, 2009).

Thus, the activity of the ICWC and Central Asian governments in pursuance of the 1992 agreement was a practical response to the criticism of the International Crisis Group (ICG 2002) that generalized the many negative opinions regarding interstate relations and pointed to the "lack of required data on river runoff, lack of transparency in the decision-making process and lack of the political support to basin organizations".

The next stage of interstate relations was an attempt (although not among the best) to co-ordinate the interests of the irrigation and hydropower sectors within the framework of the 1998 agreement. After 2003, these efforts were supported by the ADB RETA 6163 Project, under which a draft of a new agreement for water resource management in the Syrdarya river basin was prepared. This document specifies water quotas, sharing water resources between riparian states. It scopes the national responsibility and the rights to river flow management as well as defines a mechanism for compensating for the economic losses incurred by upstream countries due to additional summer water releases from the storage reservoirs. The final harmonization of this document, together with an attachment setting out operational rules of the Naryn/Syrdarya Cascade, should, to a large extent, cool things down and advance the development of sustainable mechanisms for implementing the water release plans.

There are two more shortcomings mentioned by critics of the ICWC activities. First, that there is insufficient representation of riparian countries. This situation should improve after putting into practice the ICWC Statute, the procedure for rotation in the ICWC executive bodies, and the draft agreement on the organizational development of the ICWC (the procedures for rotation of the directors and formation of joint executive bodies by interested parties) that were approved at the 51st ICWC Session on September 18, 2008. Second problem is a lack of representatives from other sectors in the joint bodies.

These decisions cannot be considered as final in spite of their considerable achievements, and there is further need to improve the organizational structure of the regional bodies, to take confidence-building measures and to eliminate many duplications of responsibility. Indeed there is already another stage in the process. However, the current institutional framework for water resources governance at the interstate level still suffers from some "bottlenecks" that need to be removed. Although the ICWC and its executive bodies are directly responsible for the management of water-sharing and improved water use, there are three parallel structures at the interstate level that are also involved in water governance to some extent. These are the IFAS and its national branches, which are responsible for providing funds for implementing the ASBP-2 and the preparation of new agreements and other documents aimed at improving water management (rules, procedures, etc.). They are, with a few exceptions ineffective, but due to often over optimistic ambitions, they create tensions and additional work for the basin authorities. There is the Regional Hydro-Meteorological Center that was established within the framework of IFAS and is tasked with improving the reliability of gauging flows and forecasts. This organization unfortunately also does not operate in a way that actually allows it to support improvements in water management. At some distance from the others exists the ISDC, which should be improving the monitoring and management of transboundary water quality, as well as initiating regional measures for sustainable development. Although representatives from the national conservancy agencies participate in the national working groups for improved management of water quality that were established by the ISDC, their participation is only seen outside of the ISDC program. The Regional Ecological Center with its national branches also acts in the same field.

Hydropower production, which in our opinion is insufficiently co-ordinated by the Central Asia Energy Council and UDC Energy, exerts very important impacts on the river flow regimes. Representatives from the national energy ministries and from the United Dispatch Center (UDC) are also involved in the activities of ICWC working groups, but little progress results from this.

The Eurasian Economic Community (EAEC), which established a special group for reviewing water and energy resource issues, also intermittently participates in discussions related to water resource management (these issues are also discussed at the meetings of the Shanghai Co-operation Organization (SCO)). Its regional influence on the system of water governance (particularly the organizational aspects) creates a troublesome and instable situation in the water supply from transboundary sources, which is reflected in water availability and sustainability and uniformity of water supply. This situation was explained in Section 4.4.

There are two options for a modern organizational structure suitable for water resource management in the Aral Sea basin. One was suggested by Jutzchak Alster (2001), a consultant of WARMAP. He suggested reorganizing the ICWC into a Water Commission for the Aral Sea Basin (WCASB), which would be responsible for all aspects of transboundary water-sharing, including hydropower production, irrigation, ecology, biodiversity, and the urban and rural water supply. This commission should not only be an executive body established by the water management ministries and departments but also be an instrument for implementing regional water strategy and developing strategic plans for the Central Asian republics. A main advantage of the WCASB idea is the principle of planning all aspects of water use in concordance with the annual availability of transboundary water resources. Thus, the WCASB would not be limited by the interests of irrigated farming but would take into account the needs of hydropower plants, industry and other users of transboundary water resources while taking into account the need to control water pollution and other environmental issues. The WCASB would also play the role of "trustee" for two additional non-majority users of transboundary waters, which are the Aral Sea and the delta areas of the Syrdarya and Amudarya rivers.

The WCASB would perform its duties through a number of political and organizational bodies. The supreme body would be the council or commission (hereinafter "commission") of prime ministers, those responsible for water resource management and the power to solve problems in all sectors of water use in their countries. This commission would hold regular sessions and its members should be authorized at the national level to represent the interests of all economic sectors. To strengthen the role of this commission, its autumn session should be headed by the deputies of the prime ministers from all republics. The bureau of the WCASB would be in charge of the execution of all the decisions of the commission and would include the Amudarya and Syrdarya BWOs, a secretariat and the Scientific Information Center of the ICWC as well.

The BVOs will be strengthened in order to become real regional bodies accountable only to the WCASB rather than certain riparian states. This approach requires BVOs to be financed at a regional level and to be given the ability to service the water infrastructure at the level of the transboundary rivers, as well as formal recognition of their interstate status by the member states of the commission. To emphasize their regional nature, it is proposed that a rotation system is introduced for the top managers of the BVOs and a supervisory council is established for each BVO. Each supervisory council will consist of five members (four water officials from each riparian country and one representative from the WCASB); and one of these members will be appointed as a BVO monitor. The supervisory council will allow representatives from each riparian country to monitor the work of the BVO, and it should be able to assist the director of the BVO to withstand the pressure exerted by member states. The supervisory councils will also promote integrated water resource management, and support the efforts undertaken by the basin water organizations and other executive bodies responsible for river basin management.

Developing the idea of improving the institutional frameworks, one of the co-authors of this book, Professor Dukhovny, has proposed a slightly different approach. He suggests establishing a regional organizational framework with a clear distribution of rights and duties, which could provide sustainability of operation, including all financial aspects, based on good co-ordination with the respective national authorities for the water sector and on mutual trust and openness of activities.

A second option that is sometimes proposed for an organizational structure for interstate water governance, that will avoid overlap in operation and specify clear rights and duties of entities, is the organizational structure of the Mekong River Commission. This may be taken as a prototype, but obviously should be developed by taking into consideration the peculiarities of the existing organizations in the region (see Figure 4.22)

Under this structure, it is proposed to set up an Intergovernmental Committee for the Aral Sea Basin (ICASB) headed by the prime ministers of all the basin countries, who will take charge, in turn, of the ICASB sessions. These meetings should be held two times a year, prior to and right after the growing season. The committee should consist of the ministers (or heads of relevant national departments) for water resources, hydro-meteorological services, conservation, energy and the economy as well as the deputy ministers of foreign affairs. The committee's meetings should be held exactly on the day specified in the regulations without any preliminary co-ordination (the ICWC experience shows that the process of gathering the plenipotentiaries led to a very lengthy co-ordination procedure going through the sequence host country → all ICWC members → national governments → cross-sectoral co-ordination → repeated co-ordination to reach a consensus at the full meeting). In the last five years, there were four occasions when the ICWC members did not attend the sessions and signed agreed documents after the event. The proposed committee has to replace the IFAS board, which today has a rather low status (the representation level is only vice prime ministers and deputy ministers).

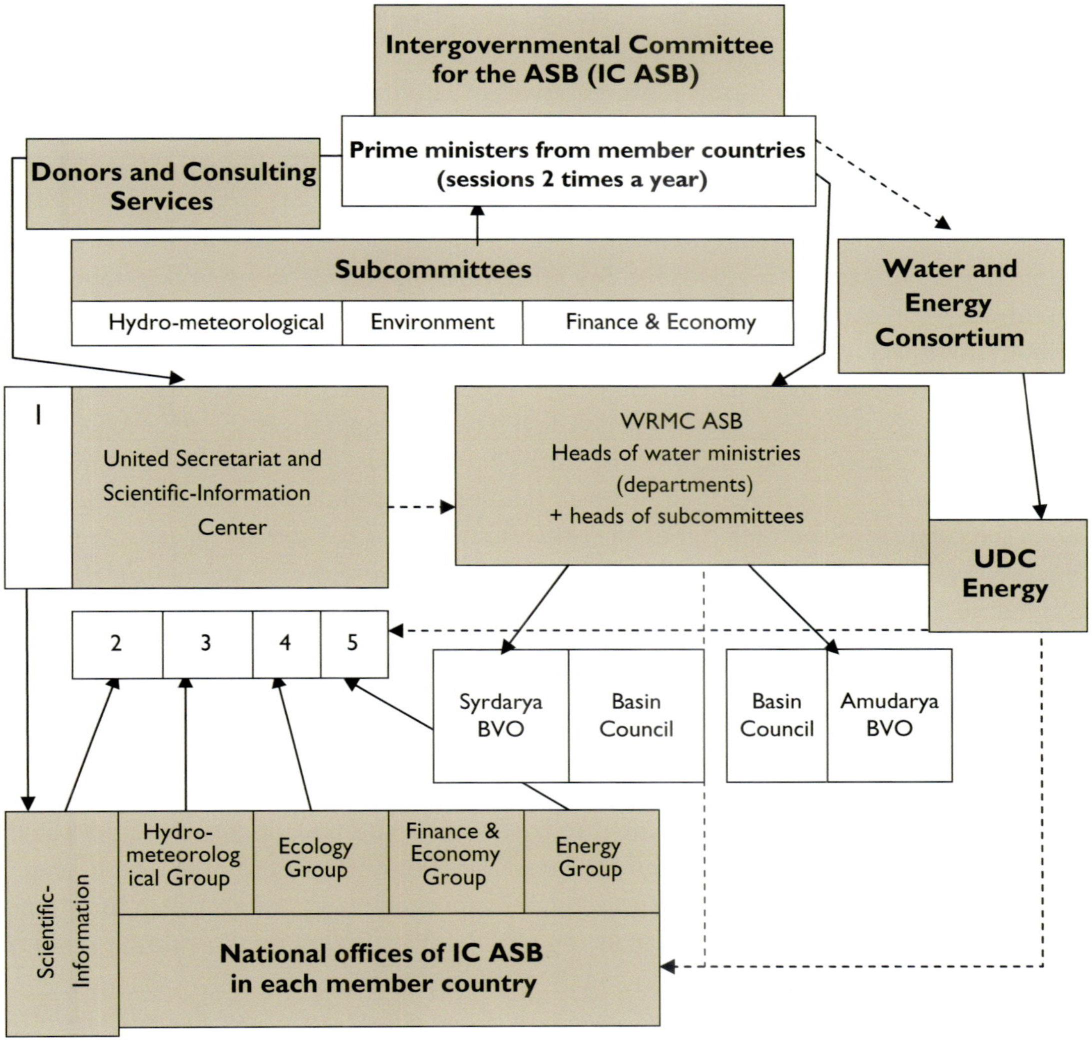

Figure 4.22 Proposed regional organizational structure for water governance in the Aral Sea basin.

The ICASB has to establish national offices in each riparian country. These offices should replace the national branch offices of IFAS, ISDC, SIC ICWC and REC. The Water Resources Management Commission for the Aral Sea Basin (WRMC ASB) can become an executive body consisting of managers of water management departments from the member countries, the Regional Hydro-Meteorological Center, the subcommittee on water resources protection (formerly ISDC) and other subcommittees (finance, investments, energy), and UDC Energy (or a representative of the Central Asian Energy Union). All the heads of the subcommittees take turns each half year (in alphabetical order).

National offices are to include specialized subdivisions acting on behalf of the relevant ministries and departments. Their function is the execution of the measures adopted at the ICASB sessions for improved transboundary water resource governance and management. The co-ordination of specialized subdivisions through relevant subcommittees should take place at half year or quarterly meetings. Staff will be supplied through appropriate national ministries and departments. This approach will provide permanent participation by the country representatives in activities related to regional problems including transboundary water issues and guarantee continuity of policies. The frequent replacement of sector representatives in working groups, with the consequent shifts in attitude and policies, often hinders the preparation of decisions, agreements and operational procedures.

A united secretariat together with the Scientific Information Center will be established as a single executive body for planning, co-ordination, financing and management of water resources. This executive body will perform those functions which at present are implemented in a very fragmented way by the SIC ICWC, the Regional Hydro-Meteorological Center, the SIC ISDC and the executive committee of the IFAS. An energy group, which will have representatives from the UDC Energy (one or two people), will be established additionally in the framework of this united secretariat. Following the experience of the Mekong River Commission, the united secretariat should be headed by someone who is not a national from the region, but the other staff should be citizens of the member countries. Given that the President of Kazakhstan, Nursultan Nazarbaev, and the President of Uzbekistan, Islam Karimov, have suggested many times that the Aral See Basin Commission come under the aegis of the United Nations, it seems logical that a UN representative should head the united secretariat. The secretariat and the WRMC ASB should have full diplomatic status. In this set up, the united secretariat will work in close co-ordination with the consulting services of the donors, who also are working within a UN framework. This proposed organizational structure will:

- concentrate all governance issues related to basin water and hydropower resources at the prime ministerial level, including the development of the key aspects of annual and long-term planning, and will also facilitate faster development of a united legislative platform for interstate relations, decision-making on issues of financing and the distribution of expenditures, and improved co-operation by removing many barriers between sectors;
- preserve the well-functioning management system at a national level for the water-related sectors, and control and co-ordinate this system through regional rules, and limitations and requirements set by the united secretariat and IWRMC;
- involve the representatives of other economic sectors and departments, such as hydropower specialists, hydrologists, economists, ecologists and others, into the decision-making process, as well as provide additional status to the united secretariat by involving representatives of the national ministries of foreign affairs (through a real transformation of ICWC into the IWRMC);
- eliminate overlapping responsibilities, latent competition and dissipation of financial resources allocated by donors and national governments, by directing implementation of clearly specified measures at the level of the basin committee;
- establish the Water & Energy Consortium as the mechanism for co-ordinating the interests of the energy sector together with irrigation practice;
- involve hydro-meteorological and conservancy agencies in a united system of water management in order to provide more reliable water monitoring data and forecasts, as well as improving operations for data collection and the development of clear regulations for water quality control and waste water discharges in view of environmental improvements.

One special aspect related to the introduction of an improved legal mechanism is the need to develop joint measures for water conservation in the basin. This was proposed as far back as 1998 at the nineteenth ICWC Session in Chimkent, with a suggestion that all riparian countries develop measures to improve water productivity and reduce actual water supply [22]. However, these good intentions were not followed up with relevant commitments with regard to material and financial resources. Any actual reductions in water consumption by various sectors of the economy (especially irrigated agriculture) was driven more by natural water shortages, by the introduction of water charges and by reducing the cultivation of crops that require much water (especially rice).

It seems, nevertheless, practical to develop incentives based on payments for water withdrawals in excess of allowed ecological levels. It has been suggested that allowed water withdrawals are established in the basin in line with natural cycles and in an environmentally sustainable way. This level can be specified as close to the 1960 water withdrawals (about 78 km^3). SIC ICWC expert V. Prikhodko has defined this level based on:

Table 4.23 Distribution of available water resources in the Aral Sea basin, km^3 (SIC-ICWC).

Assumption	Afghanistan	South Kazakhstan	Kyrgyzstan	Tajikistan	Turkmenistan	Uzbekistan	Total
Historic level of water consumption in 1960	1.60	8.50	2.10	9.10	10.60	43.50	75.40
In proportion to the population in 2030	3.60	3.70	3.82	10.70	12.17	43.80	77.81
In proportion to the population in 2030 at a rate of water consumption below 1500 m^3/hectare	3.60	4.80	3.50	10.40	13.10	42.90	78.27
In proportion to the area under irrigation in 2030	5.48	6.40	3.66	7.32	16.46	38.70	78.15
Average consumption	3.57	5.85	3.27	9.31	14.08	42.24	78.31

- historically established levels of ecologically sustainable water withdrawal;
- population growth predictions for 2020 and taking into account limits for water consumption;
- forecasts of the area of land under irrigation in 2020.

These estimates were later revised to take into account the latest forecast figures for population growth to 2030.

These figures can be used to specify the size of financial losses for breaching the ecological equilibrium in the region. If we take into account that losses resulting from a water supply less than the necessary level for environmental needs are of the order 0.3 cent/m^3, the price for water withdrawals in excess of these limits (quotas) over the whole basin will be some US$88 million ($29.6 \times 10^9 \times 0.03$)!

It is necessary to specify an average level of water withdrawal, taking into account future trends, in order to distribute this amount among the riparian countries (this could be called the contribution to a fund for upgrading the water sector).

Now, let us assume that in 2010 the riparian countries will agree to an ecologically acceptable maximum level of water consumption up to 2030, and that they agree to use a proposed dual approach to reach this target. First, annual payments for water consumption in excess of the established limits will increase from US$4.4 million in the first year to US$84 million by 2030. Second, each of four riparian countries that currently exceed the water consumption limits will invest these funds into implementation of water-saving measures in order to gradually achieve the planned levels.

This approach should not be considered too far-fetched. Under the first agreement to establish the IFAS, each riparian country was presumed to contribute 1% of its gross national product. The suggested contributions in Table 4.24 seem to be quite commensurable for all the riparian countries with the exception of Tajikistan.

A regulation concerning limits for discharging pollutants (mainly salts) into the rivers should be introduced in a similar manner in order to maintain the ecological quality in the basin. Proposals were considered in the draft agreement on ecological well-being for the Aral Sea basin, but this has not so far been followed up. Following an analysis of salt disposals into the Amudarya river, it was proposed to introduce limits to drainage water disposal into the river for each riparian country. Such commitments were already assumed in the Agreement on Collaboration on Water Management Issues signed by Turkmenistan and Uzbekistan on January 16, 1996 (Article 9): "In accordance with the earlier conventions, the parties have agreed to cease drainage water discharges into the Amudarya from both riverbanks from 1999 onwards. In case of urgency, the parties will settle issues in the framework of separate agreements."

The first steps are currently being made towards establishing a monitoring mechanism with the Central Asian water community under the aegis of the UN. The International Fund for the Aral Sea (IFAS) has received the status of an interstate observer according to a UN resolution dated December 11, 2008. However, until now, there are no clear answers as to how this status will be fulfilled and what advantages it would bring to the IFAS, although this issue should be settled in due course.

Table 4.24 Estimated payments into a fund for improvement of the water sector (SIC-ICWC).

Criteria	Afghanistan	South Kazakhstan	Kyrgyzstan	Tajikistan	Turkmenistan	Uzbekistan	Total
Water consumption in 2007, km^3	1.960	7.528	2.706	13.238	26.854	55.348	107.905
Limit allowed for ecological balance	3.57	5.850	3.270	9.305	14.080	42.240	78.3
Difference	–	1.698	–	4.085	12.774	11.108	29.605
Annual decrease necessary to achieve a balance in 2030	–	85.0	–	200.0	638.0	555.0	1480.0
Input into the fund, US$ millions	–	0.255	–	0.600	1.900	1.665	4.400

The establishment of the UN Regional Center for Preventive Diplomacy (Regional Peace Center) in Ashgabat (Turkmenistan) can be considered a very practical decision in this respect. The roundtable meeting held at this regional center on July 16, 2009 with the participation of Yan Kubish (UN Under-Secretary-General) has shown that donors have definite in this organization, as well as demonstrating the center's intentions to work with the executive committee and all other agencies of the IFAS. In the final document agreed at this meeting, the UN Regional Center for Preventive Diplomacy declared its readiness to support the development of a mutually agreed mechanism for the use of water and energy resources and the protection of the environment in Central Asia. This document considered the interests of all nations in the region and received the consent of all representatives from the riparian countries.

An important result of this meeting was the recognition of the need for co-ordination between donors and international organizations in getting agreement for the Central Asia Water Sector Initiative as an efficient instrument for implementing projects and work in the water sector [45].

Efforts to create a single program for all donors in order to overcome duplication, competition and waste of money were undertaken from the beginning of international donor involvement in Central Asia. However, so far these efforts have not produced results, as was shown in a special survey into the efficiency of the work of the donors (Dukhovny 2007). In recent years, no new initiatives have been undertaken to achieve better co-operation with the donor community.

Some donors are trying to support local beneficiaries in a way that provides them with an enabling environment for their aspirations. They facilitate development of strategic approaches and adaptation to new economic conditions, and fund programs that train local specialists in advanced practice and technologies including study tours to developed countries. Examples of efficient assistance by international donors include the IWRM Fergana Project (organized between ICWC and IWMI under financial support of the SDC), the Central Asia Regional Water Information Base Project (funded by the Swiss Agency for the Development and Cooperation and implemented jointly by SIC ICWC, UNECE and UNEP/GRID-Arendal), and the activities of the ICWC Training Center (with financial support from CIDA). These projects help to create a sound basis for the sustainable and effective functioning of the water sector. In these projects, international donors work to benefit developing nations and try to meet the needs of the beneficiaries without political, economic and other preconditions and with serious regard to local experts as equal partners. Donor countries such as Switzerland, Canada and The Netherlands, as well as some international programs implemented by NATO SfP, ADB and the European Union (such as FP 5, FP 6, INTAS), should be referred to within this context.

Other groups of donors may dictate their own priorities to beneficiaries and tend to distrust local specialists, and sometimes experience excessive delays in implementation. Sometimes their assistance comes with conditions, where 70% to 80% of the donor funds flow back to the donor nations through payments to their own consultants, purchase of equipment, etc. These projects are often not result oriented, because the assistance in itself seems considered more important than the efficiency of the proposed program.

One specifically dangerous and harmful recent tendency is for assistance to be provided through bilateral agreements that support one riparian country to the detriment of other riparian countries and groups. These actions are only "pushing" certain geopolitical interests in Central Asia. At present, these interests are closely interwoven, relating as they do to the use of water, fuel and energy resources, and with actors like China, Russia, the USA, European Union, Iran and many other countries have joined the struggle for "fuel and resources hegemony" in the region. There are many attempts to exert a definite influence over many different spheres of activity through the use of water resources. An article by Ben Slay "Water risks in Central Asia: who governs upper watersheds govern the whole basin", is quite instructive in this respect.

If the UN Regional Center for Preventive Diplomacy succeeds in establishing sound co-ordination and can ensure that donor funds are applied to overcome real regional problems, then donor efficiency will be highly increased. Another role for the center is to build confidence among riparian countries and create mechanisms for objective assessment that will be needed to mitigate the current situation. The center should become a regional agency involved in monitoring and act as an independent arbitrator and mediator able to recommend measures for settling contentious situations in the basin. Preventing damage to any country in the region must be the key criterion when resolving disputes. Given its status as a UN agency, the regional center could efficiently suppress the "hydro egoism" and illicit actions that may disturb sustainable water supply, giving priority to irrigated farming as this is basis for food security and the fight against poverty.

The shortcomings in water resource management in both Central Asia as a whole and in each riparian country separately described earlier in this chapter are not exceptional when seen from a global perspective. Already by the end of the twentieth century, there was a clear notion that to improve water resources management it would be necessary to first establish a firm common platform for *water governance* that defines the rules of play and creates the right enabling environment.

In 2002, Tadao Chino, President of the Asian Development Bank, stated that "the water crisis in Asia is, in fact, a crisis of water governance". This statement was cited in many documents of the Fifth World Water Forum, including in the letter of Mr. Brokman, the President of the UN General Assembly.

The concept of *water governance* is now officially used by the World Bank, the UN, the International Institute of Administrative Sciences and many others. However, many references actually offer a very narrow interpretation of the word "governance", defining it as an activity only related to the higher levels of the water management hierarchy. This is reflected in the UNDP projects related to the introduction of integrated water resource management (IWRM), which are limited to the national and basin level as, for example, shown in the draft national IWRM plan for Kazakhstan. This situation is similar to the Central Asian anecdote that poses the question "what progress in reform do we observe?" and gets the answer "the situation is as in the taiga, where the wind sways the crowns of the trees, but under the crowns there is not a breath of air". Fortunately, an awareness of the necessity for good governance at all levels of the water management hierarchy is gradually growing. Interesting work in this area has been done by Wouter Lincklaen-Ariens (2003), who first presented a comprehensive analysis and a broad platform for the improvement of water governance.

In Dukhovny, Sokolov et al. (2009), it is argued that the SIC-ICWC has shown that *governance* in general forms the political (legal, institutional, financial and economic) basis of the system of state and internal social relations. With respect to water resources and IWRM, the authors formulate the role of *governance* as the framework of incentives and limitations within which *management* is responsible for the realization of principles of sustainability in planning, use and protection of water and natural resources. They consider *governance* as another tool to overcome "hydro egoism" in its many guises.

Developing Falkenmark's (1998) ideas concerning *water solidarity,* we have formulated the following key requirements, which should be seen as counter-measures against progressive hydro egoism:

- present neutral and true information to all stakeholders and decision-makers;
- establish the required institutional framework to seek solutions through compromise;
- note that public participation is a required social instrument in planning and decision making;
- focus on the social, as much as the economic, values of water resources;
- do not allow water use that could result in damage to anybody;
- recognize "hydro-solidarity" on a national level and establish a National Water Council;
- establish a fair legal system with emphasis on mechanisms to provide end user rights;
- as a rule develop future water visions on the basis of sound forecasting and scenario development approaches.

Table 4.25 Governance as a tool for enhancing collaboration and joint use of transboundary waters.

Water management hierarchy level	Vulnerabilities and opportunities for ignoring the interests of other riparians	Measures for overcoming the existing vulnerabilities:	
		Political will and decisions	Legal measures
1. Global level	• Inaccurate formulating of some provisions of the IWL, UN and EU conventions;	• Recognizing hydro-solidarity as the only approach;	• Preparation of a World Water Code (or contract) accepted by the UN;
	• UN convention not put into force because ignored by leading states;	• Transferring matters for water security to the jurisdiction of the UN Security Council	• Preparation of proceedings for clarification of the provisions of the water code
	• Lack of a model national water code and rules for transboundary water use		
2. Regional level: transboundary waters or (interstate) basins	• Water rights of riparians not specified in terms of quantity and quality, and possible deviations under different conditions;	• Provisions of a regional water strategy agreed by the heads of state;	• Basin agreements on water allocation, regime of river flows and protocols;
	• Competition for water resources between sectors: hydropower, irrigation, water supply and environment;	• Political platform for meetings and decision-making by national leaders;	• Manual for transboundary water management and use;
	• Attempts to create water pressure on downstream countries by countries located in the upper watersheds	• Agreed action plans on transboundary waters;	• Clear demarcation between national sovereignty and regional commitments;
		• Regular reporting to national governments;	• Agreements on free access to information;

	Institutional measures	Financial measures	Social measures
	• UN Water transformed into a UN permanent agency responsible for monitoring water rights and recording water contracts (agreements)	• World Bank and other donors commitment to financing projects on the themes of water for food, water for the poor and water for the environment	• Develop organizations for water education in NGOs, parliamentary groups, and public awareness groups in general, promoting the principles and values of water use
	• Interstate bodies created and functioning according to principles of participatory approach;	• Procedures for public participation in transboundary water management;	• Agreed hydro-solidarity;
	• Equal representation of all riparian states in elected bodies;	• Procedures for financing the new infrastructure for transboundary water management;	• Mobilization of regional unity;
	• Arbitration commission established;	• Procedures for compensating for damage inflicted;	• Establishing and supporting a Basin Water Council with equal representation of provinces, large cities, hydropower stations and irrigation districts;
	• Combined Water and Energy Consortium co-ordinates the regime of water releases from reservoirs;	• Procedures for benefit sharing;	• System of public participation present at all levels of water management hierarchy

(*Continued*)

Table 4.25 (Continued)

Water management hierarchy level	Vulnerabilities and opportunities for ignoring the interests of other riparians	Measures for overcoming the existing vulnerabilities:		
		Political will and decisions	Legal measures	
		• Heads of state should affirm interstate water rights taking into account possible changes;	• Agreements on ecological flows	
		• IWRM adopted everywhere		
3. National level	• Administrations separated over different countries;	• Riparian states consider water management issues top priority;	• National Water Code based on the IWRM principles;	
	• Mechanism of controlling water rights absent or does not work;	• Governments recognize the water rights of all users and guarantee mechanisms for implementation;	• Clear commitments agreed between the central government and local authorities;	
	• Departmental approach;	• Water management organized from national level to individual water body;	• Clear commitments agreed between the government and public-private organizations;	
	• Water management not sustainable;	• System of water licenses is working;	• Environmental regulations allow an improvement in the ecological situation at local levels	
	• Vulnerability of existing water laws;	• State support for introduction of IWRM as the principal policy;		
	• Financial mechanisms do not work;	• State has the National Water Strategy developed, approved and put into action;		

Institutional measures	Financi al measures	Social measures
• Organizing equal access to hydrometeorological data and forecasts	• Financing of regional and basin organizations	
• Water organizations established based on hydrographic principles;	• System of water charges and payments for water pollution exists;	• National Water Council established and in operation and headed by one of the key politicians of the State;
• National Water Agency responsible as the leading organization and co-ordination unit for all water-related sectors;	• Large water projects also supported by governments in terms of financing;	• National Water Strategy discussed and accepted by the national water community, including representatives of all water users and water-related sectors;
• Water education and training under the umbrella of national water organizations;	• The state supports a system of penalties for users exceeding environmental limits and causing pollution;	• National public awareness campaign for promoting IWRM and broad public participation;
• Developing a program of preparing future water leaders is a national priority task;	• The government participates in financial support of basin organizations	• National competitions for achieving highest water productivity ongoing;
• Establishing water co-ordination committees and water user associations supported by governments;		• "Water guru" or national water manager is a most respected person in the country
• Special agencies for development and support of WUAs established within the structure of the Ministries of Water Resources;		

(*Continued*)

Table 4.25 (Continued)

Water management hierarchy level	Vulnerabilities and opportunities for ignoring the interests of other riparians	Measures for overcoming the existing vulnerabilities:	
		Political will and decisions	**Legal measures**
	• Lack of water information or does not meet requirements;	• Riparian states consider water management issues top priority;	
	• Lack of water resources accountability;		
	• Water management system is vulnerable and is not adapted to climate change or climate variation		
4. Basin, provincial and local level	• Water planning and allocation can't guarantee secured water supply in terms of water quality, quantity and required regimes of water delivery;	• Provincial and local leaders recognize and promote IWRM principles and participate as stakeholders in the process;	• Water users and water management organizations have agreed to a plan of water allocation and delivery, as well as to obligations concerning water charges;
	• Water delivery is not stable;	• Local authorities create the enabling environment for rational use and stability of water resources;	• Agreements between water user associations and water management organizations have been signed and are implemented accordingly
	• Different organizations mismanage water resources at provincial and lower levels	• Provincial plan of water-saving measures and adaption to climate change adopted and approved	

Institutional measures	Financial measures	Social measures
• System of extension services in operation in all riparian states		
• Establish a Basin Water Council; a Commission for Water Co-ordination (CWC) and water user associations as regular public-private bodies;	• Water users pay for water services to water user associations. WUAs pay the water coordination commission	• Water users create network for permanent collaboration with WMOs to improve the current situation;
• Establish a Committee of Water Users of the downstream and river delta areas	• Local government bodies support the water management organizations (WMO);	• Introducing IWRM and water demand management;
	• Financial system (including penalties) for exceeding water limits and encouraging water-saving practices is in place	• CWC and WUAs initiate social mobilization of water users on a regular basis

Hydro-solidarity can be only developed through well-established governance and capacity building in the water sector. Taking into account the multi-level hierarchical structure of the water sector, it is very important to organize a specific governance framework that allows for developing rules and procedures at each level of the water management hierarchy, and can link all these levels in both top-down and bottom-up fashion. This is one way to prevent malfunctions in water resources management at the "interfacial joints" between the different levels of the water management hierarchy.

The governance system should be the template for all levels of the water management hierarchy, management mechanisms should be developed or strengthened accordingly (Falkenmark 1998). A program approach to the development of governance systems is set out in Table 4.25. A closer look at this program demonstrates the versatility of this approach to *governance*. Governance should establish an enabling environment for the successful application of IWRM principles and mechanisms at all levels of the water management hierarchy, from the global management system down to interest groups and water user associations. At the international level, for example, the complexities and weaknesses of the international water laws create opportunities for the followers of the Harmon doctrine. Detailed analysis of the opportunities for improved governance will allow the step-by-step development of clear standards and rules for all participants of the water management hierarchy, ensuring equitable, fair and sustainable water use based on water-saving practices at a national and international level.

Of course, these proposals can be implemented in the future only if the governments and leaders of the riparian countries aim at long-term prosperity in the region. It is clear that any reform of the existing relations will be painful and will require effort and the mobilization of powerful public forces. Therefore public participation in transboundary water resource management is very important, as shown in the process organized in the framework of the IWRM-Fergana Project, which took place on a combined inter-provincial, inter-district and grass-root level.

Establishing basin councils under the BVOs can make an essential contribution to progress in raising public awareness and developing an overall aspiration towards sustainable water resource management. Results will be more sustainable and can be achieved much faster if all water users of a river scheme are represented. For example, when considering this for the Syrdarya, representatives of the lower reaches (for example the Federation of Delta Water Users) and representatives of the large hydropower facilities upstream should be regularly gathering to discuss current problems in a similar way to the water councils in France.

4.6 Capacity Building in the Water Community

One key element in introducing new water governance and water management practices is sector-wide capacity building. In this process, from needs assessment to development and implementation of enhanced water management capacity, a similar step-wise approach to that shown in Table 4.24 can be followed. Table 4.25 shows how the different technical and institutional elements that play a role at different levels in the water hierarchy operate.

Capacity building has many different elements, ranging from education and training for professionals and even non-professionals active in the sector to the development of management and monitoring systems. Integration of all these elements is necessary to put IWRM principles into practice in Central Asia. The following elements make up a non-exhaustive summary of the main activities that should be put into practice (and has already been partly put into practice) for capacity building in Central Asia:

- technical, scientific and skills training for water sector and water sector-related staff (education, training, research);
- development of hardware and software to support analysis, management and decision-making;
- use of devices and agreed operations standards for water monitoring such as the SCADA system (joint monitoring);

Table 4.26 Technical and institutional elements in capacity building.

Level of the hierarchy	IS	Com	MIS	SCADA	HMZ DM	WM	BP WCH	PP	Tr
Transboundary waters	x	x	x	x		x		x	x
National level	x	x	X	x		x		x	X
Sub-basin (a national basin)	x	x	X	x		x		x	X
Irrigation system	x	x	X	x	x	x	x	x	X
WUAs	x	x	X		x	x	x	x	X
End-users	x	x	X		x	x	x	x	X

IS – Information systems/decision support.
MS – Management system (including water and climate forecast technology).
SCADA – Data control and acquisition for monitoring.
DM, HMZ – Demand management, water allocation based zoning.
WM – Water quality/quantity measurement.
BP, WCH – Business planning, water charges.
PP – Public participation/planning.
Tr. – Education and training.
Com – Communication and information.

- water management information systems, and concepts and methods to establish accountability practices;
- build public participation practices and communication capacity at all levels of society;
- internet-based databases and information capacity such as the CAREWIB system;
- a sector-wide institutional, legal and financial framework for the operation and management of the (transboundary) water sector.

While proper water governance based on agreed political, institutional, legal and financial principles is needed for sustainable water use and management, the water management hierarchy in charge of routine management of transboundary water resources can be a problem in itself. Therefore development and strengthening the potential for co-operation should be a key task and constant concern of the sector, with hydropower, environment and agriculture interests as the key stakeholders. In Central Asia, the central role and responsibility for regional co-operation and development is vested within the ICWC, which with the support of various international donors, such as the Swiss Agency for Development and Co-operation (SDC), Asian Development Bank, UNESCO-IHE and the Canadian International Development Agency (CIDA), has managed to initiate and organize many capacity building activities:

- establishing the Central Asia Regional Water Information Base (CAREWIB Project) together with the SDC, UNECE and GRID Arendal;
- establishing a training system for top water managers and senior water professionals, WUA specialists and water users in general in partnership with McGill University Mount Royal College (Canada) and UNESCO-IHE;
- introducing integrated water resource management in Central Asia in partnership with IWMI, SDC and the GWP CACENA – this has included establishing and enhancing the involvement of public-private organizations such as WUAs, CWUC, etc.;
- introducing the SCADA system in the Syrdarya basin and the automatic monitoring of the water distribution systems;
- developing participatory approaches by establishing water user groups in various projects while taking into account gender aspects.

All these activities are conducted in joint projects, with specialists from the riparian countries sharing experiences in conferences and seminars and researching new methods for co-operation in, and improvement of, transboundary water resource management. This is important because it provides a basin-wide common framework for analysis and management of water and creates mutual understanding of the current situation and each others' needs. Eventually this achieves the consensus that is the essential condition for joint management. In this respect, it is important to note that consensus among water professionals is often reached quite fast because there is a shared understanding that water relations are closely linked and therefore require well-balanced management solutions. Concessions and compromises are, therefore, ever present conditions. Water professionals are, and must be in the future, "the catalyst" in the process of establishing inter-sector consensus and understanding, and they should try and involve selected decision-makers in this process. The ultimate goal is to promote "hydro-solidarity" as the basis for future progress.

4.6.1 Management Information Systems

Information and control is the basis for any successful management process, and much information needs to be provided when we deal with integrated water resources management. Mistakes in water resource management threaten not only the water supply but may also cause floods, droughts, water-borne diseases, loss of crops, famine and other hardships. Therefore a well-designed management information system (MIS) should be the backbone of IWRM, encompassing all the interrelations with the outside world. Figure 4.23 shows the links between water resources on the one hand and the impacts by internal and external factors to the state and some aspects of the operational regimes adopted for these resources. This is a very complex system and, in the process of establishing a management information system for IWRM, certain rules and principles must be followed. One principal requirement is that it must link to practical results of work based on ideas developed over a long period in Central Asia. This was clearly formulated in the book, *Integrated Water Resources Management: Putting Good Theory into Practice – the Central Asian Experience* (Dukhovny, Sokolov et al. 2009).

The core of such a management information system consists of thematic data sets, covering water resources, water consumption, social environment, land resources, environment (biodiversity and bio-productivity), climate, infrastructure, hydropower, irrigation and drainage, water supply, industry, micro-economy, and so on. Information on key participants in the water management process (history, scope of activities, duties, legal and institutional basis, staff capabilities, financial support, logistics, etc.) should be included in the structure of information blocks. These blocks encompass both water management agencies and water user organizations, holding actual and factual information on relevant

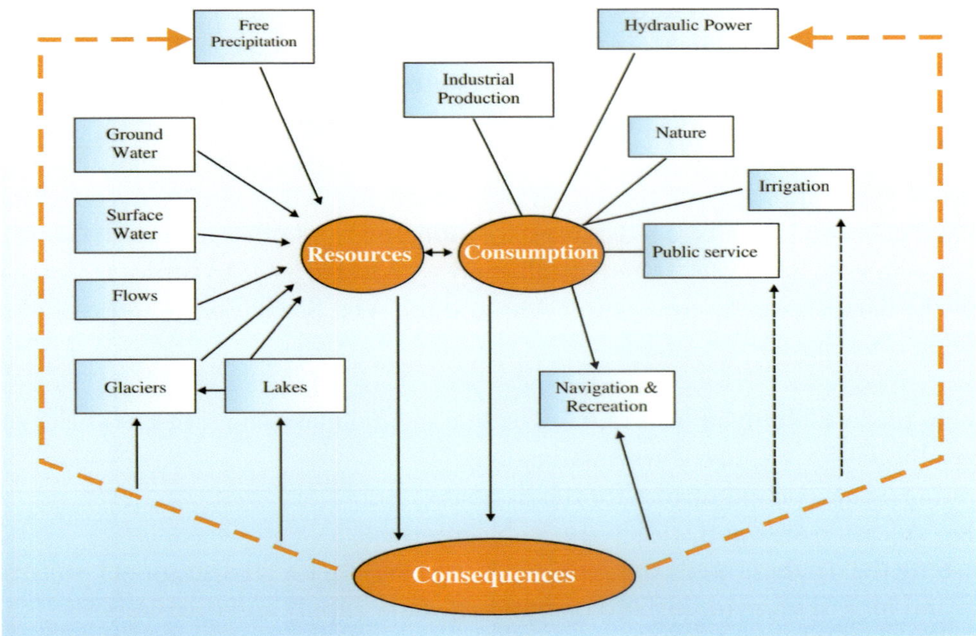

Figure 4.23 Resources to consumption, and consequences for interrelations (SIC-ICWC).

parameters and indicators. The database is built on certain general principles in order to provide a well-adjusted structure for connection to user models and for continuous updating.

1. There are three time bases – a historical time series, current information and future outlook. Time series include all hydrological and climate data, status of land resources, vegetation cover, soil fertility, etc. Socioeconomic indicators go back 25–30 years, in order to allow for reliable socio-economic analysis and forecasting. Because of the long implementation period for water management projects, time series should be at least twice as long as the average planning period.
2. Current time series are usually based on daily intervals, but for some data even one-hour intervals may be required. After a certain time has elapsed, these daily data can be either dumped, or aggregated and archived. Long-term time series may consist of a series of 10-day intervals or monthly average data.
3. Spatial division of water management objects (hydraulic structures, rivers, canals, etc.) is contained in the form of GIS (geographic information system) based thematic layers that hold data in terms of absolute coordinates and link IWRM components and interactions over an area or linearly. One example of important present use of GIS is the detailed monitoring of the formation of snowmelt runoff, which depends on land gradients, soils, precipitation etc. Information on soil conditions, hydro-geological cross-sections, crop patterns and other components is a reliable basis for calculating water requirements over irrigated areas. The GIS also allows the system to link formation of return water with a drainage network and with water supply over a certain area including the source zones of different pollutants.
4. Each database consists of a general area and an area focused on a group of certain of requirements, such as analyzing, reporting, planning (operational, annual and long-term planning) and management. A general database of water resources is often built top-down according to hydrological principles rather than from the source, using a "distribution tree" towards water consumers. However, a database to hold information on consumption of water resources, land reclamation or economic indicators should be built in the reverse order, going from the consumers back through the "distribution tree" towards the source.

Because of the complexity of the physical situation as well as the many different institutions involved in data collection, inputs to a database are numerous and often disordered. Setting up a database should therefore be done very systematically and with a sharp eye to future maintenance and use. In the case of the database set up for Central Asia and the Aral Sea basin the scope of work was based on the following elements and priorities:

- analyzing the availability of information, length and detail of data series, evaluating data validity and possibilities for improved systematization;
- elaborating a concept and format for the database;
- selecting an easy-to-use and widely applicable interface;
- establishing the database on the basis of providing a service for all operational processes, including annual and long-term planning and operational water supply and distribution;
- monitoring water quality and management;
- analyzing and adjusting the water management process;

- providing transparency of water management information exchange and building trust among water users;
- assisting water users to adopt economical water consumption practices and achieve full potential productivity;
- preparing analytical reports for improved management methods and providing information to decision makers and stakeholders;
- assessing current trends and adjusting future water use strategies.

First steps in the development of a computer-based regional information system were made between 1997 and 2002 within the WARMAP project, which was run by the SIC ICWC with the assistance of European specialists. This work was later continued in collaboration with the UNECE and UNEP/GRID Arendal Office in Geneva with financial support of the SDC in the Central Asia Regional Water Information Base Project (CAREWIB). This information system is now a powerful tool that is widely used both in Central Asia and internationally. The system consists of the Central Asia Water Information Portal and the information system. The portal is available through a central CAWater-Info website (www.cawater-info.net) giving access to a group of official websites showing projects, online databases, a knowledge base and the "Water World" website. In addition, traditional services such as daily updated news editions, a calendar of events and a forum are available on this portal.

The overall scheme of information exchange and generation is such that, apart from central information, considerable data come from various organizations and projects that are international partners of SIC ICWC. The CAWater-Info Portal is accessible free of charge for all users, and the information block contains daily updated information from all countries in the region as well as periodical press releases. The knowledge base contains processed structured knowledge, with the electronic library of the SIC ICWC, publications of the month, Central Asian newspapers and magazines, a photo library, a thematic knowledge base and glossary, and a bibliographic database. The large electronic library contains background material on matters related to water, land and environmental laws adopted in the riparian countries of the region as well as analytical articles, reviews, abstracts and publications prepared within the framework of ICWC projects.

The second part of the CAREWIB project is a regional information system that covers a very large volume of constantly updated information on water and land resources of the region, which is for official use only (each ICWC member has their own copy). The database contains water-flow records for all rivers of Central Asia (with data for some rivers going back to 1911), data of weather stations from the very first days of observations, catalogs and data on hydraulic work on rivers, main irrigation and drainage canals, as well as data from the water gauging stations. In addition, the database contains data for each country and province in accordance with four selected data blocks:

- water resources and water consumption;
- land resources and land use (information on sown areas, crop patterns, crop yields, profitability, etc.);
- socioeconomic indicators (including demography);
- efficiency of land and water resources use.

This data have been collected since 1980, and data on water use are given on a monthly basis. Part of this portal contains analytical reviews and online data on water use. Two related websites (the Amudarya BVO and Syrdarya BVO online databases) present data on water withdrawals in the basins of the Amudarya and Syrdarya rivers. These databases also contain information on basin morphology, water resources, water infrastructure (water intakes and reservoirs) and the technical status of these structures. Water management aspects, such as water consumption and actual 10-day water withdrawals (updated every 10 days with comparison against forecasts and planned withdrawal), are available here as well. The website has a "analysis of the water-related situation in the river basins of Amudarya and Syrdarya", where one can find data on the operational modes of the reservoirs, planned and actual water releases from the reservoirs, planned and actual water-sharing between riparian countries, and water diversion into the main irrigation canals (see Figure 4.24).

As well as maintaining the portal, project staff is using the site to publish periodicals, reports and monographs in order to assure wide dissemination of information on the basin. The website allows everyone to know to what extent the forecasts of the Hydrometeorological Service regarding, for example, inflows into the reservoirs has proved to be correct, how the owners of the reservoirs have implemented their water release schedule as well as how the BWO distributes water resources between the riparian states as based on the ICWC plan. For further capacity building and knowledge sharing, the national cells of the CAREWIB project are being transformed into "hubs" around national databases, which are based on the same principles and structure as the CAREWIB project.

A specific type of database was developed in the Rivertwin project for the Chirchik-Akhangaran-Keles sub-basin in the Syrdarya basin. In this area, the basins of three rivers are linked by connective canals. The structure and thematic information of this database was built along the same lines as

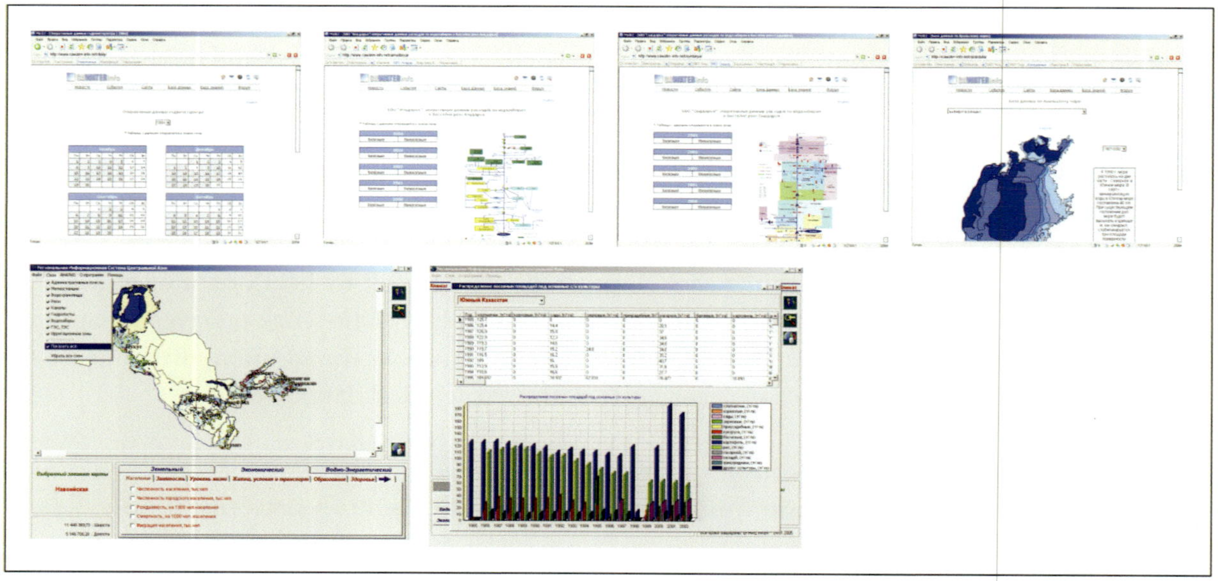

Figure 4.24 Overview of the CAREWIB website, 2010 (SIC-ICWC-CAREWIB).

the regional database using similar principles, but it has a more in-depth set of data on socioeconomic aspects and the information required for elaborating scenarios of future water resource development. These data will allow models to take climate change and other necessary information into account in order to research optimal variants of joint development of this sub-basin together with the Syrdarya basin as a whole. One new and essential component of this database is a GIS with information drawn from an analysis of satellite images and detailed soil and hydro-geological surveys. This allows detailed zoning in accordance with crop water requirements, and specific water consumption calculations for irrigated crops in these areas.

The database includes information on the combined network of watercourses (rivers, irrigation and drainage canals, water supply system pipelines, etc.), waterworks (regulators, distribution structures, hydroschemes and reservoirs) and facilities for water resource use (hydropower plants) and consumption (water supply) as well as monitoring stations (gauging stations). This network is linked with surface water resources, represented by flow rates at water gauging stations around the catchment areas, and with data on groundwater resources, represented by clusters of facilities for water abstraction.

The database used for the actual management of the irrigation systems has the same blocks as the larger system, but with more detailed components and parameters for water consumption and water distribution, local water resources (groundwater or return water), land reclamation conditions and land productivity factors, and with less information on macroeconomic parameters and hydrological parameters of the water resources.

In the water sector, all information systems in principle should be developed and built using a uniform logical scheme (see Figure 4.26) that allows not only a common "information interface" for the entire water management infrastructure of a region, country or basin to be built but also provides the potential to unify database structures. This is especially important for improving the consistency of data through comparison of the data identity at different levels of the water management hierarchy. In addition, unified interfaces largely facilitate wider database use and make it more cost effective to provide advanced training to a large number of specialists.

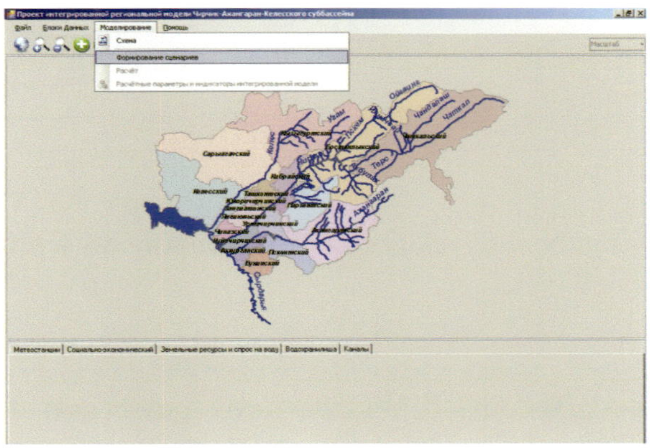

Figure 4.25 Geographic information system (GIS) for the Chirchik-Akhangaran-Keles sub-basin (SIC-ICWC).

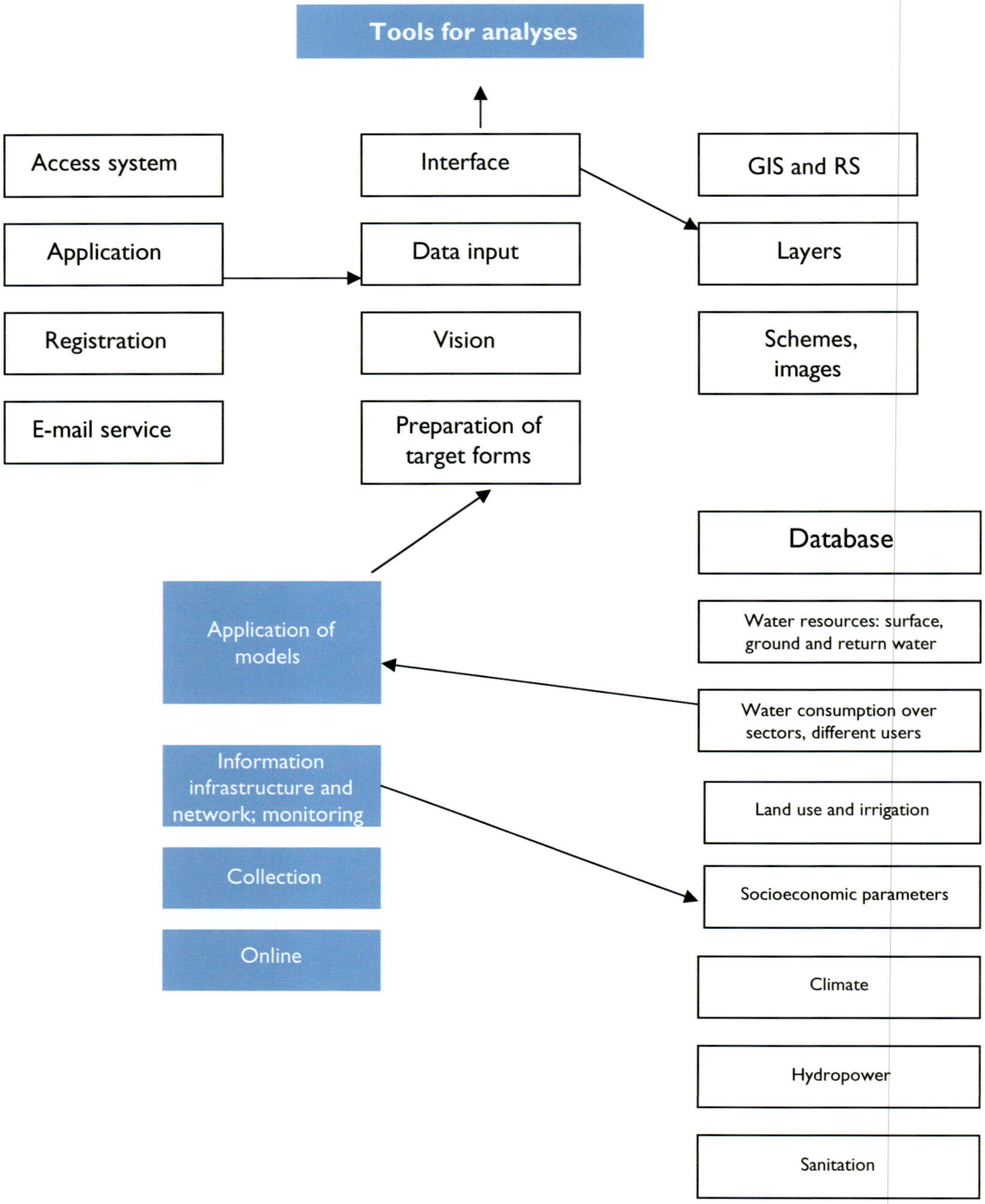

Figure 4.26 Diagram showing the interrelations of the components within the management information system (SIC-ICWC).

It is the strong conviction of the SIC-ICWC that development and implementation of information systems at all levels of the water management hierarchy, including strong provisions for openness and transparency as well as regular data updates, is of primary importance for achieving the goal of improved co-operation between riparian countries in Central Asia and, indeed, anywhere else in the world.

4.6.2 Program for Training and Qualification

In the past, and dating from the Soviet period, capacity building was a task of SANIIRI. It conducted regular scientific and practitioner seminars for the advanced training of water professionals from developing countries as offering technical assistance under programs developed by the UN Economic Commissions for Africa, Asia and the Pacific, Latin America and the Caribbean. The regional branch of the USSR's Institute for Advanced Training of Water Professionals from Central Asia, Kazakhstan and the Caucasus was also the responsibility of SANIIRI. In the course of the transition towards a market economy, and in a situation where the scientific and technical potential of the water sector in all Central Asian countries was set back, first SANIIRI and now SIC ICWC has continued to work in this field of human resources development. Training is provided to young specialists, who are introduced to subjects including informatics, management, economy and legislation. Relations are maintained with the International Commission on Irrigation and Drainage (ICID), UN Economic Commissions, UNESCO, FAO, CIDA, USAID, NATO SfP and other international agencies. This combination of regional experience and international scientific and technical co-operation has enabled ICWC to implement a large number of projects that have included components for training and study tours for water professionals and decision makers, exposing them to IWRM principles. These activities also helped to build the necessary political support and raise public awareness of the looming water crisis.

The urgent need for capacity building for joint transboundary water resource management and development is obvious when one considers the crucial importance of developing and implementing a regional water management strategy for the Aral Sea basin. This idea led to the decision (taken during the ICWC session held on October 24, 1998) to establish a training center with financial support by the Canadian International Development Agency (CIDA). The Central Asian Water Resources Management Training project was launched in partnership with the McGill University and Mount Royal College (Canada) in 2000. The ICWC Training Center started with a mandate to organize advanced courses for senior water professionals and it aimed to raise awareness of IWRM principles with senior policy makers and key stakeholders involved in projects at various levels of the water management hierarchy.

The training activity is very much based on the findings of research from joint interstate programs and regional projects, such as the IWRM-Fergana project, the IWRM Strategic Planning project and a project for transition to IWRM in the lower reaches of the Amudarya and Syrdarya. The research is used as case studies in lectures and presentations. Co-operation with colleagues from leading international universities and institutions, such as McGill University and Mount Royal College, UNESCO-IHE and ILRI (The Netherlands), Bonn University and Stuttgart University (Germany), contribute to the diversity and quality of the programs. Higher efficiency is achieved through decentralization and establishing training center branches. The ICWC Training Center in Osh trains specialists from the Fergana Valley's provinces within the boundaries of Kyrgyzstan, Tajikistan and Uzbekistan, and in Urgench training is provided to specialists representing provinces in the Amudarya's lower reaches within the boundaries of Turkmenistan, Uzbekistan and Karakalpakstan.

As part of the IWRM-Fergana project, a pilot training center was established in the Kuva District of Fergana Province in co-operation with the Akbarabad Water Users Association, where training is organized for farmers and representatives of other WUAs, community government officials and village committees. These local training activities show that in order to improve water use practice and water productivity the dissemination of positive experiences is best organized through extension services. This requires establishing special training locations close to the demonstration fields. In these units trainers can be trained at higher levels, and they in turn will be able to train farmers in up-to-date methods of water measurement, water accounting, water application technologies and other technologies related to increasing water and land productivity.

Another important feature of the training activity are seminars on specific themes involving representatives of the related sectors. Some basic topics covered in these training seminars have been:

- integrated water resources management;
- co-operation in transboundary river basins;
- water, governance, legislation and policy;
- improving irrigated farming practices.

At first, only specialists from ministries and departments of nature protection and energy and NGO representatives were invited, along with water professionals from the riparian countries. Later, representatives of ministries of foreign affairs and ministries of justice from each country were also invited. These people were engaged in the preparation of interstate agreements and national legislation in the field of water economy and nature protection, and they were specifically invited to attend he seminars on water legislation and policy that were held with participation of experts from Dundee University and the Israel Center of Negotiations and Arbitrage.

The interactive training format, the dissemination of tutorial papers and other training documents, and the dialogues and discussions on current problems faced by the water sector held under the leadership of experienced moderators, have facilitated a common framework for the analysis of problems and create an atmosphere of openness and trust. Almost each regional training seminar turns into a roundtable of different countries and economic sectors, where the brainstorming sessions (supported by moderators) contribute to promote consensus in the region at all levels. The minutes contain collective recommendations drawn up at the end of each seminar, and these are sent to all ICWC members and contribute to developing measures for upgrading and improving existing systems.

Establishing an enabling environment for friendly contacts between specialists from different countries and economic sectors engaged in solving water problems is an important side effect of these regional trainings. Relationships are developed during joint exercises and even during leisure time. This is especially important because participants in these training courses are mainly young people who, in the near future, will be local or national leaders or be active as policy makers in other sectors. Up to 2010, more than 4500 specialists were trained in topics including IWRM principles, water legislation, good irrigated farming practice, participatory approaches, etc. (see Table 4.27). The ICWC Training Center

Table 4.27 Distribution of trainees, by country and year of participation (SIC-ICWC).

Country/ Year	Kazakhstan	Kyrgyz Republic	Tajikistan	Turkmenistan	Uzbekistan	Other countries	Total
2000	9	11	11	6	12	–	49
2001	63	45	55	27	99	7	296
2002	176	171	160	128	204	5	844
2003	141	189	164	130	259	1	884
2004	173	145	144	80	427	1	970
2005	149	107	126	56	392	9	839
2006	96	85	84	28	286	15	594
2007	67	59	71	15	226	–	438
Total	874	812	815	470	1905	38	4914

has established a special database with current information on each training course participant in order to support feedback to these professionals and this alumni database is updated annually.

Thanks to the activities of the training center and the links it has developed, there is growing understanding of the need for consolidation based on collaboration and readiness to restore common activities in the basin. Acknowledgement of the importance of international commissions is increasing, due to the widening circle of like-minded people in appropriate ministries and institutions in each riparian country. There are the similar positive developments related to the advanced courses for improving irrigated farming, including the growing awareness of the need to introduce water-saving technologies. As a result, opportunities for achieving higher water productivity and increased agricultural outputs are growing with (perhaps 10%) less water consumption. This has encouraged the governments to invest in a network of demonstration sites that play a role in the extension services to farmers, WUAs and water management organizations and can disseminate up-to-date methods of water conservation. Figure 4.27 shows the number of participants in this initiative.

The seminar series organized by the Water and Education project aims to establish a generation of future water consumers that care for water resources. IWRM basics are explained to school teachers in order to facilitate integration of IWRM principles into school curricula.

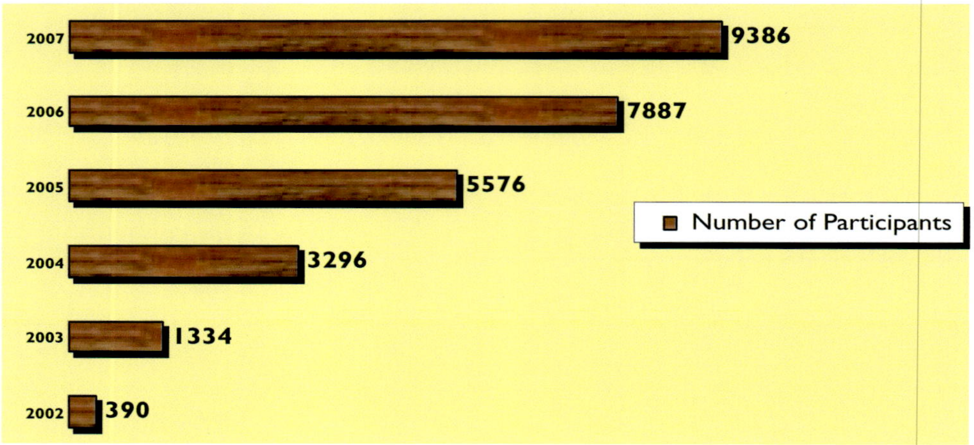

Figure 4.27 Seminar trainees within the IWRM-Fergana project (SIC-ICWC).

Regularity is a key factor in maintaining the effectiveness of the training system. This is important not only because of the permanent staff turnover at all levels of the water management hierarchy, but also because of the need to update water users with the latest knowledge and with information on the methods tested in pilot systems. Advanced courses on IWRM for heads of departments and divisions of the ministries of agriculture and water resources (committees, departments, central administration) are therefore be held on a regular basis. In addition, similar advanced courses are organized for heads of provincial water authorities, heads of water user organizations, representatives of local government and WUA representatives all over the region.

4.6.3 IWRM as the Leading Concept for Water Sector Development in Central Asia

Implementing joint projects that unify member countries around common targets is the most essential contribution to strengthening collaboration between riparian countries in a river basin. The ICWC started to implement such projects since 1998. For example, the WARMAP project including a research component on irrigated farming potential that covered 40 farms in different climatic zones and countries of the region (SIC-ICWC 2001). These studies, and a project related to dissemination of best practices, laid a practical foundation for the campaign for water-saving in irrigated farming initiated by all member countries of the ICWC. A main result of this work is that four riparian countries in parallel pilot sites created not only the enabling atmosphere but also the team spirit that is required to make the experts from different countries work together.

The most significant progress in this area has been made during the regional IWRM-Fergana project, which was implemented by specialists representing the agriculture and water resources sectors of Kyrgyzstan, Tajikistan and Uzbekistan under overall co-ordination of the SIC-ICWC and IWMI with SDC financial support. The aim of this project was to improve the efficiency of water resource management by means of introducing IWRM principles into the Fergana Valley. The wider objective of the project was "*to ensure better food security, environmental sustainability and social harmony through improved efficiency of water resource management in the Fergana Valley*". The project has aimed at obtaining maximum participation from all water users and water management organizations right from the start, and selected pilot sites for introducing IWRM approaches in line with this objective. In the first stage of the project, field teams consisting of specialists from three republics and all provinces covered by the project were engaged in selecting these pilot sites based on the principles and technical requirements elaborated by the regional working group. These IWRM principles, that were discussed with all project participants globally, were developed within this definition:

> *"IWRM is a management system, based on considering different kinds of water resources (surface water, groundwater and return water) within their hydrological limits, that links the interests of different socioeconomic sectors and hierarchical levels of water use, that involves all stakeholders in the decision-making process and promotes effective use of water, land and other natural resources to the meet requirements of ecosystems and human society under conditions of a sustainable water supply."*

This means that IWRM is to be based on these key principles:

- water resource management is implemented within the hydrological limits and in concordance with the geomorphology of the drainage basin under consideration;
- management takes into consideration all kinds of water resources (surface water, groundwater and return water) and the climatic features of the region;
- there is close co-ordination between all kinds of water use and with organizations involved in water resource management, including cross-sectoral (horizontal) co-ordination and co-ordination between the hierarchical levels of water governance (basin, sub-basin, irrigation system, WUA and on-farm end user);
- there should be public participation not only in the water management process but also in the planning, financing, operation and maintenance and development of the water infrastructure;
- the priorities of sustainability of ecosystems in relation to water requirements (environmental flows) should be recognized and be part of the practice of water management organizations;
- there should be active participation of water management organizations and water users in activities related to water saving and monitoring and control of unproductive water losses (water demand control along with resource management);
- there should be information exchange, and openness and transparency in all dimensions of the water resources management system;
- the operational circumstances of water management organizations should be economically and financially sustainable.

From an analysis of the proposed pilot areas in the IWRM-Fergana project, the project participants selected three pilot canals: the Aravan-Akbura Canal in Osh Province in Kyrgyzstan, Khodja-Bakirgan Canal in Soghd Province in Tajikistan, and the South-Fergana Canal that crosses the Andijan and Fergana provinces in Uzbekistan. After seven years of project activities, and taking into consideration hydro-geographical boundaries, a participatory approach and the democratic principles of water management, the IWRM concept was developed and was submitted to the national water authorities. The IWRM concept was evaluated and was eventually approved by all relevant water authorities in Uzbekistan, Kyrgyzstan and Tajikistan in May 2003.

A comprehensive approach for social participation was developed along with the preparation of a training program for social mobilization and capacity building at the level of the WUAs along each irrigation canal. Regular training seminars and sociological surveys have been conducted by the project, and these have provided new opportunities for stakeholder involvement in water sector reform in the Fergana Valley. Thanks to the project, new water user associations were established and the previously established WUAs were restructured. Since July 2002, the project has operated a monthly planning schedule (project agenda) and organized ad-hoc training seminars for water management specialists, water users and members of NGOs in the Fergana Valley. Much attention was paid to the dissemination of the IWRM concept and ideology. An internet-based communication network linking all key project participants (SIC-ICWC, national departments, provincial water management organizations, pilot canal administration and WUAs) was developed. The project has also established a management information system (database, mathematical models, GIS), operating online, which has become a powerful tool for planning, analysis of operations, and improved water allocation and distribution.

Alternative organizational structures for water management at the level of the WUAs and main irrigation canals were specified, discussed and co-ordinated by the project partners and stakeholders. Based on these agreements, the water authorities of Uzbekistan, Tajikistan, and Kyrgyzstan created new departments such as the canal administrations (on the Aravan-Akbura Canal in Osh Province in Kyrgyzstan, Khodja-Bakirgan Canal in Soghd Province in Tajikistan and South-Fergana Canal in Uzbekistan). New activities to involve water users in the decision-making process for improved water governance were initiated in December 2003. Pilot canal water user unions were established and officially registered on all pilot canals, and started to participate in the joint governance of the canals. Agreements related to joint water governance were eventually signed, and canal water committees consisting of representatives of the state water management organizations (WMO) and canal water user unions (CWUU) were set up.

The first steps were therefore taken towards establishing procedures for water resources planning, record-keeping, reporting and monitoring at each level of the new water management hierarchy. It is expected that more activities will be implemented at all levels of the water hierarchy through establishing these canal water committees. An effective factor in the transition towards IWRM is participation by representatives of the wider civil society in the governance process, and this also has now obtained a legal basis.

Many technical aspects of water management equally depend on public participation. It is not easy to provide guaranteed and equitable water distribution over an irrigation system as a whole. When water is delivered in accordance with planned volumes, a related increase in the productivity of water and land resources may be expected. However, water users themselves should participate more in the specification of the command areas of each irrigation canal, in the assessment of their water demands, and in the

accounting of additionally available water resources (groundwater and return water). They also have a role in adjusting water supply by crop rotation and adjusting water use in relation to weather and economic (market) conditions, as well as in improving hydraulic measurements and record-keeping at all levels of the water management system. To tackle issues and problems that may arise, it is necessary to establish extension services that assist water users in the introduction of new technologies, in the advanced practice of planning and production, and in new approaches to solving water distribution problems. The IWRM-Fergana project has developed and brought into use model regulations for canal water committees, as well as produced recommendations for each of the three pilot canals.

When the project team realized that the existing national legislation in the region cannot be a platform to support necessary reforms in the water sector, it prepared recommendations for change that were submitted to all the national water authorities in the region. The law has to be amended to specify the role and duties of governments and water management organizations with regard to water resource development, use and protection. There is an obvious need to clearly specify socioeconomic and ecological values of water, and to set out water rights and the role and responsibilities of water user associations and other water consumers. There is, for example, a need to regulate the relations between water authorities and conservation agencies, agricultural authorities and local governments. Financial mechanisms in the water sector should also have a clear legal basis. The IWRM-Fergana project has paid much attention to settling disputes at the level of WUAs and irrigation canal administrations. Sociological surveys have been conducted, and consequently recommendations have been presented and discussed at on-site seminars.

The project has rendered technical assistance in the form of inspections and provision of additional equipment for water flow gauging in structures and in pilot irrigation canals. This has allowed proper water records to be kept for the pilot canals within WUAs, making the water distribution process more transparent. The project has started real-time management of the water delivery process in the pilot irrigation canals and within the pilot WUAs. This has been done by closely monitoring and updating the planned water supply schedule based on actual water user applications and taking into account the weather conditions during the growing season. This is to be considered a first step towards equitable and rightful water distribution and, at the same time, towards reducing unproductive water losses.

By preparing dedicated files (passports) for the demonstration plots within the pilot farms, an instrument was created that allowed an analysis of the farmers production reserves, and provided the means to assess the potential for improving the productivity of land and water resources. An instrument for forecasting water consumption in line with weather conditions is being tested in real time, and the intention is to introduce this device on a wide scale during a next phase of the project. Analysis shows that on nine out of ten pilot plots, land and water productivity notably improved. Productivity only actually decreased on one pilot plot located on the South-Fergana Canal, where the farmer did not follow project recommendations. Many women were also involved in discussions on the management of land and water productivity and other water resource management problems in the Fergana Valley. Based on the outcomes of these activities, an enabling environment has been created for the wider introduction of extension services to the farmers. The results of this process to introduce IWRM has shown that the involvement of water users (through community initiatives) considerably improves the

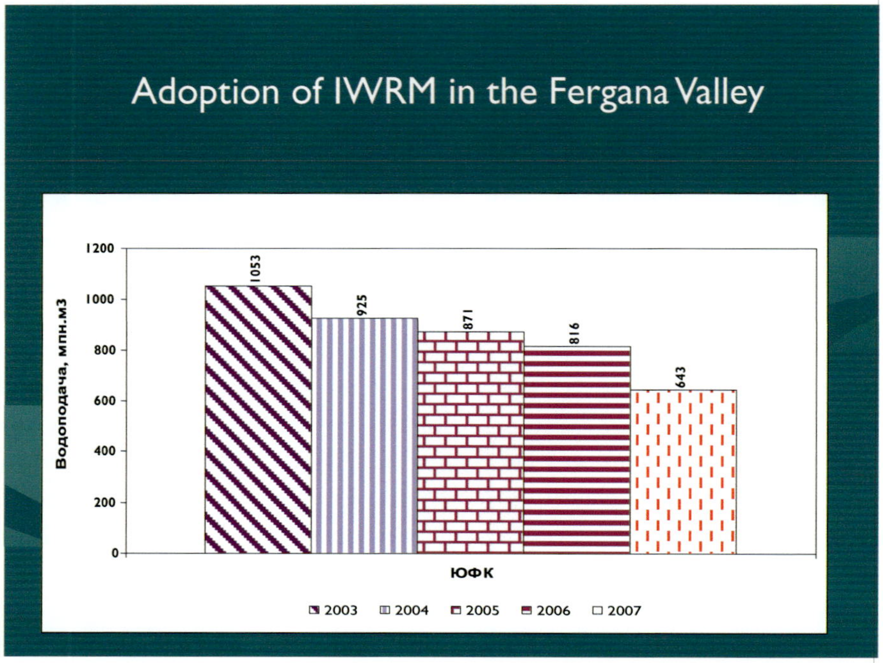

Figure 4.28 Reduced water diversion into the South Fergana Canal (SFC) due to introduction of IWRM (SIC-ICWC).

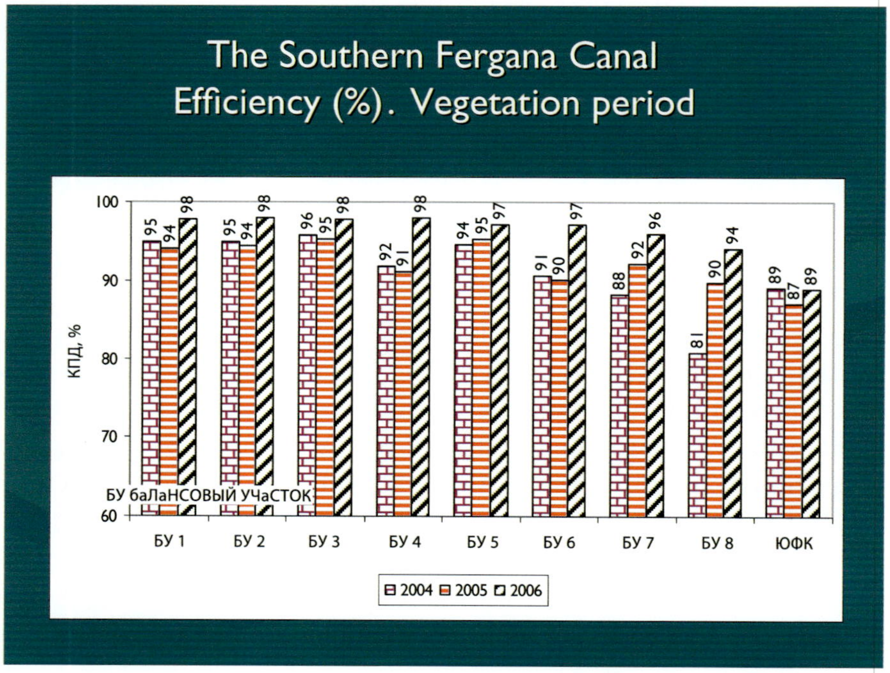

Figure 4.29 Dynamics of the irrigation efficiency at water-balance sites and in the SFC as a whole (SIC-ICWC).

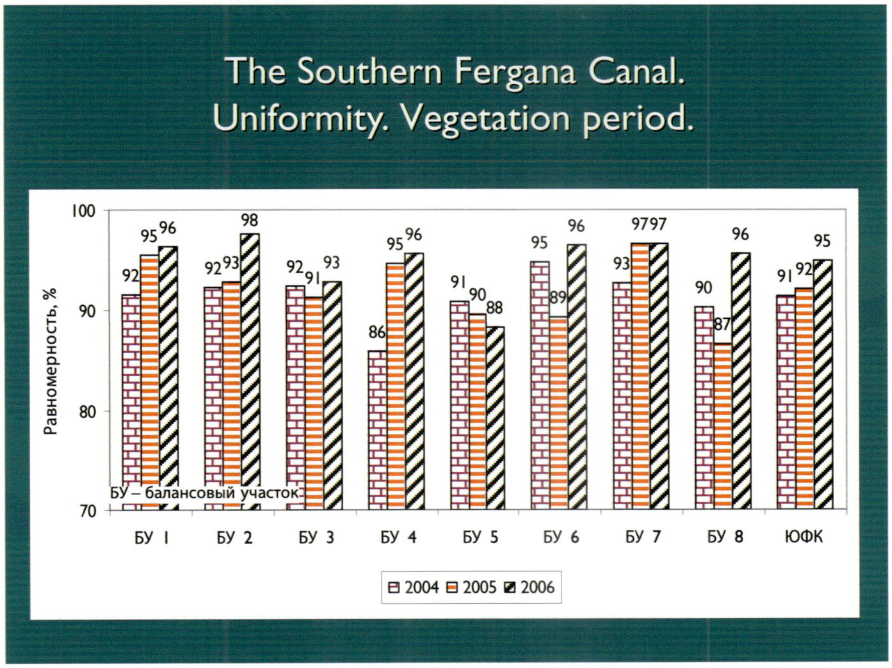

Figure 4.30 Uniformity of water supply at water-balance sites and the SFC as a whole (SIC-ICWC).

efficiency of water use and decreases water withdrawals from near the head intake by more than 25% at some places. The IWRM approach allows the creation of a system of fair and equitable water use, which, by its very nature, is close to the system of traditional water use in accordance with the ancient Sharia canons in Central Asia.

4.6.4 Development of a Monitoring System as a Technical Means for Improved Water Use

It is incorrect to assume that institutional improvements and the involvement of water users in the decision-making process alone can decrease water consumption. There needs to be a parallel application of the appropriate technical means. Real progress to a large extent depends on the proper combination of institutional, legal and technical measures as demonstrated by various projects implemented in Central Asia in the last decades. The detailed description of these technical tools is described in *Integrated Water Resources Management: Putting Good Theory into Real Practice – the Central Asian Experience* (Dukhovny, Sokolov et al. 2009). A summary of relevant conclusions is given here.

Close monitoring of water resources and water use practice is the only basis for an evaluation of the efficiency of an irrigation system. Construction (or rehabilitation) of a system of water gauging stations and water measuring facilities that cover the main irrigation canals and the distribution network is crucial in this respect. Equipping irrigation canals and water off-takes in the Aral Sea basin has been done using a hydro-geographical ("top-down") approach, and most of the project funding allocated for monitoring was spent for this purpose but some training was also organized for mirabs and specialists in hydraulic measurements. When establishing WUAs, much attention was paid to ensure that the off-takes into private farms were equipped with water measuring facilities. Where private farms were too small to be equipped separately, they were joined in groups to canals. For the purpose of improving the accuracy of water measurements, the SCADA system was introduced. This was successfully completed on the Aravan-Akbura Canal in 2008 and on the South Fergana Canal in 2009 (see Figure 4.31).

The installation of water measuring facilities was combined with the introduction of special regulations for water measuring and a system for transferring water resources from the upper water use levels towards the lower levels.

Another key element of establishing an advanced monitoring system is the introduction of a fully computerized management system and the building up of a dedicated database. Computerized systems with special software for water distribution planning (drafting water use plans, and making day-to-day plans for water distribution with adjustments depending on changes in weather conditions, hydrological conditions or water management) have been introduced for the main irrigation canals and distribution canals within connected WUAs.

Software developed for the irrigation canals and for use by the WUAs is a valuable tool to facilitate calculations and obtain updated management indicators. The canal water user councils (CWUC)

Figure 4.31 Water gauge near a weir in a secondary canal (SIC-ICWC).

Figure 4.32 All the participants in the water distribution and use system are connected through the internet (SIC-ICWC).

and the WUAs are directly involved in water distribution planning and monitoring. For this purpose, both the offices of the canal administrations and the sites where the equipment is actually used and also the offices of the WUAs are equipped with computers. However, due to the remote locations of some of these offices, there have been regular telecommunication problems between the water users and information offices and therefore methods to improve this situation had to be investigated.

4.6.5 Evaluating Water Demands and Adjusting the Irrigation Schedule Based on Zoning of the Crop Water Requirement

Estimating the water demands of private farms should be undertaken both by calculating the crop water requirement for different planning zones and by using the data collected at representative demonstration and prototype sites. One key component predetermining the required water allocation is clear information on actual crop water requirements, taking into consideration time-dependent hydro-geological conditions, and soil and climate conditions. Therefore, within the framework of the IWRM-Fergana project, an analysis has been made of the applicability of out-of-date standards and crop water requirements based on planning zones established more than 20 years ago for the whole region within the South Fergana Canal command area using available data in the Fergana Province.

Wide application of zoning related to crop water requirement has facilitated better water requirement figures and reduced the volumes of water supply over the command area as a whole. Water

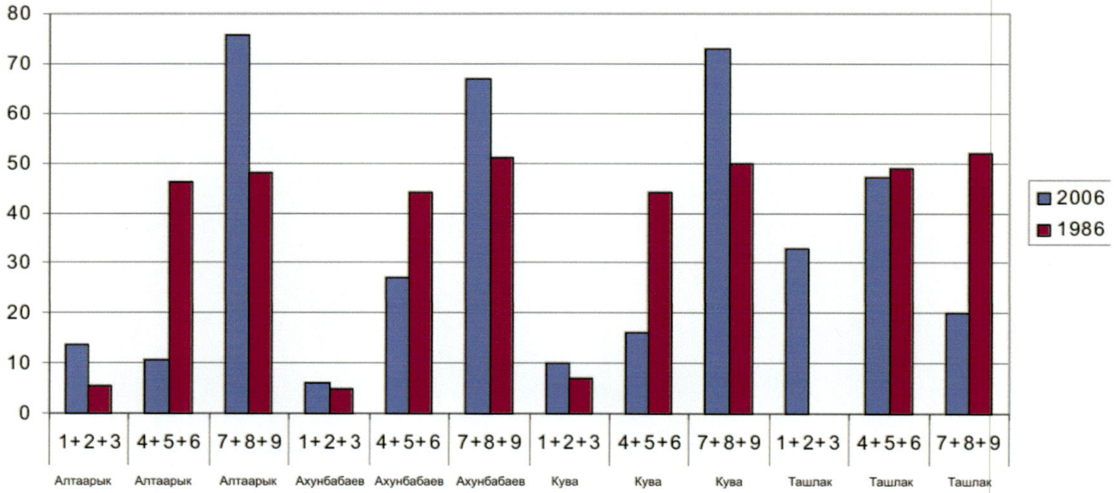

Figure 4.33 Changes in distribution of irrigated area per crop water requirement zone in Fergana Valley (SIC-ICWC).

distribution among primary water users is quite a complicated process that requires the co-ordination of crop water consumption design in combination with the performance of the primary canals (carrying capacity, available water resources and procedures for water rotation) and the preparedness of the farmers themselves. It is important to use a method of daily planning for large farms and a method known as *varabandi* for small farms (under 5 hectares), which is used in Kyrgyzstan and everywhere where grapes and other fruit are produced.

At present, the many small farms that have replaced former brigades in the collective and state farms create considerable difficulties for the organization of water distribution for new water users. If a water use plan is drafted for continuous water delivery, with a certain flow rate for each water user or for a number of small irrigated plots, then unproductive irrigation water losses and erratic durations of water applications will considerably increase due to the very small flow rates in the irrigation canals. It is also very difficult to specify a required flow rate for a 10-day irrigation period for a large number of small water users, if that water use plan is to be implemented by these former brigades.

However, independent of the size of their irrigated area, all water users want to receive their required irrigation water for each water application in a short time period (of one to five days). This causes problems as the existing irrigation networks were designed for a specific water demand specified by the cropping pattern at the time (rotation of cotton and alfalfa with irrigation intervals of 10 to 25 days). In these circumstances, it was proposed to use daily water distribution planning figures (within 10-day periods during the growing season) to ensure uniform and equitable water distribution among water users within the WUAs. This approach makes it possible to reduce irrigation water losses and enhance water use discipline.

It is difficult to describe the financial and economic mechanisms underlying integrated water resources management in brief, but in our view these should include:

- payment for water services by WUAs and WMOs, returning a specified share of these payments to the government depending on state-organized market demand and the related planned sown areas;
- payment for water services should not exceed 5% of the total net income of a farmer and should be differentiated depending on the actual level of services rendered;
- the system of payment should encourage water-saving practices at the level of WUAs and WMOs;
- agricultural water users have to pay a contribution for the maintenance of the main irrigation canals, which are state property;
- non agricultural water users have to cover all operating costs (made by the state and WUAs) related to water delivery;
- land reclamation and services related to maintenance of the drainage network should be paid for by the state;
- the state should encourage all initiatives by WMOs aimed at cost recovery in excess of the government financed part of their activities;
- because related sectors receive a considerable part of the income generated from water use in general, they should contribute to the operation and maintenance costs of the water infrastructure.

4.6.6 Concluding Remarks

The activities of government administrations and water management organizations in the period after independence deserve both attention and respect. Based on their former experience and collaboration in water resource management, the five Central Asian countries have managed to establish a rather unique form of co-operation and joint operation in support of sustainable operation of the water use system, water sharing and water supply to the various economic sectors and ecosystems. All this took place under very complicated and uncertain conditions during the development of these newly independent states.

The interstate co-operation in Central Asia is to be considered unique, especially when taking into consideration the number of riparian countries, the acuteness of the water use problems in these highly arid regions, and the progress necessary to achieve regional goals of food security, labor opportunities and economic prosperity. Central Asian countries have developed and put into practice a number of advanced concepts, tools and methodologies that support and promote a closer integration of the interests of both the nations as well as the individual water user organizations.

International institutions (the World Bank, Asian Development Bank, European Union) and some bilateral donors (Switzerland, The Netherlands, Canada and others) have made a valuable contribution to improved regional co-operation. Nevertheless, attempts to commercially develop hydropower gives rise to the concern that both national governments and water professionals will not be able to deal with the related water allocation dilemmas. These events have already sown their seeds of discord in the Central Asia water management arena. It is thought that only by the joint efforts of the riparian countries, the international donor community and the world community at large, represented by the United Nations, will the region be capable of overcoming these troubles.

References

1. Alster, J. 2001. The Aral Sea basin – the organizational structure for managing transboundary water resources, Collected articles: "Water Security in the World and Region." The SIC ICWC, Tashkent, pp. 34–40.
2. PA Consulting Group. 2002. Central Asia Natural Researches Management program, Transboundary water and Energy project, USAID.
3. Dukhovny, V.A. 2007a. "The ICWC – achievements and future challenges: water collaboration for sustainable development." Tashkent. SIC ICWC. (in Russian).
4. Dukhovny, V.A. 2007b. Globalization of water in Central Asia. Journal "Irrigation and Drainage" No. 56. Moscow (in English).
5. Dukhovny V.A. 2002 "Big challenges and unlimited opportunities – what are constrains for co-operation?" Proceedings UNESCO IHE, pp. 119–124
6. Dukhovny, V., Sokolov, V. et al. 2009. Integrated Water Resources Management: Putting Good Theory into Real Practice – The Central Asian Experience, Tashkent.
7. Weinthal, E. 2002. State Making and Environmental Cooperation, MIT Press, Cambridge, Massachusets Institute of Technology, p. 13.
8. Falkenmark, M. 1998. Forward to the future conceptual framework for water dependence – Volvo Environmental Prize lecture, Ambio 28.4; 360.
9. Stulina, G.V. (ed.). 2006. Gender Aspects of IWRM (in Russian).
10. Le Moigne, G. 1994. Co-operation between countries in the Aral Sea basin and their partners in the field of managing jointly-used water resources: problems and opportunities, www.worldbank.org/servlet/WDSP/IB/11.01
11. Petrov, G.N. 2009. Problems related to using water and energy resources of transboundary rivers in Central Asia and ways of their resolve. The Institute of Water Problems of Tajikistan, Dushanbe (in Russian).
12. Hodgson, S. 2004. Land and water – the rights interface, FAO, Rome, p. 9. (in Russian), http://www.fao.org/legal/pub-e.html
13. ICG. 2002. ICG Asia Report, No 34, Central Asia: Water and Conflicts, 30 May 2002, p. 1.
14. Sultanov, B. (ed.) 2008. "Kazakhstan in the modern world: reality and prospects." Almaaty, pp. 90–93 Krasnova, I. 2003. "Water problems in Central Asia and their settling." Newsletter of Ecological Development Programme (in Russian).
15. Lange, K. 2001. Energy and Economic Security: the Syrdarya Crisis in Central Asia. Presentation at 2nd International Conference of the IWHA, Bergen, Norway, August 10–12, (in Russian).
16. Mkrtychan, L., Halnepezova, A., and Ataev, J. 2000. "Agricultural policy reform and food self-sufficiency in Turkmenistan", Food policy reform in CA, IFPRI, pp. 164–175. (in Russian).

17. Mamatkanov, D., Asanbekov, A. 2000. The economic mechanism for water resources management and basic regulations of the inter-state water-sharing strategy. The CAEG International Institute, Bishkek (in Russian).
18. SIC ICWC. 2001. Means of water conservation – the findings of works in the frame of the sub-project WUFMAS and Component A2 of the GEF Project "WEAMP". The SIC ICWC, Tashkent, 2001, p. 148.
19. Nazarbaev, N.A. 2000. "We build a new state", Moscow,
20. Bilen, O. 2009. "Turning and water issues in Middle East".
21. Wouters, P., and Vinogradov S. 2002, International Law of Water Resources: Evolution, IWLRI, Dundee, www.cawater-infor.net/bk/water_law/pdf/wouters-vinogradov_presentation_1-pdf (in Russian).
22. Protocol of the 19th ICWC Session of May 15, 1998. Collected articles: "The Regional Inter-State Coordination Water Commission for Central Asia." Tashkent, 1992/2007, p. 108. (in Russian), 2007.
23. UN-IFAS. 2008. "Review of Donors' Assistance in the Aral Sea Region", Tashkent, p. 69. (in Russian).
24. Khamidov, M. 2008. "Reservoir operation is going to response to short-term demand response operation", pp. 2–18, SIC ICWC, Information proceedings, p. 213. (in Russian).
25. SIC ICWC under edition of Dukhovny V.A. 2008. Rivertwin – the IWRM model for twin river basins, Tashkent.
26. Risbekov, Yu. 2009. Collaboration on transboundary rivers: problems, experience, lessons, and forecasts of experts. Editor-in-Chief V. Dukhovny, the SIC ICWC, Almaty, p. 202. (in Russian).
27. Sorokin, A.G., 2002, Problems of Amudarya Basin Management. Web: www/cawater-info.net (in Russian).
28. Siegfried, T. and Bernauer, T 2008, Measuring International Policy Performance: Collected articles: "Water Resources Management in Central Asia", Tashkent.
29. Sorokin, A., Averina, L. 2002, Evaluating the Consortium's activity under different scenario of managing the water-fuel-energy resources. Presentation at the Training Center's seminar "Simulating the regimes of irrigation and power generation" (in Russian).
30. ECOSOC. 2002. Substantive Issues Arising in the Implementation of the International Covenant on Economic, Social and Cultural Rights. Advance Unedited Version, UN Economic and Social Council, the Committee on Economic, Social and Cultural Rights. General Comment No. 15 (2002). Geneva, 26 November 2002. p. 2.
31. JICA. 2009. "Study on intra-regional cooperation on water and power for efficient resources management in Central Asia", February 2009, JICA, Tokyo, Japan.
32. The World Bank. 2000. The Aral Sea Basin Program Review, the Report 17590-Kaz.
33. The World Bank. 2003a. The World Bank Report, Tajikistan.
34. "The Land Code of Turkmenistan", Article 27, 1 November 2004, Proceedings "The Water and Land Legislation in Turkmenistan", 2004. www.cawater-info.net/RETA/library/rus/land/turkmenistan.pdf (in Russian).
35. UN SPECA. 2001. Diagnostic report for the regional strategy on rational and effective use of water resources in Central Asia, prepared for Rational and Efficient Use of Energy and Water Resources in Central Asia project. The UN Special Programme for the Economies of Central Asia. www.cawater-info.net/documents/pdf/final_report_e.pdf
36. Sredazgiprovodkhlopok. 1984. The Revised General Scheme of Water Resources Use and Protection in the Amudarya River Basin (in Russian).
37. www.cawater-info.net/library/agreem.htm
38. World Bank. 2003b. "Water Resources in Europe and Central Asia", Vol. 1, WB, May 2003, p. 36.

39. Wegerich, K. 2008a. Models of WUA accountability under the local reality in South Kazakhstan. Proceedings "Water Resources Management in Central Asia", Tashkent, pp. 5–62
40. WUFMAS. 1998. Annual report, 1998 year, TACIS project, ENVKYR 9601, p. 128.
41. World Bank 1999. World Bank: Uzbekistan Review of policy social and structure report, Washigton. (in Russian).
42. Linklaen-Ariens, W. 2003. "Improving water governance in the Asia and Pacific region: drawing a map of challenges and approximates" Material of Conference "New thinking in water governance", Singapore, July 2003, p. 23.
43. "IWRM in Lower Reaches", 2005, REPORT FY2003 OESI WP, http://www.cawater-info.net/library/rus/iwrm_lower_ru.pdf (in Russian) under Dukhovny V.A., Khorst M.G.
44. Wegerich, K., 2008b, Passing over the conflict. The Chu Talas Basin agreement as a model for Central Asia? Collected articles: "Water Resources Management in Central Asia", Tashkent, p. 88–95.
45. http://sic.icwc-aral.uz/releases/rus/164.htm (in Russian) 2009.
46. Ben Slay: Water risks in Central Asia: who governs upper watersheds govern the whole basin http://www.centrasia.ru/newsA.php?st=1247636580 (in Russian).
47. Rakhmon, E. 2001, "Tajiks in the mirror of history", "Ifron", Dushanbe, p. 153.
48. Akaev, A. 1999. "By thinking about the future with optimism", Moscow, The International Academy, 1999, p. 240.
49. Nyazov, S. "Rukhname", 2001, Government publishers office of Turkmenistan, p. 86. (in Russian).
50. Dukhovny V.A., Sorokin A.G., 2007, Evaluation of the Rogun reservoirs' impacts on the Amudarya runoff regime, Tashkent, SIC ICWC, p. 98.
51. Rene Kagnat, 2002, Dark turnarounds of Central Asia, Paris, Payot Voyageurs.

Ensuring future water management in Central Asia through negotiations

5

Water and the Future for Central Asia

5

Water and the Future for Central Asia

Although the prospects for the region as a whole are still unclear and to some extent pessimistic (mainly for geo-political reasons), Central Asia is demonstrating positive examples of survival and development under conditions of water scarcity. For some time after independence forecasts on the water future were not noticed by the new governments, but gradually the prospects of the region became clearer mainly thanks to the persistent activities of the World Bank and for example the preparation of the Water Vision 2025 for the Aral Sea Basin produced by UNESCO for the World Water Forum in 2000. The main aim in the vision for the Aral Sea basin is to achieve food security under conditions of equally meeting ecological requirements and ensuring water security. Already at that time the vision took notice of potential conflicts between countries and sectors and made a start considering alternative scenarios and strategies for the future.

Scenario development and the development of integrated models for socio-economic development and water management on the basis of data bases developed in the last decades, has become a core activity of water science institutions in the region. Three base scenarios prevail, which are business as usual (present day situation), national (assuming that countries take national priorities as a guideline for their policies) and regional (assuming that countries take regional, transboundary, priorities as a guideline for their policies). The chapter describes the outcome of these scenario's for a number of key parameters including for example irrigation area development, water productivity and labor opportunities. Special attention is given to external scenario drivers that may impact the water and environment sector such as climate change, global finance and markets and the development of regional energy policies.

One specific area of interest is the influence of global policies with reference to the geopolitical future of Central Asia. After the Caspian Basin, Central Asia has become the scene of competition between global forces mainly because of its energy potential (hydrocarbon fuel and hydropower resources) and because of its location at the cross roads of east and west. The situation in the region itself can be characterized as a misbalance between three countries rich in hydrocarbon fuel resources (Kazakhstan, Turkmenistan and Uzbekistan) and two countries located in the upper watersheds (Kyrgyzstan and Tajikistan), which have a huge, but underutilized, hydropower potential. The regional energy debate is most likely going to dominate the debate for regional cooperation and this chapter is providing outlines for how to deal with this issue. A final factor will be the future developments in Afghanistan, which country is equally considered a riparian of the Aral Sea Basin and will play a role in the future water scenarios of the region.

As a main lesson learned the development and strengthening of user-focused, balanced regional water management is key to a future sustainable use of the resources. Therefore decision structures should be simplified, by for example establishing a single water governance body which would govern the executive water management agencies. Afghanistan should be included (even if only as an observer) from the start. The mission and mandate of this council should be to issue solicited and unsolicited advice on key water and environment management issues and it should be facilitated with all the means necessary to research the background and consequences of this advice. The operating costs of the council should be the joint responsibility of all six countries of the Aral Sea basin and it should preferably operate under external, international chairmanship. This is not to say that this proposal will solve the problems of basin wide water management in Central Asia, but it pays tribute to the principle that only by accepting neutral advice and guidance the countries and institutions of the region will progress towards a solution.

5.1 Geopolitical Perspectives

The vast territory of Central Asia, with its richness in natural conditions, mineral resources and human resources, should become a prosperous region again in the future. This prosperity should be based on an integrated governance system, which is politically and economically sound and stable and which allows people to live in harmony with their habitat and makes life attractive for the inhabitants of neighboring countries as well as for the indigenous population. In recent times, only a few local people have left Central Asia for reasons such as new jobs, a better life, fear of extremism or pure speculation. The main reason why people don't leave is a shared nostalgia for this remarkable and diverse land that is so strong felt by its inhabitants. This feeling of nostalgia for the good times is also often shared by the foreigners who have worked here and who have become inspired by these lands and its people. Central Asia attracts the attention of its neighbors as well as many global players because of its geopolitical and economic potential. This potential is due to many different factors. First, this region is the location of the interface between Europe and Asia. If Turkey can be considered a narrow and short bridge between these two continents, Central Asia is a thousand kilometers long border area where both Europeans and Asians have crossed in the past and will keep on meeting in the future thanks to the humanitarian and trade relations that have been in existence for centuries. According to some, Central Asia's location is actually the reason that the future of the region will be mainly shaped by military priorities (Kazakh Institute of Strategic Research 2007). We should remember that the great invasions (Alexander the Great, Genghis Khan and Tamerlane) are firmly in the past, and in a world dominated by nuclear weapons and ballistic missiles, even an event such as the battle of Stalingrad would be irrelevant. The world leaders, who govern mega forces, must now be controlled and responsible to maintain the equilibrium, if mankind in general and Central Asia in particular is to survive.

The strategic location of the region is very important owing to:

- concentrations of very large deposits of natural gas, oil and uranium, which, in the short term, contribute to guarantee the future energy supply of the world;
- a very extensive overland road network, which has linked East and West since ancient times. (This is now supplemented with air corridors, gas and oil pipelines and a regional power grid. In the past, the East and the West exchanged goods along the routes of the Silk Road. Later, this was trade replaced by the North South exchange of timber and industrial products for cotton, vegetables, fruit and mineral resources. Today, oil, gas, minerals and hydroelectricity dominate the trade from Central Asia with many different countries of the world.);
- important human resources, with a hard-working background and with cultural traditions dating back centuries, now combined with a high level of education and a strong aspiration for development;
- industrial capacity combined with an ample supply of power from existing plants and a well-developed electricity supply system.

5.2 Visions, Forecasts and Realities

The prospects for the development of water resources were systematically evaluated during the Soviet era by drawing up schemes of integrated water resource use in the Amudarya and Syrdarya river basins. As explained in Chapter 3, these schemes unfortunately were not adequately followed up. Rather than focus on several key development indicators, principle consideration was actually only given to one indicator—the area of new lands brought under irrigation. Only in the 1980s, after strong criticism in the wider Soviet society, was attention also given to the volumes of water abstractions. This, in turn, generated aspirations for water conservation that eventually even resulted in some decrease in total water withdrawals in the Aral Sea basin as compared with the earlier plans. However, the main cause for the differences between the official schemes with the realities on the ground were the actual rates with which the irrigation systems were improved and the lack of adequate water resource management practice. The challenge of achieving the planned levels of land productivity and the problems of reusing on-farm drainage waters were also not considered by the planning and water management organizations. These practices were disclosed and analyzed in some scientific publications of that time (Poslavsky 1983). Many alternative plans were developed, and these can be found in the documents of the Borovoy Commission and in the feasibility study for the transfer of water from the Siberian rivers into Central Asia, on which their forecasts of future water resources use and availability were based. Eventually intense agricultural development, based on advanced technologies, on irrigated farming and on the skill of traditional local farmers, made the region self-sufficient from the point of view of food supplies. In a World Bank report on the future development of the region, Central Asia was mentioned as one of the few regions in the world, along with Russia, that would supposedly remain a supplier of foodstuffs to neighboring countries. However, successful and sustainable development of water resources and greater water use efficiency is impossible to achieve without the close co-operation of all riparian countries. Under conditions of shared transboundary surface and groundwater resources, co-operation between users is the key condition for water resource preservation and for strengthening the strategic potential of the region. In Central Asia, where water is the real guarantee of life, development and prosperity, damage to the hydrological cycle can cause a chain reaction in many, if not in all, spheres of life. Therefore the regional economy should be considered from this point of view first.

Although the prospects for Asia as a whole (as appraised by many scenario developers) are still quite hazy and rather pessimistic (Koh et al. 2009), Central Asia is demonstrating positive examples of survival and development under conditions of water scarcity and it may even become a "beacon" for others. For an initial period after independence, these long-term forecasts did not draw the attention from the new governments in the riparian countries, but gradually the prospects of the region became clearer mainly thanks to the persistent activities of the World Bank from 1994 to 1997.

A first and rather significant step was the preparation of Water Vision 2025 for the Aral Sea Basin—a large-scale research work produced by UNESCO under the leadership and initiative of the well-known scientist Professor J. Bogardi (UNESCO 2000). UNESCO established five national

working groups, which together with representatives from SIC ICWC and the UNESCO Division of Water Science co-operated within the Scientific Advisory Board for the Aral Sea Basin (SABAS). Using data and proposals from the national groups, the board produced an analysis of the current situation and possible scenarios for water resource development in the basin. The basic ideology that was represented in this vision is different from that in the future development scenarios which we use at present.

A vision is to be considered a practical picture of the future we seek to create. It is an image of a future that can be achieved (in all likelihood) and is worth achieving. Just as a thought stirs action, the vision generates our prospective world. The vision can contribute as well as respond to trends. It can help to create new trends that we want to happen and to prevent those that we do not want. A vision provides a sense of mission and a guideline for future strategies and actions.

Planning from a vision requires having a different mindset to that adopted when planning from the reality of today towards tomorrow. Planning from a vision requires working backwards and learning from history before developing new strategies. A vision concentrates on "where we want to be" as a starting point rather than on "where we are". Therefore it helps to identify the global changes needed to make this future possible. Defining a vision also raises the question about which changes in attitude and which approaches to the future are required to accomplish what is expressed in the vision. These changes, in turn, need to be transferred into specific goals, which become the basis for future strategies, plans and actions. This is the "how to get there". Schematically, as shown in *Vision 21: A shared vision for hygiene, sanitation and water supply and a framework for future action* (WSSCC 2000), the process is:

vision > changes > goals > strategies > plans > actions

The main aim in the vision for the Aral Sea basin is to achieve food security (a minimum of 3000 calories per person per day by 2025) under conditions of meeting the ecological requirements and ensuring water security (preventing floods and droughts). At the same time, there should be an increase in the well-being of the people of the region. Even at the time of drafting this vision, the authors had foreseen the possibility of conflicts between the competing hydropower and irrigation sectors. The document states: "*As long as the food requirements can be met in the region, the use of water in winter for hydropower generation is fully acceptable. When this is not the case, the food requirements must have priority and the energy needs in winter have to be met by using other sources*" (UNESCO 2000, page 24). Scenario analysis software was used to test the reality of all possible options for the Aral Sea basin as discussed in Chapter 4. The Globesight model developed by a group led by Professor Mesarovic from Case Western Reserve University, Cleveland, Ohio, USA, was used for this purpose. This model shows the relationship between different economic and social indicators. It is relatively simple to use, which made it possible to develop and apply a local version of the model in a relatively short time. Three main scenarios were tested when preparing the vision: business as usual; priority to agriculture and rural development; and emphasis on industry and services, with a modest increase in agricultural productivity.

5.2.1 Business As Usual (BAU)

This scenario assumes that the economy, infrastructure development, budget allocations and societal behavior follow the trends established in the previous period. Water conservation is not a priority, and growth in land productivity is spontaneous rather than regulated. Wheat production is assumed to remain at the former level, but rice production will drastically decrease due to water scarcity and the higher water requirements of the population, which will grow to 50 million people by the year 2025. The status of water supply and sanitation infrastructure will remain at the same level. Supply of water for environmental needs will decrease and the Aral Sea level will continue to drop.

5.2.2 Priority to Agriculture and Rural Development

This scenario was adopted from an Australian model prototype. This model stipulates a scenario in which most of the population continues to stay in the rural areas, where development is assumed to be intense and financed from investment made possible by the income generated by the oil and gas sector. The region will develop new and diversified processing plants for agricultural outputs and a very sophisticated infrastructure to service the agriculture and water sectors. "*The priority of agriculture will be such that bright young people will be attracted to professional careers in the rural areas and stay there*" (UNESCO 2000, page 83). It is recommended that the region builds greenhouses, grows vegetables, practices horticulture and cultivates medical herbs and flowers. According to the authors of this scenario, combining the skill of local farmers with advanced technologies and innovations (as in Israel) will produce the highest levels of land and water productivity. This approach will also reduce, to a considerable degree, cotton exports, and export instead fresh and processed vegetables, fruits and other agricultural products.

5.2.3 Priority to Industrial Development

This scenario puts the emphasis on industry and services, and envisages a modest increase in agricultural productivity. Although the authors consider this scenario as most economically profitable and rational, it leads to a deficit in agricultural production and insufficient employment opportunities for the rural population as well as a dependence on imports for food security.

Table 5.1 compares all three scenarios. However, if we look at a combination of development indicators, a combined option of industrial and services development and agriculture development should actually be ranked above the three scenarios. Consequently, Vision 2025 proposes a combined scenario, which is not described in detail, but Table 5.2 presents an outline of the performance of this combined scenario against the development objectives.

This work became the basis for further in-depth study of the prospects for regional development (including the water sector), by developing a combined and integrated model for socioeconomic development and water management in the Aral Sea basin, the Aral Sea Basin management model (ASBmm). The model and interface were developed by a team from SIC ICWC (V.A. Dukhovny, A.G.

Table 5.1 Indicators of different scenarios in the Water Vision for the Aral Sea basin 2025.

Indicator	**Scenario**		
	Business as usual	**Priority for agricultural and rural development**	**Priority for industrial development**
Population, million people	50	70	70
Wheat productivity, ton/ha	2.0	4.3	3.4
Cotton productivity, ton/ha	2.5	3.3	3.1
Water use for wheat, m^3/ha	5,000	4,000	4,500
Water use for cotton, m^3/ha	10,000	6,000	8,000
Net water consumption:			
for food, km^3	55	34	42
for cotton, km^3	14	9	12
Total net water supply for irrigated agriculture, km^3	69	42	54
Irrigation system efficiency	0.60	0.65	0,62
Total gross water supply for irrigated agriculture, km^3	115	65	87
Total to industry and municipal water supply, km^3	3.47	4.85	6,0
Aral Sea and adjacent areas, km^3	3	23	6
Food supplies for the population, calorie/person	>3000	>3000	≈3000

Sorokin, M.T. Ruziev, V. Prihodko with the participation of D. Sorokin and A.I. Tuchin) in collaboration with P. Kouwenhoven, J. de Schutter and M. de Groen from the Dutch research and consulting company, Resource Analysis. Further information about this model is presented in (Ruziev and Prihodko 2007).

The model consists of hydrological and socioeconomic modules that cover the entire Amudarya and Syrdarya basins combined and is subdivided into sub-basins and planning zones. The model runs

Table 5.2 Development indicators of the combined scenario in the Water Vision for the Aral Sea basin 2025.

Possible goals in the water-related long-term vision for the Aral Sea basin	Targeted thresholds for 2025
Health	
Mortality under 5 years (number of children per 1000)	<30
Life expectancy at birth in years	>70
Nutrition	
Average availability of food in calories per inhabitant per day	>3000
Environment	
Water available for the environment in cubic km per year	>20
Wealth	
Increase of income per person. Purchasing power in urban areas as a factor	>2.5
Increase of income per person. Purchasing power in rural areas as a factor	>3.5
Agriculture	
Average water use in cubic meters per ton of wheat	<1000
Average water use in cubic meters per ton of rice	<3400
Average water use in cubic meters per ton of cotton	<1900
% of irrigated areas salinized (middle and highly salinized)	<10
Drinking water supply	
Coverage of piped water supply in urban areas, in % of people	>99
Coverage of piped water supply in rural areas, in % of people	>60
People served by good quality water by biological standards, urban, in %	>80
People served by good quality water by biological standards, rural, in %	>60

on the detailed SIC-ICWC databases and regional databases, and it allows data exchange between the calculated results from the hydrological module and the socioeconomic modules.

The interface of the model allows users to develop various scenarios (including analyses of climate change impacts) and save various combinations of scenarios and development strategies (cases) which can be compared and prioritized on the basis of preferred trends for development indicators (socioeconomic, water supply, irrigated agriculture, industry, environment), as shown in Figure 5.1. Five possible base

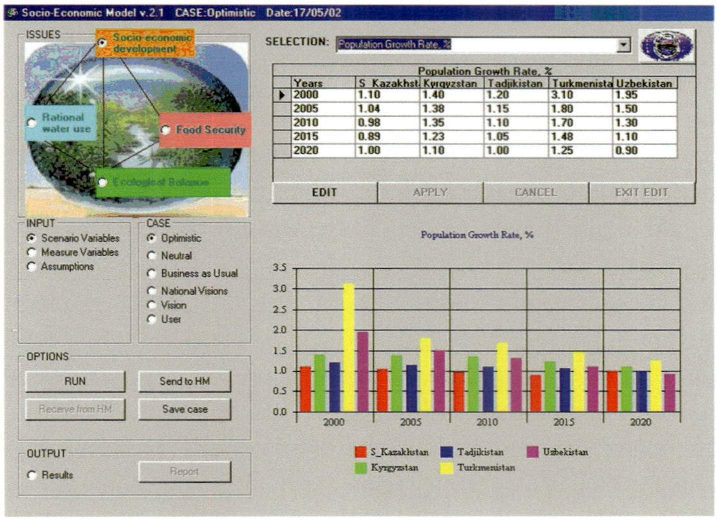

Figure 5.1 User Interface of the ASBmm (SIC-ICWC, Resource Analysis).

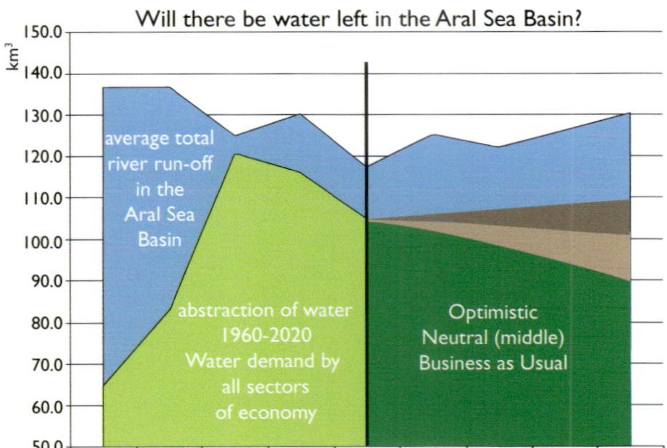

	1906	1970	1980	1990	2000	2005	2010	2015	2020
Total available water resources	136.6	136.7	125.0	130.3	117.0	124.8	122.3	125.5	130.2
Business as Usual scenario	64.7	83.5	120.7	116.1	104.6	106.2	106.7	107.4	108.4
Neutral scenario	64.7	83.5	120.7	116.1	104.6	104.6	104.2	103.0	101.8
Optimistic scenario	64.7	83.7	120.7	116.1	104.6	103.2	101.4	96.1	91.1

Figure 5.2 Water resource availability and abstraction scenarios in the Aral Sea basin as used in the ASBmm.

scenarios for the Aral Sea basin have been simulated to assess the indicators detailed in the Tables 5.1 and 5.2, and to calculate the impacts for the different countries in the region.

The base scenarios (optimistic, neutral, business as usual) tested and used in the first version of the ASBmm are briefly discussed below.

5.2.3.1 Optimistic Scenario

- The region will be developed under conditions of a strong transboundary integration process, as opposed to the current situation where each national government plans and prioritizes its own development scenarios. The developments under this scenario include:
  - mutually beneficial sharing of all transboundary water resources, based on the principles of water conservation and common environmental approaches;
  - mutually beneficial development of the agriculture sector with a focus on combined regional production, particularly with respect to the most favorable cropping patterns depending on soil conditions and water availability;
  - co-ordinated processing of agricultural products and co-ordinated development of infrastructure.
- Power supply will be developed mainly on the basis of existing hydroelectric power plants so as to provide sustainable and environmentally friendly power supply.
- Water-saving policies are adopted at a national level in order to achieve the highest possible water use efficiency.
- Significant increases in GNP will occur due to rapid industrial growth. Under the assumed conditions of industrial development, including recycling and saving of industrial water, industrial water use should reach 3.3 billion m^3/year as against 1.9 billion m^3 in 2000.

5.2.3.2 Neutral Scenario

- Integration processes in transboundary water management progress at a much slower rate compared to the optimistic scenario. In addition, the neutral scenario does not assume regional specialization for crop production and co-ordinated processing in the agriculture sector.
- Population growth rates will only slightly decrease.
- Development of newly irrigated lands is restricted not only by water availability and soil quality but also by a lack of financial resources for investment. This scenario assumes low rates of economic development and restricted funds (and results) for implementing water-saving measures in all economic sectors.

5.2.3.3 Business As Usual (BAU) Scenario

- The region evolves in line with current trends in transboundary water use, and in the field of regional agriculture integration, both in production and processing. Key efforts of riparian

states will be mainly directed at saving local water resources rather than towards transboundary co-ordination.

- Population growth rates will remain in line with present trends.

Table 5.3 presents the results for key parameters of national (priority) and regional (priority) strategies, visions and long-term development plans as used in the model calculations. Table 5.4 provides a comparison of the key development parameters assumed in these scenarios with those from a final (combined) scenario proposed by UNESCO.

The tables show the differences between the various scenarios and allow analysts to recognize the differences in indicators according to the different approaches and calculations used. Total water consumption ranges from 91.1 km^3 to 109 km^3, depending on the degree of water use efficiency. National (water use) priority scenarios show absolutely unrealistic results because the estimated water resource requirements exceed available water resources by more than 20 km^3 per year.

By 2010 almost ten years have passed since the model was developed—half of the period being considered to 2020—it is interesting to compare actual data, for example for the year 2008, with calculated indicators. These results are shown in the final columns in Table 5.3 and Table 5.4. It is interesting to note that the population in the Aral Sea basin has almost reached 50 million, outrunning the assumptions in scenarios I and II of the ASBmm and the assumptions made in the UNESCO vision but the numbers are still below the assumptions of the national vision scenarios.

The population assumptions show increases ranging between 12 million and 25 million people, which is 30% and 60% growth over the 20-year period. The actual population growth over the first ten years was about 6 million people, and therefore it is likely that the average rates of population increase will be similar to the numbers of scenario III and will be higher than the assumptions in the UNESCO vision and in scenarios I and II. Turkmenistan has the highest population growth rates. In the period to 2008, GNP has almost doubled, and this is approximately in line with the target numbers of the national vision mainly due to the indicators of Kazakhstan and Turkmenistan, whereas the indicators of the three other countries are closer to the neutral scenario. All riparian countries apart from Kyrgyzstan will achieve the GNP per capita target numbers of the optimistic scenario or even of the national scenario by 2025.

In 2007, which was an average water availability year, the total water consumption was rather higher than the target number in the neutral scenario although water consumption for irrigation almost approached the target number in the optimistic Scenario. In 2008, a year with lower water availability, the total water consumption was much less than all assumed values. However, one has to take into account that this occurred not only as a result of more thrifty water use (as envisaged in all scenarios) but also because of the obstacles placed on irrigation water releases by the owners of the reservoirs in the countries of the upper watersheds. In addition, there has been some decrease in the total area under irrigation in the Aral Sea basin (with up to 230,000 hectares being lost, 150,000 hectares in Uzbekistan, 60,000 hectares in Kazakhstan, and 20,000 hectares in Kyrgyzstan). As a result, the target numbers for the area of irrigated land in all scenarios will be hard to achieve. This is especially the case for the

Table 5.3 Key development parameters of the Aral Sea basin states (SIC-ICWC).

Scenario	Indicator	Unit	S. Kazakhstan	Kyrgyzstan	Tajikistan	Turkmenistan	Uzbekistan	ASB	Actual 2008
2000	Population	Million people	2.61	2.30	6.12	5.32	24.79	41.02	47.19
I			2.98	3.03	7.61	7.51	32.13	53.26	
II			3.01	3.06	7.82	7.68	32.98	54.56	
III			3.05	3.10	7.77	9.80	36.48	60.19	
National Vision			4.81	3.50	8.90	11.68	37.48	66.37	
UNESCO Vision			2.98	3.03	7.98	7.07	32.52	53.57	
2000	GNP per capita	$/capita	1306	637	307	617	729	681	1239
I			3012	1397	908	1883	1610	1615	
II			2391	1212	599	1059	1155	1133	
III			1895	1048	406	469	766	743	
National Vision			3595	1344	715	5490	1738	2394	
UNESCO Vision			3341	1549	891	2003	1765	1741	

2000	GNP	Billion $	3.2	1.5	1.9	3.3	18.1	27.9	58.5
I			9.0	4.2	6.9	14.1	51.7	86.0	
II			7.2	3.7	4.7	8.1	38.1	61.8	
III			5.8	3.2	3.2	4.6	27.9	44.7	
National Vision			11.3	4.2	6.2	64.1	65.2	151.0	
UNESCO Vision			10.0	4.7	7.1	14.1	57.4	93.3	
2000	Contribution of agriculture to GNP	Billion $	0.9	0.4	0.4	0.5	5.5	7.7	13.54
I			2.9	1.4	1.7	3.3	12.4	21.6	
II			2.2	1.1	1.0	1.6	10.4	16.3	
III			1.7	0.9	0.6	0.7	8.5	12.4	
National Vision			3.6	2.1	1.9	9.9	15.6	33.1	
UNESCO Vision			3.2	1.6	1.7	3.3	13.8	23.5	

Table 5.4 Key indicators of various regional 2020 scenarios compared (SIC-ICWC).

Indicator	Scenarios						
	Optimistic	Neutral	BAU	National	UNESCO	Actual 2007	Actual 2008
Irrigated area, '000 ha	8,504.00	8,451.00	8,008.00	11,185.00	8,507.00	7,828.58	7,809.80
Irrigated area per capita, ha	0.16	0.15	0.13	0.18	0.16	0.169	0.165
Total water consumption of all sectors, including	91.10	101.80	109.10	145.30	93.0	105.18	88.29
• *irrigation*	80.10	90.90	96.80	133.0	87.0	82.62	66.31
• *industry*	3.29	2.55	3.05	7.63	2.50	2.80	2.70
• *water supply*	4.39	4.92	5,88	5.24	3.4	3.99	4.88
Average daily calorie intake, '000	3.59	2.77	1.83	5.05	3.90		
Export/import of calories, million/year	11,418.00	–4,519.00	–25,787.00	48,000.00	19,020.00		
Population growth, %	0.98	1.23	1.90	2.30	1.00	2.90	1.97
GNP growth, %	308.2	221.5	160.2	8.00			
Water consumption per hectare, '000 m^3/ha	9.40	11.0	12.00	12.90	11.10	10.55	8.49
Water consumption per capita per year, in m^3	1,710.00	1,910.00	18,18.00	2,416.00	1,740.00	2,272.6	1,870.7

national priority vision scenario, which exceeds the other scenarios by some 2.5 million hectares, and will actually require a further 3 million hectares of land to come under irrigation by 2020 if its projection is to be achieved.

The conclusion to be drawn from the results produced by the model and scenarios should be that in spite of the fact that all numbers are estimates, within a "scale of uncertainty", the results are rather informative, although a detailed analysis of all the indicators has not been done.

One more outlook for the region, although one that was only prepared for Uzbekistan by the Regional Office of the International Water Management Institute (IWMI) should also be mentioned here (Yakubov and Manthrilake 2009). Two scenarios (BAU and "best case") were analyzed using the PODIUM model and a trend forecast under the BAU scenario was produced for 2025 for a number of indicators, including water use, GNP, etc., in relation to various projections for population growth. The model shows that under the "best case" option, there would be a considerable rise in the level of food security. Assuming that the BAU scenario does not differ much from the scenarios considered above, the "best case" scenario shows more progress in national development, and is therefore worth considering in more detail.

Given its assumptions, this scenario can be considered moderately optimistic and quite achievable. The key assumptions underlying this scenario are based on improvements for most development factors, including an improved diet (3000 calories per capita per day) for a moderately increasing population and a minimization of water losses by all economic users as a result of enhanced efficiency. In fact, better results than those calculated under the "best case" scenario seem possible if higher water efficiencies and crop yields are realized. With only a 10% increase in overall irrigation efficiency and an average 30% increase in crop yields, the gross irrigated land area required for all crops in 2025 will not exceed 4.1 million hectares, or only 5% more than in the base year. At the same time the gross area under crops will only need an 8% increase, going from 4 million hectares in 2000 to 4.3 million hectares in 2025. Thus, the 4.4 million hectares under crops today in Uzbekistan, including 4.3 million hectares under irrigation, would be sufficient to meet the future national needs for crop production to guarantee self-sufficiency and exports. This is a very different result from the business as usual scenario discussed earlier in this section, which would require almost one million hectares extra under crops by 2025. One of key assumptions under this scenario is enhanced water use efficiency in all economic sectors, especially in the agricultural sector, and with Uzbekistan's water demand in 2025 for all major economic users (agriculture, public water supply and industry) estimated at 54.3 km^3. This figure represents the total diversions from all water resources (surface water, groundwater and return water). It is 8% less than in the base year 2000 (58.5 km^3). This water demand includes an estimated 48 km^3 requirement for irrigation, almost 4 km^3 of water for drinking and other domestic uses, and slightly over 4 km^3 for industry. Compared with the business as usual scenario, improved efficiency should save almost 20 km^3 of water in the agriculture sector, about 2 km^3 in domestic services, and should make available more than 11 km^3 to meet environmental needs.

When making a comparison of these scenarios, it seems clear that some indicators are actually achievable or have already been partly achieved. This should allow some optimism for the water management and socioeconomic future of the region. The following chapters provide more detail.

5.3 Globalization and the Future of Central Asia

Globalization, as a process characteristic of the second half of the twentieth century and the beginning of the twenty-first century, can be traced back already to the Bretton Woods Conference in 1944.

The framework of the UN agencies and the International Monetary Fund established after the Second World War has provided a legal basis for a "new world economic order" by peaceful means and opened the "gates" for this globalization process. The presence of two global spheres of influence (Eastern and Western blocks) somewhat hindered this process, but in spite of the existence of the Iron Curtain globalization was developing in both political blocks, be it at different rates. There were very powerful socioeconomic movements leading towards integration in the fields of energy supply and industrial development and, later on, in monetary, financial, institutional and finally political spheres, which resulted in the setting up of the European Union as an example of real unification of national resources and capabilities. The disintegration of the USSR, along with the subsequent expansion of the European Union, has strengthened this geopolitical development and it has also opened the borders of the formerly rather isolated countries of the Eastern block, including Central Asia, where global influences have started to increase.

This global process has influenced politics, economics, technology, environment, culture, ideology and even religion since the 1950s. Each particular aspect of globalization has produced an effect on the water sector in countries and regions throughout the world, even during periods when the sector was independently developing within individual countries. The various aspects of globalization that impact on the water sector play different roles at each stage of the development of a nation and region. The degree of penetration of globalization influences at regional and national levels varies, depending, for example, on the extent of the management response and the culture of governance. The dynamics of these processes have been quite intense in Central Asia, which despite being behind the Iron Curtain for a long time, could not withstand these global influences. The influence of "globalism" on the water sector in Central Asia has been analyzed in a paper by Dukhovny (2007).

> *Liberalization of foreign policy, where international migration, transfer of goods, services, labor resources and funds is becoming more free, has become the historical prerequisite for economic globalization in the shape of liquidation of antagonistic economic systems.... This tendency enhances the interdependence of nations to each other and ties them up into a closer economic community.*
>
> (Stitienko 2009)

The conditions created to integrate Central Asia into the international community and economy were described in detail in Chapter 4 of this book and further analyzed in our research work. The open attitude of this region has largely facilitated its further entry into the global political arena.

It is rather strange that globalization was mainly driven by actions of international financial institutions, which skilfully linked financial assistance to certain political conditions and recommendations, rather than by the diplomatic activities of the newly established embassies and missions. This political approach—one of "Greeks bearing gifts"—had several official objectives, including to prove the disastrous nature and total bankruptcy of the socialist system, and sought to impose the institutions' own vision on future regional development under the promise of democracy and progress. However, one hidden aim was to transform the region from an appendage of the Soviet monopoly to one that produced raw materials into a market for their competing economies and as an important source of fuel

and energy resources. Central Asia has a quite powerful industrial, agricultural and human potential. In order to be able to develop these potentials anew, some of the existing infrastructure and institutions would first have to be destroyed, and quite favourable local conditions were actually created for that purpose. First, there was the breaking up of many economic ties with Russia, including the loss of federal subsidies, leaving the Central Asian national governments unable to generate the financial resources needed to develop a system of governance that supported the exploitation of the available potential. These conditions during the first stage of independence led to economic recession, a setback in agricultural production, the disruption of scientific potential, a huge brain drain, and the deterioration of the education system.

What main objectives should the nations of the region set under such conditions? One key requirement made by all international financial institutions was privatization. However, self-sufficiency is a prerequisite to economic stability—this is actually only a new version of the slogan: the rescue of a drowning man is in his own hands—and privatization actually led to industrial retardation at first, and then to the liquidation and substantial theft of economic stock.

The complete focus on full liberalization and privatisation of irrigated agriculture, coupled to some extent with a rejection of existing forms of co-operation, were also often rather fatal decisions. The regional irrigated agriculture system, which was designed for large-scale mechanized forms of production, was literally degraded, and water and land productivity was lost to a considerable extent. Under the situation of large-scale agriculture that existed at independence, managers and workers were well organized at all levels, with tasks based on the specific skills and background of all involved. It takes time and effort to transform a farm worker into a skilled independent farmer. Besides, the Japanese approach may be the most appropriate for Central Asia. It is oriented towards small-scale farming but incorporates cooperative and regional forms of ownership and responsibility. However, these alternatives have been disregarded and not tried out in the region.

Although the major world grain suppliers, such as the United States, Canada and China, and the major world cotton suppliers, such as the United States and China, use large-scale farming with a high level of mechanization, It is interesting that the recommendations for Central Asia were for privatization to produce small-scale farming. As a result, the mean plot size of arable land was reduced to one hectare in Kyrgyzstan, to 4 to 6 hectares in Kazakhstan and to 10 to 15 hectares in Uzbekistan. Highly efficient production of crops such as cotton, wheat and corn is practically impossible under these conditions. Thus, in due time, this process will be reversed, as can already be observed by a consolidation of plots. For example, in 2005, in the Southern Kazakhstan Province, the average area of the plots increased up to 20 hectares through a process of sub-tenancy, transfer of title to tenancy, etc.

All the riparian countries have come to understand that privatization and leasing of small plots is only profitable for cultivation of high-value crops, such as truck crops and orchard crops. Cash crops, such as grain, cotton and many others, should be cultivated on large-scale and in highly mechanized farms. Therefore all countries, and especially Uzbekistan and Kazakhstan, have reversed the trend for smaller farms, and have sought to optimize the size of plots under cultivation of commercial crops, with farm sizes up to 60–120 hectares and more. Another manifestation of globalization is the constant

fluctuation of world market prices for commercial crops, which is a very critical factor that affects production. These fluctuations are often so abrupt that producers have neither the time nor the flexibility for proper decision-making with regard to crop selection. The opinion expressed by Joel K. Bourke (2009) in this respect is quite typical when he is notes that the huge growth in the prices for foodstuffs in 2008 aroused and confused our planet. Between 2005 and 2008, prices for wheat and corn tripled, and the price of rice increased by 400%! Approximately 75 million people found themselves below the poverty line, and all this took place when there were record grain yields. However, due to the fact that consumption exceeded production in the last decade, there are only 61 days of global food reserves.

Although the prices for agricultural products can fluctuate from one region to another, any price movements will affect the efficiency of irrigated farming owing to the openness of the markets. The World Bank has forecast that taking into account the price jump for cotton over the first four months in 2008 (30% over the same period in 2007) and the subsequent gradual decrease in prices by 15%, it should be assumed that the cotton prices will drop to 1 US$/kg because of the expanded area under cotton in India together with the growth of cotton productivity in China. In 2008, the prices for wheat almost doubled. The World Bank supposes that by 2010 prices will stay at a level of 300 US$/ton and gradually reduce to around 180 US$/ton by 2020.

However, these are not the only aspects of globalization that present specific risks to the region. Similarities with other regions of political instability, such as Afghanistan, Pakistan and Middle East, plus the presence of pan-Islamic tendencies in neighboring countries, cause many other threats. Moreover, the production and trafficking of narcotics create global risks and challenges that attract the attention of the leading world forces including the European Union (Sultanov 2008). Although the interests of all global leaders should coincide in this region, it is strange to see that everyone has their own line of behavior, takes their own interests into consideration, and sets political targets that are surreptitiously combined with specific economic and trade objectives in Central Asia.

After the Caspian Basin, Central Asia has become the scene of competition between global forces because of its energy potential (hydrocarbon fuel and hydropower resources). Some analysts from Central Asian countries predict that foreign actors, primarily the United States, Russia and China, intend to project their own future interests as a "system formation" framework. It is a cause for grave concern that energy co-operation is developing in the region as a new round of geopolitical maneuvers are being initiated by these foreign actors—maneuvers which are focused on the strengthening their own positions in the region and bringing Central Asian nations into their own sphere of interests.

The situation in the region can be characterized as a misbalance between three countries rich in hydrocarbon fuel resources (Kazakhstan, Turkmenistan and Uzbekistan) and two countries located in the upper watersheds (Kyrgyzstan and Tajikistan), which have a huge, but underutilized, hydropower potential. As the saying may go "nature is not a credulous or easily fooled person, and he wisely distributed natural resources, granting "black fuel" deposits to the downstream countries and a huge potential of "white fuel" to the upstream countries". All global forces hold an interest in gas and oil resources, which are easily accessible and profitable in the short-term (investments may start showing returns after

only three years), but, from the long-term point of view, possession of hydropower plants is a better key to future prosperity. However, this approach is more costly, with a long and uncertain period of cost recovery in the present conditions. Nevertheless, those who participate in the present development of the fuel resource reservoirs in Central Asia are trying (sometimes taking "three step forwards and two steps backwards") to develop the infrastructure for long-term water and energy resource management.

In August 2006, Russia, China, the Republic of Korea, Malaysia and Uzbekistan established a consortium for developing the oil and gas deposits on the dried Aral Sea bed (Vasileva 2007). At present this territory, where once the waters of Adjibay Bay washed, is covered with a network of exploration and production wells drilled by the Uzbekneftigas, an Uzbek company, together with the Chinese State Energy Company, Petronas (Malaysia), Lukoil (Russia) and KNOS (South Korea).

The future will show what will remain (or what will be left) under the former waters of the Aral Sea. An interesting notion is that although all these countries have joined the consortium, there is a question of who is competing against who in the global arena, as was the case on many occasions in the past. It might be Russian and Chinese interests against those of the United States and Europe, but other combinations are possible. Evidence of the new geopolitics in the region include the construction of the Atasu-Alashankau oil pipeline, with a carrying capacity of 10 million tons a year (to increase to 20 million ton in the future), in a project implemented by Kazakhtranskoil and the China National Oil Development Corporation in 2005, and the purchase of US$4 billion worth of shares in Petro Kazakhstan by the China National Rehabilitation Corporation. We should note that there are also plans to transport Russian oil through the Atasu-Alashankau pipeline (KISI 2008).

Partly in response to these actions, and within the framework of the Big Central Asia program, the United States is proposing to construct new power systems with the purpose of redirecting energy flows, starting from Kyrgyzstan and Tajikistan and working towards Afghanistan, Pakistan and India. As a consequence, in 2006, the governments of Turkmenistan, Kyrgyzstan, Afghanistan and Pakistan signed a memorandum to develop a regional power market (KISI 2008).

There is information that the United States, through USAID, is initiating efforts to create a "water bridge" between Afghanistan and Tajikistan for the development of hydropower facilities on the Panj River. This in particular concerns the Dasht and Dzune Hydropower Plant (design capacity 3 million kW). Power supply projects initiated by the United States that reorient power supply lines towards Afghanistan should be also considered in this perspective.

However, an advantage of Russian and Chinese involvement in Central Asia is the presence of the Shanghai Cooperation Organization (SCO), in which, with Russia and China, all Central Asian countries participate with the exception of Turkmenistan. Although Kyrgyzstan and Tajikistan try to create a checks and balances on the work of the SCO, because these countries are focusing on hydropower development, the USA position in the region is weakened by this organization. When looking at the history of investments in the hydropower sector, in seems at first sight that Russia has had some clear successes, but the truth in fact is different:

- April 1994: the governments of Russia and Tajikistan sign an agreement on completion of the Rogun Hydropower Plant (agreement dissolved in 2006);
- in 2003: draft agreement prepared by the RAO UES of Russia on the transfer of surplus electrical power from Central Asia to Russia;
- October 2004: agreement on long-term co-operation between RUSAL (an open joint stock company) and the government of the Republic of Tajikistan concerning the Rogun Hydropower Plant construction phase I project;
- From 2004 to 2009: the RAO UES of Russia undertook construction of the Sanguta Hydropower Plant 2.

It has also been reported that Dimity Medvedev, President of Russia, made a promise to Kyrgyzstan concerning participation in construction of the Kambarata Hydropower Plant 1 (capital share almost US$2 billion). Russia has a two-faced policy here, combining promises to invest in construction of the Kambarata Hydropower Plant 1 (Kyrgyzstan) and the Rogun Hydropower Plant (Tajikistan), with assurances to Uzbekistan that constructing these hydropower plants will not be done without the independent examination and consent of the downstream countries. This has caused some dissatisfaction, with Tajikistan talking about walking out of international bodies (Mr. Rahmon has threatened that he will not participate in the sessions of the SCO and EAEC) and cryptic attacks by the mass media in Uzbekistan. The general situation is that Russia from the north and China from the south are trying to connect to the electrical power system of Central Asia—Russia via north Kazakhstan and China via Kyrgyzstan-, where it hopes to cover the power deficit in Xinjiang (Peyrouse 2007). The new Kazakhstan-Chinese project for the construction of a thermal power-station with a 40 billion kW per year production capacity fuelled by coal from the Ekibastuz coalfield can be considered in the same framework. In considering all these "energy intentions" it should be noted that the operations of present and future hydropower plants will play (and already are playing) a considerable role in the allocation of the water resources available for irrigated farming and industry. China made its first investments in the Tajikistan hydropower sector, after Sino Hydro signed the contract for the construction of the Yavan Hydropower Plant on the Zeravshan River in 2007, but so far it has only implemented design work because of objections made by Uzbekistan. The projects to construct hydropower plants on the Obikhingou and Surghab rivers (upstream from the Rogun Hydropower Plant) are at a feasibility study stage, but there are competitive interests of the RAO UES of Russia in this case.

At present, a complicated geopolitical fight that involves all riparian countries, foreign investors and international powers (Russia, United States, Iran, Japan, European Union, World Bank and Asian Development Bank) is developing around the existing and future hydropower structures. Kyrgyzstan and Tajikistan are attempting to find external investors, demonstrating their abilities to dictate their own conditions for planning and operating water release from the reservoirs without taking the interests of the downstream countries into consideration in their use of transboundary water resources. Without warning in 2007 and 2008, as explained in Chapter 4, and even during the 2009–2010 non-growing season, power engineering specialists implemented a regime of filling reservoirs without any co-ordination with other economic sectors. This regime was greatly disadvantageous for all the irrigation systems in the region. For example, during the whole of the 2009 growing season until the last ten-day period of July, Tajikistan maintained a Full Supply Level (FSL) in the Kayrakum Reservoir

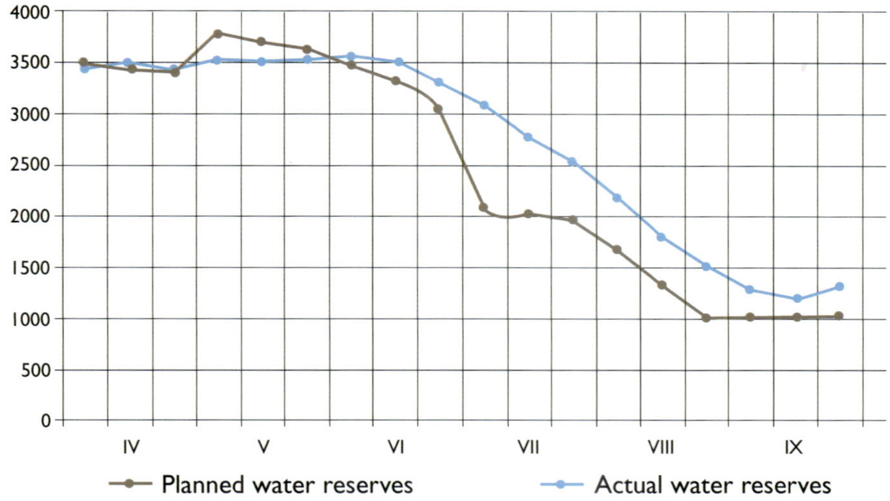

Figure 5.3 Water reserves in the Kayrakum Reservoir, 2009.

for the benefit of the hydropower sector, and only started to release water in August and September when it was already too late for irrigation purposes (see Figure 5.3). About 1,500 million m^3 of water was not supplied from the reservoirs according to the seasonal regulation for the peak growing season. In November of the same year, when water was needed for the pre sowing of irrigation areas planned for wheat, the outlet gates of the Kayrakum Reservoir again stayed closed, completely ignoring all agreements on agricultural and ecological flows. The agreed operational water releases are 225 m^3/sec, with an ecological (base) flow of 100 m^3/sec, but the actual flow rate downstream from the reservoir that amounted to only 70 m^3/sec. A similar picture is seen at the Toktogul Reservoir in Kyrgyzstan. Kazakhstan and Uzbekistan, in response, have threatened to withdraw from the treaty for a united power system in Central Asia. This action could result in serious disturbance to the energy supplies in the region and completely disturb the option for system operators to regulate power frequencies based on various supply requirement needs. Adding to these tensions, the external actors not only tacitly accepted this situation but also from time stirred it up by actions that actually increase the heat in the power and water debate between the riparian countries. USAID, for example, has initiated joint activities with Afghanistan and Tajikistan related to developing their hydropower potential in the Panj River without involving the downstream countries in the planning of these operations.

In an interesting analysis by the Euro Asian Development Bank (reported in *Kommersant-Vlast* February 23, 2009), it was concluded that the joint "political will" (of the political leaders) has to be the basis for sustainable water resources management. In spite of the fact that the President of Tajikistan, Mr. Rahmon, repeats many times that "never and under no circumstances, will Tajikistan plan hydropower facilities that cause losses to its neighbors"—(he made this pledge at the SCO summit on August 16, 2007) the ministers of Tajikistan and Kyrgyzstan insist on excluding a provision with regard to "obligatory co-ordination of all projects for hydropower schemes on transboundary rivers with all riparians" in the draft agreement on the Syrdarya River. As long as there are no agreements between Central

Asian countries that regulate the procedures and conditions for constructing infrastructure on the transboundary watercourses, and while not all countries are members of the UN Convention, these counterproductive actions will go on. Under such conditions, all attempts by the SCO, Euro-Asian Economic Community and IFAS to produce an acceptable water and energy policy for the region will continue to face growing pressure from the ever more powerful hydro-energy lobby.

So far attempts by the Euro-Asian Economic Community within the framework of the High Level Group and by the ICWC (RETA 6163 project) aimed at establishing a platform to improve this situation (by seeking compromises between irrigated farming and hydropower) have not been successful. However, recently these attempts have received a new impetus as a result of instructions by the heads of state to the executive committees of IFAS and of the ICWC. The instructions require the creation of a mutually acceptable mechanism for developing a policy of integrated water resources use and environmental protection while taking into account the interests of all riparian countries (Almaty, April 28, 2008summit). The new executive committee of IFAS has started appropriate activities, involving the international donor community in the process. A decision was made to elaborate the analytical tools needed to simulate different possible scenarios and options and to search for balanced solutions acceptable to all. At that, Kazakhstan and Uzbekistan insisted on an approach under which the UN 1992 and 1997 Conventions will be the basis for all decisions, but upstream countries still want to proceed from their requirements to meet their energy needs first.

Global forces, represented by the World Bank and the European Union (which have become the leading players offering assistance to IFAS) are also participants in seeking consensus. Thus, the region is entering a new period when professionals (managers, lawyers, economists, programmers and others) should develop feasible scenarios and options that will be analyzed by taking into consideration national, regional and even global interests. Meanwhile decision makers in national and international organizations will have to wait patiently for proposals to come out of the process. It is a situation that provides the stakeholders to this process with an opportunity to not only be witnesses but to also be participants in these investigations and research. However, this can be useful only if all political forces unite around the aim of seeking a mutually acceptable consensus.

The intention to eliminate or, at least, mitigate the tensions from potential conflicts is an issue of global value. The world does not have an interest in blowing "sparks of anger" in the region, where from time immemorial peaceful coexistence has been the basis for the joint life of the many nations and people residing within the limited space of their oases. Today, the competition for markets, natural resources and spheres of influence can and should be arranged in a civilized manner.

There is hope that, while balancing their interests, East and the West will confine themselves to competition in the financial and engineering spheres without actually entering into the regional political arena. In his speech at the summit of the heads of state on April 28, 2009, Islam Karimov, President of Uzbekistan, said: "*I think that we, as the heads of state, who are responsible for our people and history, must think, first of all, about how to find common points of contact. These discussions should not be raised up to the political level. Otherwise it is possible that the interests of our countries and our people will be put aside, and the strategic and geopolitical interests and targets of third party forces will come to the foreground. However, most*

important is the prevention of any escalation in the contradictions between countries and high-ranking officials as well as the search for compromise. There is no other path for us to follow."

5.4 Basic Approaches to the Development of Scenarios and Models for Future Water Resource Development

Formulating scenarios for the long term (more than 10 years) is always a difficult business with unpredictable results, since it is concerns development in a multifactor and multidimensional future. This development can only be analyzed on the basis of a model, if it is built up modular and allowing a synthesis combination of future scenarios and alternative strategies. Moreover, it has to take account of the unreliability and uncertainty of the background data required to specify trends for the many factors that play a role and that, for example, have different relationships at the level of sectors and within different geographical zones.

For the whole post-Soviet period in Central Asia, it is almost impossible to specify relationships and trends between planning indicators due to the ongoing transition from the stagnant economies of the 1980s and the 1990s. The indicators have to be based on the planning processes and socioeconomic results achieved during a time when natural resources were heavily exploited and when the countries where moving towards independence with their national statistics in total disorder. These very unstable socioeconomic dynamics were explained in Chapter 4 of this book, which set out the basic socioeconomic indicators of the post-Soviet period. Given this situation, rather than to base scenarios on exact data and past results, best estimates have to be used to direct decision makers towards the possible options caused by the dynamics of the many external and internal development factors.

A methodology for building up future development scenarios has been proposed in publications prepared in the framework of two projects relating to the whole Central Asian region (Dukhovny 2004) and to the sub basin of the Chirchik-Akhangaran-Keles river system (tributaries of the Syrdarya river) (Dukhovny 2008b). The basic approach of this methodology consists of the integration of separate thematic scenarios (by sector or a specific focus), each of which can exist within the scope of the required indicator but at the same time should be linked to defined national interests and interests per sector within certain boundary conditions. The composition of integrated scenario variables and modeling principles is presented in Figure 5.4. Climatic, socioeconomic, agricultural, water management and ecological sub scenarios are used to develop the specific parts of an integrated model. A key driving force of all sub scenarios is the governance system that creates the basis for more or less development depending on the priorities of the different sectors, target investments, and (most important) the political environment (quality of governance and relations between riparian countries). All these elements together provide the basic prospects for water availability and will determine the indicators for sustainable development, including poverty eradication, food security and the overall well-being of nations. The complexity of developing an integrated regional model scenario consists of the fact that the various

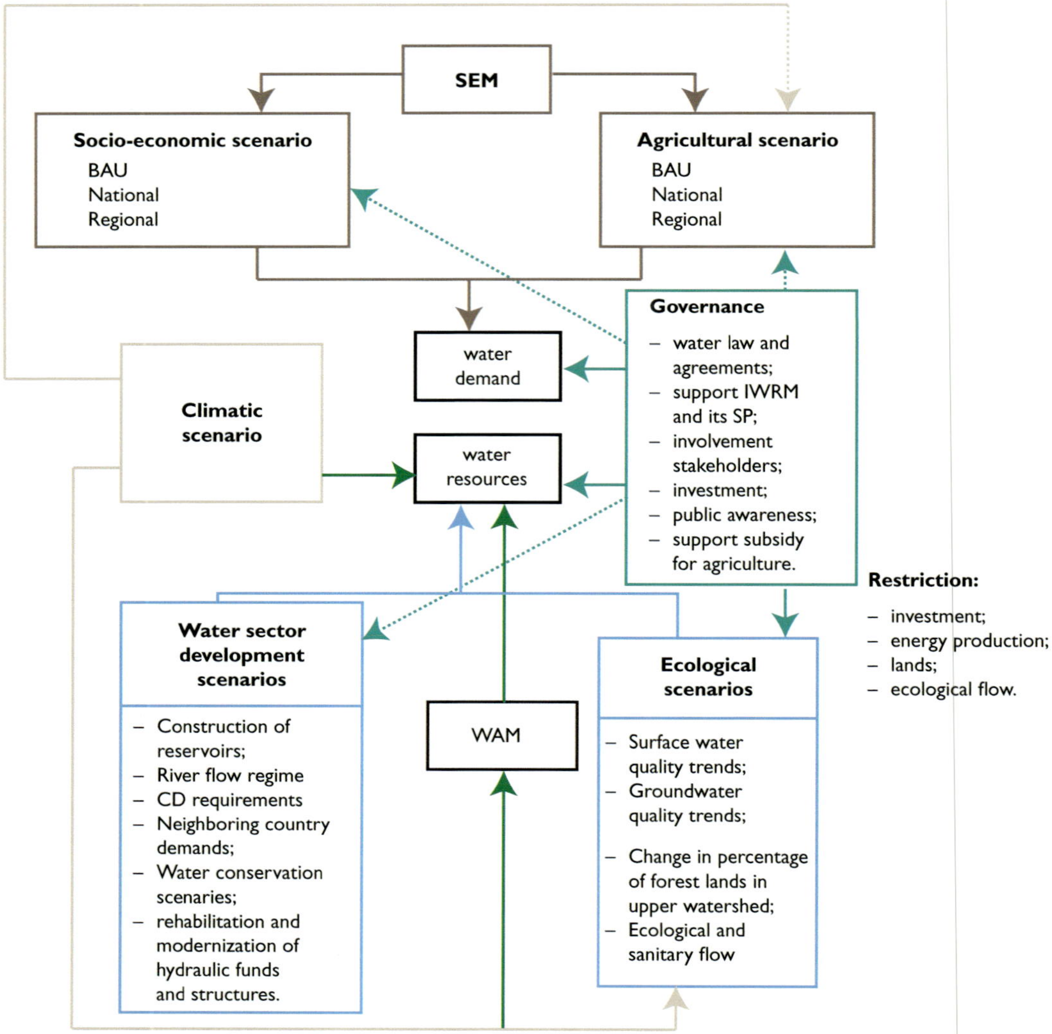

Figure 5.4 Links between specific scenarios in a river basin management model (SIC-ICWC ASBmm).

national scenarios (socioeconomic, agricultural, water management, ecological) as well as the policies of the governments can be quite different.

The system of water governance for this model is based on the experience gained in the Fergana IWRM project[1]. The countries and areas of the region which are partly located within Central Asia, such Chinese Xinjiang and Afghanistan, have different political systems with a different level of democratization, access to information and public participation in politics, as well as a different level of public security and political stability. More differences exist in the economic status of the riparian countries, which vary from very low income countries with a extremely high level of poverty and a poor education

[1]The joint IWMI-SIC ICWC project for IWRM and agricultural sector reform in the Fergana Valley.

Table 5.5 Comparative national development indicators for Central Asian countries, 2008.

Indicator	Country					
	South Kazakhstan	**Kyrgyzstan**	**Tajikistan**	**Turkmenistan**	**Uzbekistan**	**Afghanistan**
Population, millions	3.02	3.03	7.37	6.21	27.56	25.0
GDP, US$ billions	11.73	2.03	4.03	14.26	26.45	10.6
GDP per capita, US$	3884.1	670	546.8	2296.3	959.7	424
Industry, value added, % of GDP	42	20	27	41.8	33	26.3
Exports of goods and services (% of GDP)	57.6	56.5	21	70.4	42	17.3
Irrigated area per capita, hectares	0.20	0.12	0.10	0.29	0.15	0.13
Energy use (kg of oil equivalent per capita)	3412.4	603	596.9	3418	1796	14
Electricity consumption (kWh per capita)	5133.9	1842	2330.4	2021	1659	24.8
Participation in						
the 1992 Convention	+	–	–	–	+	–
the 1997 Convention	–	–	–	–	+	–
Establishing basin organizations and introducing IWRM	+	+	–	–	+	–

system such as Afghanistan, to rather prosperous and continuously advancing countries such as Kazakhstan. This difference creates "seeds of risk" that can germinate and impede peaceful national development, including development of the water sector. The unevenness in national policies also affects this potential risk situation to the some extent (see Table 5.5).

Table 5.5 assumes different storylines for the development of the water sector in Central Asia. Kazakhstan, Kyrgyzstan and Uzbekistan will aim to improve their water use practice, introducing

IWRM and supporting the water sector with the active participation of water users. Turkmenistan will adhere to the administrative (government-led) approach of water governance. The storyline for Tajikistan and Afghanistan with regard to irrigated farming is not yet clear, but like Kyrgyzstan, they will definitely try to prioritize their hydropower sector.

As was argued in Chapter 4 that preserving current trends will unavoidably lead to conflict and the loss of water resources available for useful application in irrigation (with reductions of some 15–20%) due to overall disorganization of regional water resource management practice. This storyline is typical for a business as usual scenario and for national scenarios, especially in dry years. For the long term, however, it is obviously necessary to aim at achieving reasonable consensus and progress under mutually profitable decisions based on international water laws. The unwillingness of countries in the region, apart from Kazakhstan and Uzbekistan, to sign the UN conventions might suggest a pessimistic future, with the possibility of countries "playing their own cards" probably by using one of the leading world donors with a special interest in the region. Political observers, however, have expressed doubts that the region is moving towards such an outcome. It is more generally thought that given the forces of globalization, national leaders that attempt to ignore UN conventions, where they often seem to overestimate their own significance and that of their sponsors, will eventually find themselves in partial or complete isolation (Sayfulin 2008).

Given the limited geographical space of Central Asia, moving along the "absolute sovereignty" road means to be doomed to permanent tensions with neighbors over energy, water, transport and many other issues. One may often be amazed, however, by the views expressed by those who have a complete lack of faith in a better future.

A review of some of these gloomy predictions was published in *Kommersant-Vlast* in February 2009. This did not concentrate on Central Asia, but considered a number of unfavorable scenarios affecting peace around the region. We can only hope that these scenarios will not happen, and that policies of containment will become predominant all over the world. Nevertheless, taking into consideration the existence of ill intentions or even terrorist actions, scenarios for water and especially agricultural development should take into account the need to secure the riparian countries against extreme situations that could lead to conflicts. Strengthening of regional co-operation can be a good basic tool for preventing extreme situations caused by natural disasters or by endogenous and exogenous factors. In this respect, Central Asian nations have a great advantage, which is the wisdom in ancient traditions of peaceful and co-operative water use that can still assist in providing a joint transition towards water-saving practices and developing the appropriate boundary conditions for water use at an interstate level.

Climate scenarios are independent of the economic activities of the riparian countries in the region, and can be considered exogenous variables. Without any mitigating measures undertaken by riparian states, the assumption is that the average temperatures will continue to rise, and according to all available information, glaciers will melt at rates of different intensity. There are numerous trend scenarios for climate indicators that result in different impacts on the region and which are non-across countries.

A review on how Central Asian countries could adapt to climate change was carried out by G. Stulina and presented at the FAO seminar held in Budapest in November 2009. [31]

Figures for temperature indicators over recent years are available from a Euro-Asian Bank publication (Ibatullin et al. 2009). Since the end of nineteenth century, regular observations of climate parameters have been conducted in Central Asia. Most meteorological stations in the mountain regions were established in the first half of twentieth century. The largest number of meteorological stations was operational in the 1980s, but since then the, numbers have been cut by a third with many closures in the mountain regions. According to observations, a considerable increase in surface air temperature is the basic consequence of climate change in Central Asia. For example, mean annual temperatures have gone up in the each country in the region, rising as:

0.29°C every 10 years in Uzbekistan (in the period1950–2005);
0.26°C every 10 years in Kazakhstan (1936–2005);
0.18°C every 10 years in Turkmenistan (1961–1995);
0.10°C every 10 years in Tajikistan (1940–2005);
0.08°C every 10 years in Kyrgyzstan (1883–2005).

Temperature increases are, however, unevenly distributed over Central Asia. Te greatest increases in mean annual temperatures have been observed over the plains, with somewhat lesser increases in the mountain regions.

Therefore, rising temperatures are more significant in Uzbekistan and Kazakhstan, and to a lesser degree Turkmenistan, and are less in Tajikistan and Kyrgyzstan. Almost certain, rising temperatures will cause water consumption for the irrigated agriculture sector to increase by 20–30% by 2030 depending on the geographical zone. A scenario for climate change was developed by the hydrometeorological services of the five Central Asian countries using the unified standard method. Anthropogenic climate change was analyzed for scenarios A2 and B2, applying version 2.4 (Turkmenistan) and version 4.1 (Kazakhstan, Kyrgyzstan, Tajikistan, and Uzbekistan) of the MAGICC/SCENGEN program (Model for Assessment of Greenhouse-Gas Induced Climate Change Scenario Generator). This program was developed using the IPCC's assigned methodology (including a vulnerability evaluation). To some extent it is necessary to consider both scenarios (A2 and B2) for greenhouse gas emissions because the effects (impacts) are equally probable.

An evaluation of water resources change in the region was based on the work of Chub et al. at Uzgidromet [32]. As a rule, the minimum temperatures during the non vegetation period have increased at a faster rate than maximum temperatures in summer. For example, the average rate of increase of the maximum temperatures in Uzbekistan is 0.22°C each 10 years (since 1951), but minimum temperatures have been increasing by 0.36°C each 10 years.

One of the most dangerous implications of climate change is the observed increase in the frequency of extreme meteorological events. There has been a considerable increase in air temperatures

along with a decrease in precipitation, leading to enhanced aridization (lower precipitation, lower humidity) over the desert and semi-desert plains of Central Asia. This change in humidity has been confirmed by data recorded at 60% of the meteorological stations in Kazakhstan. Analysis of damaging agro-meteorological events recorded by private farms in Kazakhstan has shown that from 2005 to 2007, atmospheric droughts (60% of the reported incidents) and soil droughts (20%) were the most frequent negative agro-meteorological events in this republic. There has been an almost negligible reduction in climate-based aridization in some mountainous and piedmont regions where temperature increases are less severe. When using the integrated scenario, two extreme value combinations have been selected as boundary conditions—a maximum temperature rise and a minimum runoff (option "climate-max") and a minimum temperature rise with a maximum runoff (option "climate-min"), between which different options for sub-scenarios can be grouped.

The next cluster in a regional model are the socioeconomic scenarios. Building this type of scenario is a most complicated task that requires very detailed base studies at the national level down to the level of individual planning zones and even farms and plots. The projection of socioeconomic development in practice depends on key indicators such as:

- demographic indicators (population growth, migration rates, education, employment, etc.);
- gross national revenue, actual and growth rate;
- gross domestic product, actual and growth rate;
- size of and changes in industrial output volumes;
- size of and changes in energy demand and supply;
- size of and changes in service sector volumes;
- size of and changes in agriculture sector outputs
- investments (national, international, loans, etc.);
- water availability for municipal and industrial needs;
- access to sanitary facilities;
- access to the gas supply network;
- meeting food requirements (calorie intake, etc.).

There are many other determining factors shaping the socioeconomic sub-scenario, such as status and trends of growth indicators for population, education level, modernization efforts, etc. National policies for different regional development priorities (stimulus programs for zones and regions) and sectors are also key elements of a development strategy. Finally, some mainly regionally influenced indicators have to be carefully evaluated, such as growth rates for national and regional income and the volume of national and international (direct and indirect) investment.

The development of a planning zone (large agriculture cluster), or of a country as a whole, depends of the relative progress of development against externally induced counter acting factors. It must take into account demographic trends and changes in the potential of the planning zone or of the country as a whole (Dukhovny 2008a). Zonal development is a function of the decrease of its available potential through the degradation (or losses) of four constituents (finance, production output, human resources and raw materials), and of an increase in its available potential through the internal resources of enterprises

(companies) or through additional investments made by the state and in (direct and indirect) foreign loans.

Foreign experts assess the future of the Central Asian economy with varying degrees of optimism. In particular, Dowling and Wignaraja (2005) speak about the rather muted prospects for economic growth. Global Insight (2006), on the contrary, gives more optimistic assessments. Table 5.6 was compiled assuming no risk factors. This means that world prices for energy resources are maintained at the current level. It should also be noted that from 2001 to 2005, the actual rates of increase of exports and imports in Kazakhstan and Uzbekistan were 1.5–2 times higher than in the best options used here, meaning that there is some "margin of safety" in the forecasts.

The GDP of a country or of individual economic zones is estimated on the basis of the outputs of the main sectors contributing to the economy. These sectors are industry, energy, agriculture, processing industry and services. The GDP of the industrial sector and of the services sector in urban areas is hardly influenced by water availability. Given that industrial water use amounts to only 7–10% of total supply, we can assume that meeting this demand is generally possible without any problems. Very large-scale industrial schemes are an exception to this rule, and they should be considered separately along with the impact of any effluent quality issues. The GDP component generated by the energy sector has to be examined separately as well, as energy production depends on the amount and the regime of water releases through the turbines of the hydropower stations. The GDP contribution from the irrigated farming and livestock sectors depend on the area under crops and on cropping patterns, which in turn depend on water availability. These sectors in turn create a certain volume of work for the processing industry and the services sector in rural areas, which can make a sometimes important contribution to GDP.

Table 5.6 Future overall growth assumptions for Central Asia under a no-risk assessment.

	Feasible GDP growth rates %			**Export growth rate %**	**Import growth rate %**
	2006–2010	**2011–2015**	**2006–2015**	**2006–2015**	**2006–2015**
Kazakhstan	8.7	7.0	7.9	14.0	11.0
Uzbekistan	4.5	4.2	4.3	8.2	10.0
Kyrgyzstan	3.8	3.4	3.6	2.2	2.5
Tajikistan	5.5	4.5	5.0	5.8	5.4
Turkmenistan	12.4	8.2	10.0	15.0	17.0

In the further explanation of scenario development and as a methodology for evaluating scenarios and cases (these are combination of scenarios and strategies for intervention), we will consider and evaluate three different sub-scenarios:

- the agricultural sub-scenario;
- the water management sub-scenario;
- the ecological (environmental flow based) sub-scenario.

5.4.1 Agricultural Sub-Scenario

Agricultural development is determined in part by development in the water sector. Agriculture trends are many and varied, and depend on many factors that, in general, are difficult to identify and difficult to assess with certainty. In our approach, it is assumed necessary to focus on these key factors:

- growth in area of land devoted to agricultural use, including all irrigated land;
- use of all available agricultural land and opportunities for reclamation (including irrigated land);
- changes in crop productivity;
- changes in land productivity;
- cropping patterns;
- livestock production;
- changes in (global and local) prices and profit margins on agricultural production;
- changes in production volumes, the profit margins in associated and processing sectors (indirect benefits), and opportunities for development;
- changes in production volumes of the fishery (and other aquatic products) sector.

Besides these primary factors, agricultural production also depends on natural conditions, current water supply, land conditions, and the provision of fertilizers, agricultural chemicals and equipment. There are also secondary factors such as the actual status of infrastructure, the possibilities for processing agricultural outputs, the institutional framework and governance structure, the availability of extension services and current skill levels of farmers, the status of mechanization, etc. In addition, these destabilizing factors could be considered:

- deterioration in soil fertility;
- changes in demand for agriculture outputs and product prices;
- deterioration of the irrigation and drainage infrastructure;
- role of imports and exports in delivering food security (including use of "virtual water" principles).

In order to be able to develop the agriculture block and to model the different scenarios, we used the methodology tested in the framework of the RiverTwin project (5.2) (Dukhovny, Sorokin, de Schutter and Maskey 2009). This methodology as an initial step, made use of a specific definition of production volumes of both irrigated and rain-fed agriculture. It calculates net present value (wages + profit + taxes + depreciation) under changing areas under crops, cropping patterns and productivity (depending on

soils, climate conditions and technologies) as a function of water availability. The livestock production volume was defined using analytical methods based on the study of national statistics for food availability. The agro-processing and service sectors were defined by assuming a correlation with the aggregate volumes of farming and livestock production and taking into account additional investment for development of agro-processing and agro-surveying infrastructure. This approach makes it possible to analyze the dynamics of the development of the agricultural sector in relation to water availability.

The resulting model shows that adopting appropriate government policies and investing in the sector are very important. Experience from Tajikistan shows that a government policy oriented to improving farmer self-sufficiency had a considerable impact on production due to exogenous factors and caused large deviations from assumed development indicators. The government guidance model, with state support to farmers (by subsidizing farmers and regulating the market) and development of agricultural infrastructure coupled with water sector improvements and with simultaneous introduction of IWRM principles, showed much better potential for economic growth.

The Fergana IWRM project (which was described in Chapter 4) shows the advantages of putting IWRM into practice. This is further confirmed by the fact that with limited investment (about US$7 million for six years), the project achieved savings of more than 20% in irrigation water use (250 to 300 million m^3). Any other approach to improve the efficiency of irrigation systems that would aim to save an equivalent amount of water would require a US$50–60 million investment. Moreover, if new water reservoirs for storing this volume of saved water were to be constructed in the region, more than US$150 million could be possibly saved. Therefore, wise government policies can reduce water demand and guarantee sufficient water availability for irrigation even in dry years.

5.4.2 Water Management Sub-Scenario

This scenario is mainly covered by inclusion of the following components:

- volumes of water resources available for irrigation and ecological flows;
- improvements in water use efficiency in irrigated farming and other economic sectors in order to save water resources for other needs and further development;
- increases in effectiveness of water distribution and use;
- additional investment directed to water saving and improved water use efficiency;
- potential for adaptation to climate change.

The specifications of the water management scenarios are similar for all riparian countries. The model has the flexibility to reflect different strategies. For example, Kyrgyzstan and Tajikistan have prioritized the hydropower sector for national development and mainly invest in this sector. At the same time, they neglect the problems in irrigated farming and the need to improve the efficiency of water use in this sector, confining themselves to pilot projects financed by international donors (World Bank, ADB, SDC and others) only. In principle, a strengthening of the hydropower sector could be supported and accepted as a leading development policy for these countries, but it should be pursued

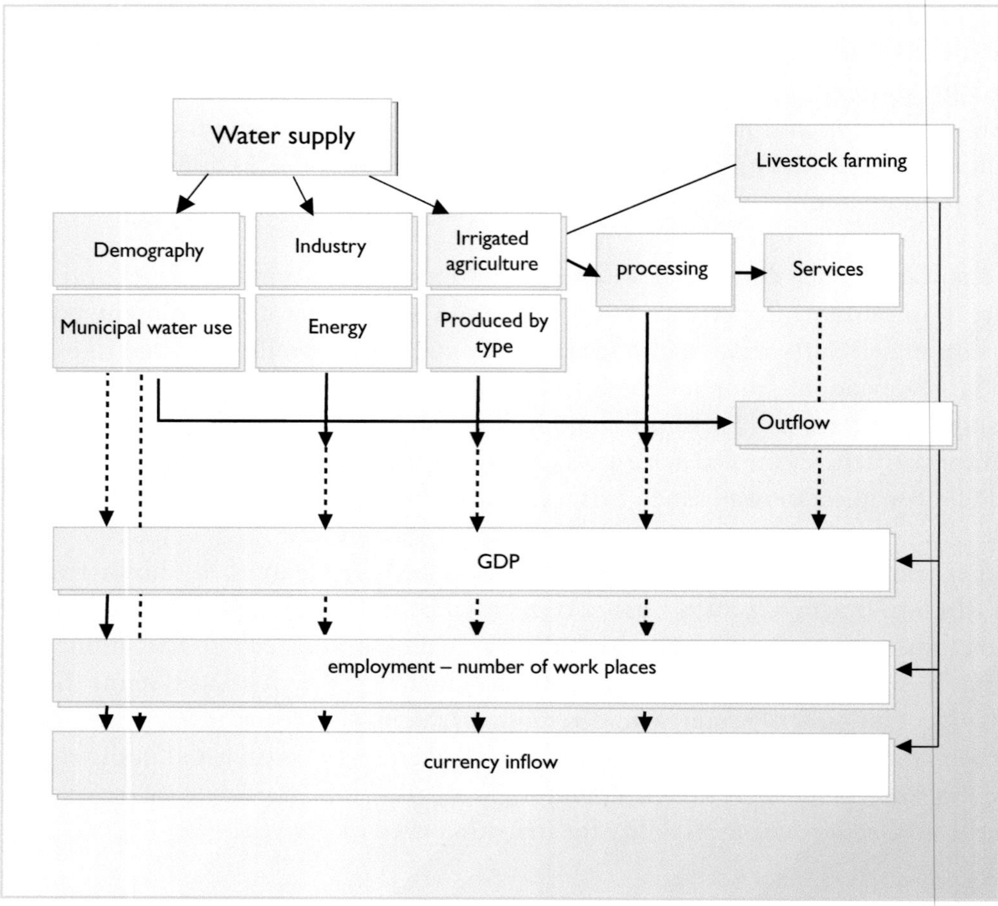

Figure 5.5 Scheme of inclusion of the socioeconomic blocks in the model and the relationship of these blocks to water resource management.

without the exclusive national focus that is underlying present policies. Again one-sided decisions about how to regulate the river flows (focus on winter releases for hydropower generation) that ignore environmental requirements and reduce available water resources for downstream water users by sometimes 10–15% cannot be accepted. Constructing new hydropower plants and reservoirs will intensify tensions between the riparian countries and will prevent required consensus for agreements in other fields of joint economic activity. A comprehensive forecast of the future situation related to the foreseen operational regime of the Rogun Hydropower Plant shows that operating a dedicated hydropower regime in the Vakhsh Cascade could cause some US$900 million worth of damage to Turkmenistan and Uzbekistan. Therefore, the socioeconomic feasibility of constructing large dams should be considered under fully integrated analysis conditions.

An alternative strategy that can be investigated under this water management sub-scenario is that of irrigation development. Although the total area of land suitable for irrigation amounts to 25 million hectares (with 8 million hectare already under irrigation), the land resources for new areas of low-cost

irrigation development are practically exhausted. The present cost of irrigation development in new areas that could meet modern water-saving standards ranges between US$6,000 per hectare and US$12,000 per hectare, making cost recovery opportunities for these projects very low. Only Afghanistan still has rather good possibilities for developing new irrigated land, but this should be reviewed separately.

One more strategy that was studied in our work (Dukhovny 2003) is the rehabilitation and remodeling of irrigation and, more importantly, drainage systems that are losing operational capacity due to a lack of operation and maintenance (O&M). This practice leads to loss of land productivity and additional water consumption for leaching operations. Rehabilitation of irrigation and drainage systems is needed in more than 3.5 million hectares of irrigated land in the region. Equipping all large waterworks on transboundary and secondary watercourses with the SCADA (gauging) system (see Figure 5.6) is a very efficient tool for preserving water resources in this respect. The introduction of these systems in the Syrdarya basin improved the accuracy of water distribution from ±10% to ±20%, and is evidence of the efficiency of this approach. A pre-feasibility study for introduction of a SCADA system on the Amudarya shows that about 4–5 km^3 of water can be saved in this basin each year. This would compensate for the possible volumes of water diverted from this river by Afghanistan in the future. The investment is estimated at US$30–35 million).

5.4.3 Environmental Sub-Scenario

The environmental sub-scenario mainly differs from the others by the assumed volumes of ecological (environmental) flows for improving the ecological conditions of the Aral Sea and adjacent areas, and the river deltas and the rivers themselves. These are to be considered as nature areas and included as registered water users. These aspects were discussed in part in Chapter 4, but the problems related to the Aral Sea will be described in this section.

The building up of integrated (combined) scenarios is done by using a single interface that allows scenario indicators to be selected for future water management development. A matrix of possible combinations of sub-scenarios (see Figure 5.7) shows that there are, at least, 36 combinations which can be considered and which are coupled to three environmental sub-scenarios (business as usual, preservation of selected water bodies, preservation of the deltas of the rivers). Under these conditions, there are more than 100 combinations of sub-scenarios. Future use of the interface, which is still in the process of design (2010), will demonstrate the feasibility of this approach. Because of the capacity of the model, pre-selection of marginal sub-scenarios will be needed (as in the RiverTwin project) as for example was done for the feasibility study for the Rogun project where 18 sub-scenarios were examined. A scheme of three integrated scenarios for socioeconomic and agricultural development, each of which are subdivided according to "climate-max" and "climate-min", and at least nine combinations of strategies are suggested in Figure 5.7.

Different combinations of water management strategies will be superimposed on these variants, allowing a practically infinite number of new hydro-schemes and reservoirs to be modeled, by grouping these schemes into one of three categories—"hydropower", "irrigation" and "combined". Each case

Figure 5.6 SCADA equipment installed at the head gate of the Large Andijan Canal (SIC-ICWC).

can be evaluated under the options "business as usual" and "optimum development", which aims to improve the irrigated agriculture sector by the introduction of IWRM, SCADA monitoring systems, etc. It is expected that further research will allow better and more concrete results to be produced, providing better decision indicators in the future.

One very specific case which requires attention is the future development of the water sector in Afghanistan. This could be a source of very complex trouble for the Amudarya basin or a driving force in the search for latent reserves in the basin. In spite of the fact that all official documents produced by the national governments, IFAS and ICWC have not been disclosed, the problem of future water consumption in North Afghanistan, in a region adjacent to the Amudarya basin, arouses the concern of the regional water professionals (Dukhovny and Sorokin 2002) and many foreign experts (Fuchinobe H., Tsukatani T. and Toderich K.N., Kyoto University, 2002). The great interest in this problem is reflected by the number of publications and reviews that show very different opinions related to this issue [see, for example, Ahmad and Wasiq 2004).

In 2002, the SIC ICWC prepared a synthesis of all documents concerning irrigation development in this region and reviewing the legal right of Afghanistan to divert water from the Amudarya. The previous agreements between Afghanistan and the Soviet Union did not really deal with the water-sharing issue (these agreements are dated February 28, 1921; June 24, 1931; July 13, 1946; January 18, 1958; October 16, 1961; and February 6, 1968). The 1965 *Scheme of Water Resource Use and Protection of the Amudarya Basin* is still in force. This states that the volume of water diversion from the Amudarya and Vakhsh rivers and their tributaries should be approximately 2.1 km^3 a year, which corresponds to the volume of water use in the 1960s. Along with water diversion from the Shirintagao, Sarykul, Balkh and Hulk rivers, this volume is sufficient to supply water to 1,079,100 hectares of irrigated land, of which

Water management and environmental sub-scenarios	Socioeconomic and agricultural sub-scenarios					
	Business as Usual		National		Regional	
	Climate-max	Climate-min	Climate-max	Climate-min	Climate-max	Climate-min
Operational regimes of HPPs:						
hydropower (a)						
hydropower (b)						
Operational regimes of HPPs:						
irrigation (a)						
irrigation (b)						
Operational regimes of HPPs:						
combined (a)						
combined (b)						

Figure 5.7 Matrix of sub-scenarios and different operational regimes.

only 617,900 hectares were actually irrigated during that period (data from the North Afghanistan Irrigation Development Plan by the Sredazgiprovodkhlopok Design Institute, 1968).

Using two different approaches, SIC ICWC evaluated a number of options with various growth figures for the area under irrigation and corresponding water diversions from the Amudarya.

The first approach follows the estimation of possible growth of irrigated lands given in the 1968 North Afghanistan Irrigation Development Plan. The level of irrigated land was assumed as the area of lands irrigated here in 1965. Taking into account that, as stated above, the irrigated area now and in 1965 is of a similar in size, we assume that the 2005 area of land under irrigation equals the 1965 level. The amount of newly irrigated land that has come into operation is 153,000 hectares over a period of 35 years (Figure 5.8). At a water consumption level per unit area according to the 1967 scheme, by 2040 the water withdrawals should have increased by 3.6 km^3 (the upper curve at the Figure 5.9). With the same growth of irrigated area but with a reduced water consumption per unit area down to 11,000 m^3/ha, an additional 1.5 km^3/year of water withdrawal will be required (the curve in the middle Figure 5.9).

The second approach is in analogy with the Hydroproject scheme of 1965 and shown as the second curve in figure 5.8. But taking into account the limited possibility that indeed a financial contribution will be made we expect the real development of irrigated land for the next 30 years to be no more than 30 thousand hectares which will require 3.1 km^3 in 2045 (the lower curve in the fig. 5.9).

Considerably higher regional tensions will be created by the development of hydropower on the Panj River, where there is great hydropower potential with an estimated total capacity of 17.720 MW and aggregate generation of hydropower in the order of 81.9 billion kWh. The Dasht and Dzune Reservoir with a storage capacity of 17.6 km^3 would be the largest hydroscheme in the region. According to informal sources, the construction of a hydropower station cascade on the Panj River was discussed at a meeting of the President of Tajikistan with the Minister of Energy and Water Resources of Afghanistan on August 3, 2007 (Verkhogorov 2008). A reference to the intention to construct this hydropower scheme can also be found in the annex to the Afghanistan Policy on Transboundary Water Resources of 2008. Preliminary estimates for the costs of this project amount to US$3 billion.

The design data for this hydroscheme are impressive. The scheme would consist of a rockfill dam of 320 m high and 1075 m long, two 1965 m long tunnels on the first level, a 1375 m long tunnel on the fourth level that would also function as a spillway tunnel, a 2630 m long emergency spillway, three

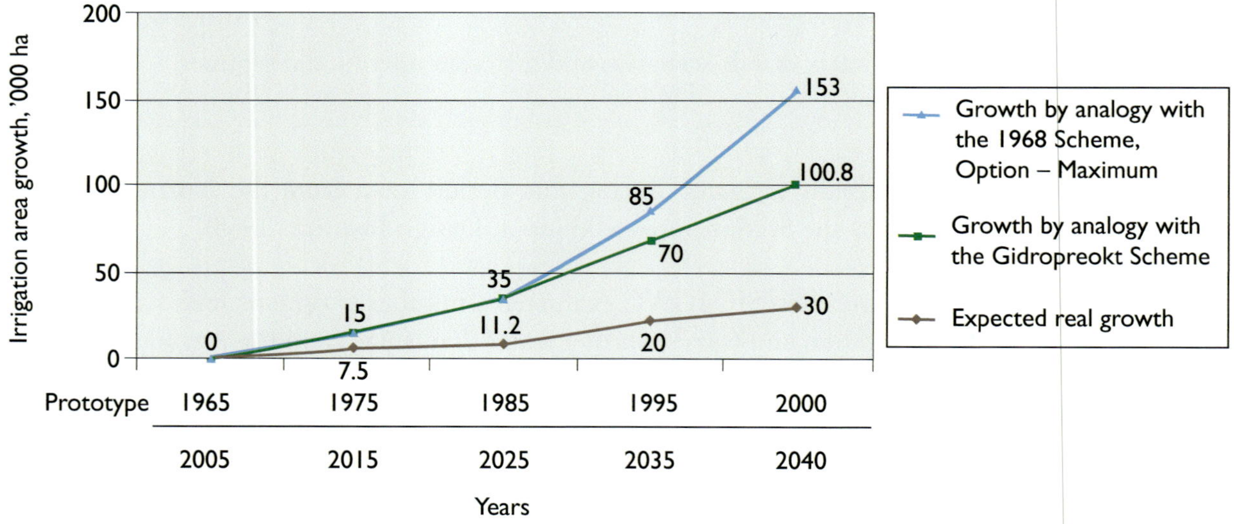

Figure 5.8 Expected growth rates for areas under irrigation in North Afghanistan (area under irrigation in '000 ha).

Note: Graphs show growth on the basis of the 1968 scheme (North Afghanistan Plan). Maximum growth according to the Hidroproekt scheme and estimated (actual) growth perspective over the years by SIC-ICWC. (Source: SIC ICWC)

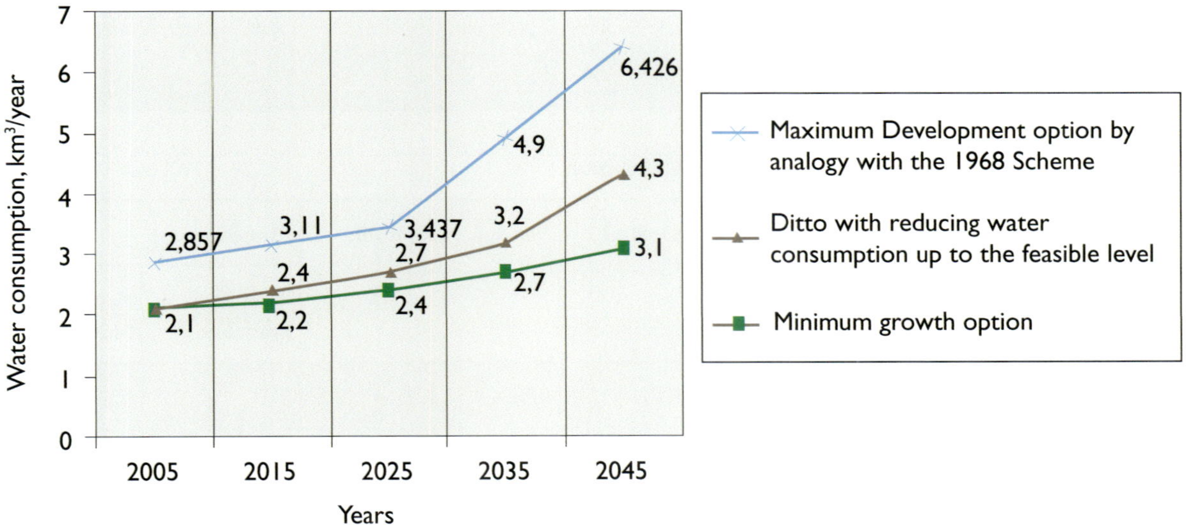

Figure 5.9 Expected growth in volumes of water diversion for irrigation from the Amudarya, km^3/year over the years).

Note: Maximum development option on the basis of the 1968 development plan. Maximum development level with reduced water consumption brought down to a feasible level. (Source: SIC-ICWC).

Table 5.7 Water use scenarios depending on future water management strategies in the Amudarya basin

Country	Water use limit, in million m^3	Actual water use, in million m^3 in 2008	Probability, %	Water deficit (in summer), in million m^3
Tajikistan	6135	5750	93.7	885
Turkmenistan	13950	10369	74.3	3581
Uzbekistan	15498	9990	64.5	5508
Total in the basin	**35583**	**26109**	**72.6**	9474

penstocks and a powerhouse. The construction work include excavation of 17 million m^3 rock soils, 98.3 million m^3 of loose rock fill, and up to 1.7 million m^3 of concrete work. Apart from the investment needed for construction, the project will require additional impacts as the reservoir will flood 14 settlements and 860 hectares of farmland in Tajikistan and 13 settlements and 1110 hectares of farmland in Afghanistan. However, the dam would solve a problem related to the possible breaching of the natural dam near Lake Sarez in Tajikistan and prevent a potentially disastrous flood there. At present, there is

Table 5.8 Trends for maximum and minimum air temperature (ΔT) *change* in Uzbekistan since 1938 [32].

Characteristics	Seasons	Regions of Uzbekistan					
		North	Central	South	Foothills	Mountains	
						Tien Shan branches	Pamir-Alai branches
T_{min}	Winter	0.97	0.77	0.51	0.85	0.03	0.02
	Spring	1.46	1.38	1.31	1.16	0.08	0.10
	Summer	2.45	2.25	2.60	1.95	0.31	0.18
	Autumn	1.86	2.23	2.17	1.72	1.50	1.42
T_{max}	Winter	0.13	0.08	0.14	0.13	0.27	0.51
	Spring	0.53	0.41	0.35	0.34	0.29	0.23
	Summer	0.38	0.42	0.22	0.31	0.07	0.00
	Autumn	0.72	0.54	0.72	0.60	0.97	1.03

only a simple horse path leading to the projected construction site (the road ends 12 km before reaching the dam site), and it is difficult to say when Tajikistan and Afghanistan will ever actually launch the construction of Dasht and Dzune Hydropower Plant.

The use of production of hydroschemes on the Panj River simply for the benefit of energy production could paralyze irrigated agriculture, both during the filling up of this large water body and afterwards in the course of its operations regime. The present flow regime of the Panj River is convenient for summer irrigation requirements, and the transition to a hydropower-based regime threatens the loss of about 10 km^3 of water for irrigation in the middle and lower reaches of this river during the summer period. This would create a permanent regime of water shortage similar to the droughts experienced in 2001 and 2008 (Table 5.7).

It is important that inter-government measures at the highest political level help prevent these developments in order to avoid major rifts and severe competition between upstream and downstream countries. Integrated scenarios should suggest a negotiable way out to tackle these challenges.

5.5 The Future is Rosy ... and Not So Rosy

The principles and model described allow the framework of a regional future to be outlined under "business as usual", "national" and "regional" (as a recommended option) perspectives and under different development priorities (socioeconomic, agricultural and ecological). If we start with an assessment of the water available for all users, these base factors will affect future availability of water resources:

- climate change;
- water use scenario (hydropower based operation, governance system, etc.);
- environmental aspects.

We can assess the extreme scenarios (minimum/maximum) according to the corresponding average values for the main indicators and possible deviations from these mean values (risk assessment).

5.5.1 Climate Change Impacts

Uzbekistan occupies the largest part of the Aral Sea basin area. It was used as the reference area to develop a climate change forecast which could be used to consider the changes to be expected towards 2035. The assessment of these climate change impacts for Uzbekistan was based on data presented in the Second National Report on Climate Change [32], which comprises daily observations since 1951 and monthly and seasonal data over longer periods. Analyzing the changes in air temperatures averaged over the various regions shows a rather significant warming-up over the whole territory. Minimum air temperatures are rising faster than maximum temperatures everywhere in Uzbekistan. Since 1951, average rates of warming (ΔT/10 years) amounted to 0.22°C for the maximum air temperatures and to 0.36°C for the minimum temperatures. The region around the Aral Sea where high rates of increase of maximum temperatures were found (and minimum air temperatures are almost stable due to the shrinking sea water area) is the only exception to this phenomenon.

Climate extremes (e.g. wet and dry periods) will intensify and estimated indexes for maximum values (T_{max} and T_{min} above 90% occurrence) show increasing trends over the whole territory. The number of days each year of heavy rains are also on the increase as well. The number of days with more than 10 mm precipitation has increased over the plains and piedmont areas. Relatively small increases in the number of days with more than 20 mm of rain were observed in the mountainous areas (up about 9). Special attention should be given to this fact because of the recurrence figures for intense precipitation events in these regions.

Indicators, such as the number of days with minimum relative humidity (less than 20%) in the hot seasons and an average humidity deficit over the summer season, have been used to analyze the trend towards a dryer climate in the region. Humidity is very sensitive to local anthropogenic impacts on physiographic conditions. If there are no local human-induced impacts, the indicators will reflect this in spite of the increase in absolute moisture content caused by climate change. The change towards a dryer

climate is considerably higher in the region adjacent to the Aral Sea during the hot season than in the mountains, plains and piedmont areas. Because there is no obvious increase in atmospheric precipitation for Uzbekistan as a whole, the high rates of temperature increase are considered the major indicator of climate change.

In order to bring down the uncertainty levels of the regional climate scenarios, it is advisable to combine and compare the outputs of different models. This was done using the results from six models: CCCM1-TR, CSIRO-TR, ECHAM4, HadCM3, CCSR-NIES, and GFDL-TR. Eventually the A2 base in ECHAM4 and B2 in the HADCM3 model have been used.

Average sensitivity values for climate change due to the increase of greenhouse gas concentrations in the atmosphere were selected for developing the regional scenarios using the emissions under the scenarios A1B, A2, B1 and B2, with a mitigation influence of sulfate aerosols for three time intervals: 2016–2045, 2036–2065 and 2066–2095 (hereinafter indicated as 2030, 2050 and 2080). Analysis showed that the differences in expected changes in mean annual air temperatures according to the A and B scenario groups by 2030 and 2050 are comparable with the standard error of calculating the mean values for 30-year time series of observations. Before 2050, the estimated changes in climate parameters under each scenario were quite similar, and therefore only one scenario (A or B) was taken into consideration for an evaluation of the vulnerabilities.

An assessment of changes in standard deviation was obtained from the MAGICC/SCENGEN model, which showed some increase in the variability of air temperatures and precipitation compared to the baseline period for the whole territory. This assessment was made by assuming the probability of occurrence of extreme drought periods during very hot weather and the probability of sudden and significant drops of temperature in the region.

The scenarios show an increase in air temperatures over Uzbekistan, especially in the winter season, with the minimum air temperatures rising more than the maximum temperatures. Another conclusion is that the amount of rainfall is likely to increase in the winter season (December to February), with an increase in number of days with maximum rainfall and in the intensity of rain. This may increase the risk of extreme floods, mudflows and landslides.

The trend for reducing snow reserves has been found from data for the period from 1950 to 2005. This is in line with the rising air temperatures recorded by the meteorological stations. An increase in air temperatures in the mountain zones deteriorates the conditions for the formation of snow reserves and so lower snow availability can be forecast. Lower river discharges will be observed in some river basins as a consequence.

This research also calculated the necessary climate parameters to evaluate the vulnerability of, and climate change impacts on, different social and economic sectors in Uzbekistan under different climate scenarios. The influence of climate change on water resources is found in impacts on river flows and increased water consumption. The first indicator (river flow) has a high uncertainty, but the latter

(water consumption) is assumed to increase proportionally to temperature growth and is coupled to changes in air humidity and precipitation in all scenarios.

The Hydrometeorological Service of Uzbekistan has evaluated the impacts of climate change on river runoff under different scenarios used and largely attributes the impact on river runoff. to the differences in expected precipitation. Because of the high natural variability of precipitation observed at the weather stations in the region and the lack of exact data for change trends, this evaluation was conducted under two options:

- changes in rainfall and temperatures assumed according to the scenario;
- changes in temperatures according to the scenario but rainfall based on current rainfall data.

A mathematic model (developed for the modeling of runoff from mountainous rivers) was used for evaluating the climate change impacts. It was used as an automated information data system to produce hydrological forecasts. This modeling of the runoff of the rivers in the Aral Sea basin (within Uzbekistan) based on climate change scenarios has shown the following results.

- Substantial changes in water resources will not occur in the Syrdarya basin by 2030 under the climate change scenarios for rainfall and temperatures. Under scenario B2, some increase in river runoff in the upper reaches is possible, but all deviations will be within the natural runoff variability. A certain trend towards river runoff reduction is observed in the Amudarya basin.
- Under the scenario that assumes air temperature rise under unchanged rainfall conditions, there will be an expected 5–8% decrease in available water resources (as against the baseline period) by 2030 in the Amudarya basin, but substantial changes in water resources will not occur in the Syrdarya basin (all deviations are within the natural runoff variability.
- In the longer term (2050), under a scenario for changes in air temperatures only (rainfall remains unchanged), there may be reduced runoff in both the Syrdarya and Amudarya rivers. The possible reduction in river runoff may amount to 6–10% for the Syrdarya basin and 10–15% for the Amudarya basin.

Table 5.9 presents some of the modeling results. These describe river flow changes for the growing seasons in the hydrological districts within the zone of the upper watersheds of the Amudarya and Syrdarya rivers (see Figure 5.10). The data in the third column that show water availability around 2030 to 2035 are most important for our forecast. Although on average the indicators for this period seem not too dangerous at present (according to Hydrometeorological Service of Uzbekistan data), an absolutely different situation may arise in dry years. It is necessary to note that the estimates by the Hydrometeorological Service of Uzbekistan and the SIC ICWC, in its calculations for the Chirchik River basin (River Twin project), point to possible deviations of up to 40% from the mean annual runoff values during the growing season. The estimate of variations in available water resources and water consumption in the Chirchik-Akhangaran-Keles sub-basin for the 2003–2030 period shows that deviations in volumes of available water resources can be in the order of 40% in the short term and more than 50% in the medium term.

Table 5.9 Assessment of river flow indicators according to the climate scenarios A2 and B2 in accordance with the second national report of Uzbekistan. (32)

River/gauging station	Flow in the growing season (m^3/sec)	Changes in river flows (in % of baseline flow) in the growing season under different climate scenarios					
		B2			A		
		2030	2050	2080	2030	2050	2080
Pskem – Mullala	128	95	91	91	96	96	90
Chatkal – Charvak	195	95	94	92	98	97	92
Inflow into Charvak Reservoir	323	95	93	92	97	97	91
Akhangaran – Irtash	36	98	96	95	103	101	91
Padshaata – Tostu	9	74	75	74	78	80	77
Chadak – Julasay	10	58	58	56	62	62	57
Gavasay – Gava	6	57	56	55	61	61	56
Karakulja – Aktash	38	96	97	98	99	101	101
Yassy – Salamalik	37	101	100	97	107	105	96
Tar – Chalma	82	96	98	95	106	108	101
Kurshab – Gulcha	25	88	86	84	96	96	87
Zeravshan – Dupuli	256	104	103	102	98	84	88
Kafirnigan – Tartki	255	99	100	99	99	93	79
Vakhsh – Komsomolabad	988	75	72	70	73	69	72
Obikhingou – Tavildara	266	80	75	62	71	67	62
Kyzylsu – Samanchi	97	91	90	87	94	89	77

Under these conditions the range of discharge fluctuations is so wide that the Charvak Reservoir would not be able to smooth out a flood load, and its capacity is also not enough to cover water deficits in extreme dry years. This situation alone would make irrigation impossible, and the situation is even worse given that environmental requirements and the needs of hydropower producers and other water

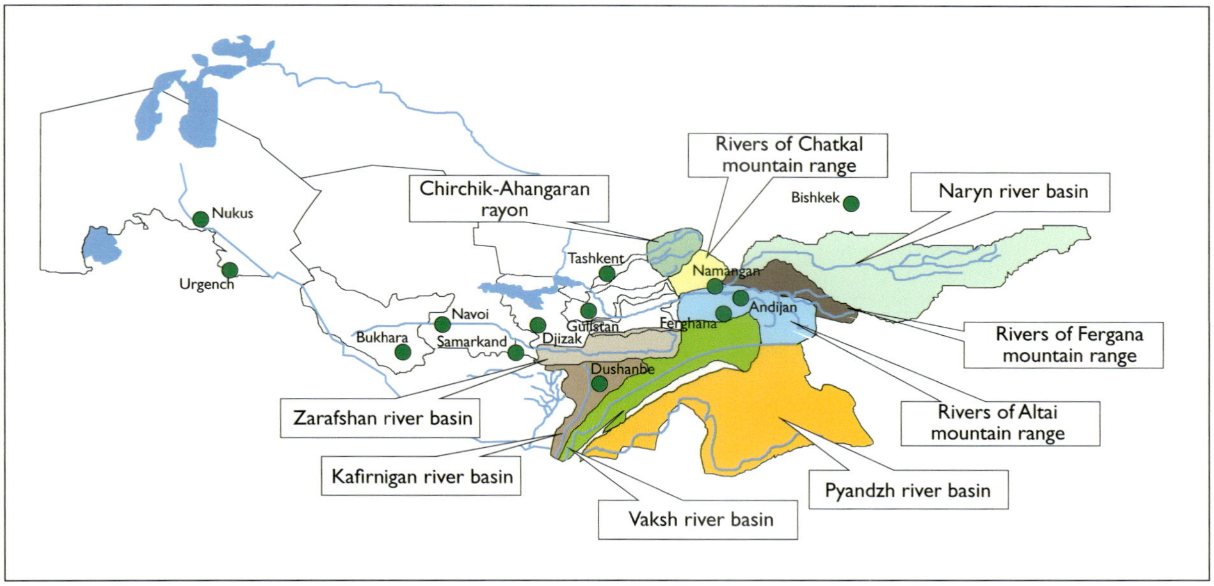

Figure 5.10 Layout of the hydrological districts in the upper watersheds of the runoff zone of the Amudarya and Syrdarya rivers (SIC-ICWC).

users have to be considered. This river basin was previously considered to be one with ample water resources for all different user functions. This situation should now therefore attract the special attention of water management organizations and government bodies because water consumption will rise by 15–20% by 2030 under all the user options presently considered!

A detailed estimate prepared by A. Usmanov, E. Cholponkulov and O. Dudko (SANIIRI) for all the regions of Uzbekistan shows that the increase in water consumption due to climate change will be around 6% in an average year and up to 14% in extreme dry years (see Figure 5.11). Unfortunately, other national reports do not show similar estimates, and therefore the findings of the Second National Report of the Hydrometeorological Service of Uzbekistan has to be used for replicating the exercise in similar river basins in the other riparian countries. Based on these assumptions, Figure 5.11 shows the estimated growth of water consumption due to climate in the basin by the year 2035.

Using these results, the available water resources in the Aral Sea basin in 2030 can be estimated for the two scenarios as shown in Table 5.11. This estimation takes into account that changes in the availability of surface water resources will affect groundwater reserves (design factor 0.8) and the amount of return water (design factor 0.6). We therefore combine these parameters for an average year and for a dry year based on these observed ratios.

It is assumed that when hydropower plants are operated according to an irrigation regime under conditions of multi-year river flow regulation, there will be no water disposal into the Arnasay Depression, into the desert sinks and into the lower reaches of the rivers in volumes that exceed agreed limits set by the ICWC. Therefore during an average year the available water resources will be equal to the rates of natural runoff, but in a dry year additional water volumes shall be provided at the expense of

Table 5.10 Indicator trends for available water resources and water consumption in the Chirchik River basin, km^3 (24).				
	Water resources		**Estimated water withdrawal for irrigation**	
	ECHAM	**HADCM2**	**ECHAM**	**HADCM2**
Baseline year 2003	9,213		4,380	
Min	5,131	5,440	4,225	4,210
Max	12,552	12,775	6,285	6,270
Average over the period 2003 to 2030	8,107	8,403	5,360	5,190

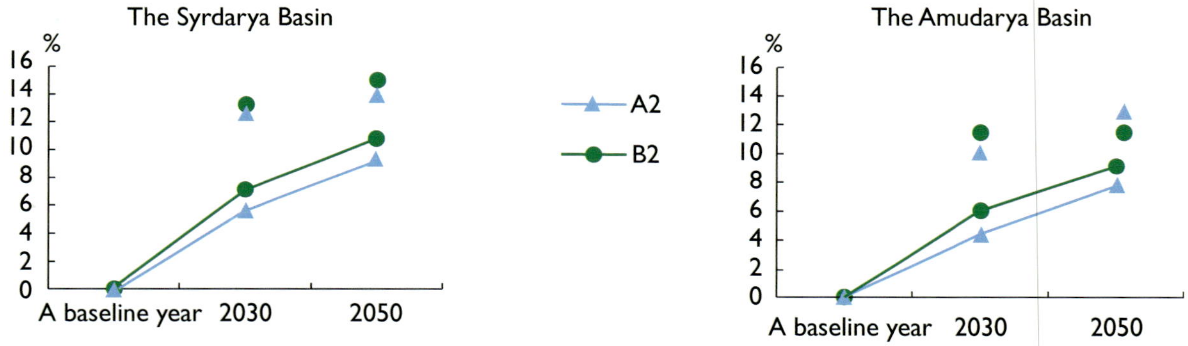

Figure 5.11 Additional increase in average and maximum irrigation water requirements in order to avoid crop yield losses due to increased evaporation in the Amudarya and Syrdarya river basins (17).

the multi-year river flow regulation up to a volume of 4 km^3/year in the Syrdarya basin, and 3 km^3/year in the Amudarya basin (Dukhovny and Sorokin 2007).

In the case of a hydropower regime the situation (of water availability throughout the year) will be worse, and in an average year the river runoff will be some 5 km^3 lower in the Amudarya basin, and 2.2 km^3 in the Syrdarya basin. Thus the available water resources will amount to a total of 126.4 km^3 (74.7 km^3 in the Amudarya basin, and 51.8 km^3 in the Syrdarya basin) in an average year, and 95.8 km^3 (56.2 km^3 in the Amudarya basin and 39.6 km^3 in the Syrdarya basin) in a dry year, which is close to the volumes recorded at present. In the worst case option, the volumes are 7 km^3 lower in an average year and almost 14 km^3 lower in a dry year! This situation—where climate change causes an abrupt decrease in water availability

Table 5.11 Estimate of the water resources of the Aral Sea basin in 2030, km^3 (SIC-ICWC).

	Average year	2008 (dry year)	Scenario B2		Scenario A2	
			Average year	Dry year	Average year	Dry year
Surface runoff, incl.:	116,483	86,762	110,933	82,600	106,695	80,021
Amudarya	79,280	59,460	73,730	55,298	71,352	53,514
Syrdarya	37,203	27,302	37,203	27,302	35,343	26,507
Groundwater:	16,891	13,573	16,472	13,178	15,747	12,598
Amudarya	5,989	4,791	5,570	4,456	5,390	4,312
Syrdarya	10,902	8,722	10,902	8,721	10,357	8,286
Return water:	32,450/21,580	12,948	20,899	12,539	20,114	12,008
Amudarya	19060/9730	5,838	9,049	5,429	8,757	5,254
Syrdarya	13,390/11,850	7,110	11,850	7,110	11,257	6,754
Water losses in open channels:	13,900	13,900	13,900	13,900	13,900	13,900
Amudarya	8,900	8,900	8,900	8,900	8,900	8,900
Syrdarya	5,000	5,000	5,000	5,000	5,000	5,000
Environmental requirements:	8,000	5,200	8,000	5,700	8,000	5,700
Amudarya	4,800	3.200	4.800	3.200	4.800	3.200
Syrdarya	3.200	2.000	3.200	2.500	3.200	2.500
Total usable water resources	*133,054*	*94,123*	*126,404*	*88,717*	*120,656*	*85,027*
Amudarya	81,299	57,989	74,645	53,083	71,799	50,980
Syrdarya	51,755	36,134	51,755	35,633	48,757	34,047

Table 5.12 Combination of climate scenarios and water management scenarios (SIC-ICWC).

Water management scenario	Climate scenarios					
	Usual natural runoff		Scenario B2		Scenario A2	
	Average year	Dry year	Average year	Dry year	Average year	Dry year
Total	133,054	94,123	126,404	88,717	120,556	85,027
Amudarya	81,299	57,989	74,649	53,083	71,799	50,980
Syrdarya	51,755	36,134	51,755	35,633	48,757	34,047
Hydropower (irrigation regime W1):			126,404	95,833	120,556	89,985
Amudarya			74,649	39,633	71,799	53,850
Syrdarya			51,755	56,200	48,757	36,635
Hydropower (hydropower regime W2)			119,274	81,264	113,996	76,386
Amudarya			69,719	45,831	67,439	43,551
Syrdarya			49,555	35,433	46,557	32,835

in the Amudarya—will immediately lead to further emphasis on national interests (hydro-egoism) where all stakeholders fight for their own water volumes.

Now we will consider the combined scenarios with regard to water demand. The socioeconomic scenarios and the water management scenarios specify the scope of water requirements, and these can be transformed into three base scenarios dealing with extreme conditions.

5.5.2 Business As Usual (Existing Trends are Maintained)

This scenario can be also called the "non-intervention scenario" or the "worsening conditions scenario". This scenario assumes that the current situation is maintained over the whole (25 year) evaluation period, including a low level of financing for operation and maintenance of the water infrastructure, a low level of modernization of irrigation and drainage systems and that state control in agricultural production

(so-called state orders) continues in the countries where this system exists. Obviously, this scenario can be used as a base for comparison with other scenarios. This scenario is based on a set of further assumptions in particular.

- Further economic reforms are not undertaken, there is a negligible increase in the number of private farms, and transition towards a free market economy is slow.
- Interstate agreements related to water resource and hydropower management remain unchanged, and the difficulties of implementing these agreements remain. This means that in the upstream countries, hydropower plants will (normally) operate according to a regime of maximum electricity production during the winter season.
- Reforms of irrigated agriculture are not envisaged in these fields:
  - *the institutional framework of water resource management*: in particular, introduction of water charges and drainage system development are not initiated;
  - *the schedule of water delivery*: inter-farm water distribution system will remain and the existing system will limit new methods of water applications;
  - *financing the water system*: insufficient funds continue to be allocated for the rehabilitation and O&M of the main and on-farm irrigation and drainage systems, and considerable wear and tear will result in failure of most part of the drainage infrastructure, and there will be low level of expenditure on the operation, maintenance, repairing and replacement of machinery and equipment;
  - *financing agricultural development*: there will be limited access to funds and credits to purchase seeds, chemical fertilizers, pesticides, new equipment, etc. and the level of supplying agricultural inputs will continue to decrease.
  - *systems efficiency*: the efficiency of irrigation systems will not exceed 60%.
- As a result, land salinization will continue and with consequent reductions in the area under crops, land productivity and profit margins will decrease. In some regions, production will probably drop considerably, in combination with partial or complete suspension of irrigation activities. Irrigated land in the upstream regions where water is supplied by gravity and there is no problem with soil salinization is unlikely to suffer damage.
- Water volumes available for environmental needs could be increased due to the decrease in water consumption for irrigation, but this will not happen because of the low management efficiencies in these countries.
- Social costs will be quite large because new jobs will not be created in the regions where water supply for irrigation is stopped.
- It is expected that the agricultural contribution to the national economy will be considerably lower, with reduced output impacting on the level of food supply and the overall well-being of the population.

A main feature of this scenario is that the water deficit in the region will grow due to the policy of each country to use water resources as much as possible for its own benefit and that there will be an enormous loss of water productivity owing to low investment levels by the government and water users in the operation and maintenance of the water infrastructure.

5.5.3 Scenario of National Preferences

The fundamental assumption under this scenario is that increased investment in the agricultural sector will be quite sufficient for the stabilization of agricultural production at current levels in all regions. The agriculture sector will remain in the hands of the government to the same or an even greater degree than today. These are other important assumptions under this scenario.

- There will an increasing gap between riparian countries in terms of indicators such as GNP per capita, financial capacity, level of market freedom, and the level of internal and external investment.
- Some more economic reforms will be implemented, including further land privatization and further movement towards a free market economy, although at different rates in each country. Transboundary water resources management will be implemented at a basin level, but it is unlikely that the basin framework agreements and the yearly bilateral and multilateral agreements will provide effective governance of water resources and hydropower production.
- In terms of irrigated farming, the following assumptions apply:
  - some institutional reforms will be implemented for improving market conditions but a water charge system will not be introduced any time soon;
  - limited investment funds will be made available and the ability of farmers to purchase seeds, fertilizers, pesticides, new machinery and equipment, etc. will be weakened;
  - reforms will be undertaken to stabilize agricultural production, and these will cover cropping patterns, size of sown areas, the structure and organization of private farms, agricultural practice and/or irrigation schedules;
  - expenditure on the operation and maintenance of irrigation and drainage infrastructure will increase to the levels that allow partial rehabilitation and proper technical maintenance, facilitating the needs of agricultural production;
  - prices for agricultural products, crop productivity and profits will remain at mainly low levels and, as a result, the agricultural sector will not be capable of financing investment in modernization.
- Water volumes available for environmental needs are likely to remain at the former and present levels.
- Social consequences will be less severe than under a "worsening conditions scenario" but the overall investment necessary for realization of this scenario must be assumed to be higher than present levels.

5.5.4 Optimistic or Regional Interest Scenario

Under this scenario, it is assumed that all five (or six, if Afghanistan is included in the future) riparian states will adapt to free market principles for economic development and management, but some measures will be also be taken to the support of social and environmental objectives. Resources will be allocated in such a way that they achieve the maximum contribution to economic development under

conditions that fulfill ecological requirements and ensure social stability. These are other important assumptions.

- Water allocation mechanisms will be agreed and put into practice over time that provide optimum and balanced use for agriculture, water supply, hydropower and the environment to the overall benefit of all riparian states.
- Under conditions of optimum water resource use, the hydropower sector will not be considered separately either from other energy sources or in each river basin, but it will serve the power production and distribution needs of the whole region. A competitive, regional electricity and energy resource market will be an essential element to achieve this objective.
- Concerning irrigated farming:
  - Farm privatization will see rapid progress, and private farms will generate increased income;
  - IWRM principles are put into practice everywhere and certain institutional reforms will therefore be undertaken—in particular, there will be involvement of stakeholders in water resource governance and management; a reorganization of irrigation systems according to hydrographical principles; the introduction of water charges that will encourage efficiency of agricultural practice and efficient water use—consequently, on-farm water resource management and irrigation methods will be improved resulting in reducing water consumption and, in turn, lowering the probability of water-logging and soil salinization;
  - limitations on funds for purchasing seeds, fertilizers, pesticides, new machinery and equipment, etc. will be avoided;
  - there will be changes in cropping patterns, size of sown areas, structure and organization of private farms, agricultural practice and investment in new machinery and equipment all driven by free market principles;
  - expenditure on the rehabilitation and operation and maintenance of irrigation and drainage infrastructure will increase considerably to levels that allow basin and national infrastructure to reach proper standards that provide for sustainable irrigated land use over a longer period;
  - the efficiency of irrigation systems will increase to at least 0.75, a level that was achieved in the newly developed irrigation schemes in the Hunger Steppe (Uzbekistan, Kazakhstan and Tajikistan) in the 1970s.

The key strategy for improving on-farm land and water use will be to increase land and water productivity to achieve water resource savings.

- It is expected that farmers will contribute to financing works where investment is required under market conditions and under conditions where the banking system and government institutions are reliable and stable. The financing of intra-farm and on-farm infrastructure will be the responsibility of private farmers.
- As a result of the measures outlined above, crop productivity and profit margins will increase, although overall production will still be regulated by market conditions.

Table 5.13 Expected water demand in the Aral Sea basin in 2025, million m³ (IWRUPS).[2]

Country	Time horizon	Economic sector						Total
		Water supply	Rural water supply	Industry	Fishery	Irrigated farming	Others	
Kazakhstan	2005	80	70	75	65	5,500	300	6,090
	2010	140	100	120	150	8,500	500	9,510
	2025	160	120	290	170	7,450	500	9,290
Kyrgyzstan	2005	95	70	50	0	3,500	0	3,715
	2010	100	85	60	0	4,500	0	4,745
	2025	150	100	190	0	4,200	0	4,640
Tajikistan	2005	500	300	400	80	8,800	750	10,830
	2010	570	450	450	100	10,380	600	12,550
	2025	670	500	580	140	9,500	500	11,890
Turkmenistan	2005	370	190	750	25	18,000	0	19,335
	2010	400	200	900	30	20,000	0	21,530
	2025	470	250	1,100	40	17,650	0	19,510
Uzbekistan	2005	2,650	1,390	1,350	1,050	56,560	0	63,000
	2010	2,700	1,400	1,390	1,320	52,400	0	59,200
	2025	5,850	1,630	1,460	2,240	48,020	0	59,200
Total in the Aral Sea basin	2005	3,695	2,020	2,625	1,220	92,360	1,050	102,970
	2010	3,910	2,235	2,920	1,600	95,780	1,100	107,535
	2025	7,300	2,600	3,620	2,590	86,820	1,000	104,530

[2]IWRUPS: Integrated Water Resource Use and Protection Scheme.

- Water volumes for environmental needs will be allocated and based on agreed water distribution mechanisms.
- It is expected that the social impacts will be positive.

Scenarios of water consumption for the next 25 years have been developed in many documents, including the IWRUPS for the Amudarya and Syrdarya rivers, the WEMP (Water and Environmental Management project of the Global Environment Facility) GEF project (2007), and the Second National Report on Climate Change. All authors have considered broadly similar scenarios, but the forecast water consumption volumes differ considerably. Table 5.13 presents data for expected water demand in 2025 under the optimistic scenario. It is important to note that the water consumption of different consumers given in this table indicates considerable growth (water usage almost doubles; industrial usage increases by nearly 25%, fishery requirements are up 1 km^3), and water consumption in the irrigated agriculture sector decreases by some 9 km^3.

When we compare these figures with the data for the optimistic scenario for some riparian countries, we find large differences. For example, the GEF Agency and the Uzsuvloyikha Institute, in their publication *Water is the life and death resource for the future of Uzbekistan* (2007), predict that by 2025, water demand for irrigation in Uzbekistan (the biggest water consumer in the region) will reach 64 km^3, rather than the 48 km^3 in the IWRUPS projection, and it even forecasts 59 km^3 according to its "minimum" option. This is 5 km^3 more than in the optimal use option and does not take climate change impacts into consideration. Taking into account temperature rise, which is expected to increase water consumption in irrigated farming by 10%, the GEF Agency and the Uzsuvloyikha Institute's estimates for water consumption by irrigated farming are 70 km^3 in the optimistic option and 65 km^3 in the minimum option (see page 64 of their report).

Table 5.14 shows the forecast for the total water consumption in the two river basins made on the basis of the WEMP study for the last year considered (2030) under the optimistic scenario.

Table 5.15 compares the results of the ASBmm model [40] with the WEMP and IWRUPS predictions. To make comparisons with the IWRUPS data given in the Table 5.13 with regard to water demand in 2025, it has been assumed that the figures forecast for 2025 will continued unchanged to 2030 and similar trends have been assumed for the other main indicators.

Only time will show which forecast will be closest to reality. Shiklomanov (2008) estimates that water demand in 2025 will increase by 23.8% as compared to the demand in 2008, for a dry year. The total water diversion in the Aral Sea Basin in 2000 (see Table 5.16) was 96.0 km^3 and was 89 km3 in the very dry 2008,, so according to his estimate a water withdrawal of about 118.85 km^3 can be expected in 2025, which is close to the BAU scenario as considered in this book.

The "optimistic option" of the ASBmm provides figures for water demand that are 18.1 km^3/year and 14.5 km^3/year lower than those in the WEMP and in the IWRUPS respectively. This difference is mainly caused by the assumed demand for the irrigated agriculture sector. It is worth to note that the

Table 5.14 Water consumption under the "optimistic scenario" over a 25-year period according to the WEMP study, million m^3.

	Amudarya		Syrdarya		The Aral Sea basin	
	Average year	Dry year	Average year	Dry year	Average year	Dry year
Total water consumption in all economic sectors						
Kazakhstan	–	–	8200	7093	8200	7093
Kyrgyzstan	100	86,5	4050	3503	4150	3590
Tajikistan	8400	7266	1900	1643.5	10300	8910
Turkmenistan	18810	16262	–	–	18810	16262
Uzbekistan	29300	25345	17900	15480	47200	40825
Afghanistan	5000	4325			5000	4325
Losses in canals	5650	4667	3500	3028	9150	7695
Use of groundwater	67260	57951	35550	30748	102810	88099
The same with climate change affects	73986	63746	39105	33822	113091	97568

water demand forecast in the WEMP analysis should be increased by 10% to take temperature rise due to climate change into account.

Combining the ASBmm optimal option with the W1-B2 option in the climate and water management scenario (see Table 5.12) could increase water supply to the Aral Sea and adjacent areas up to 39.4 km^3/year. Of course this situation actually is unrealistic, but it would be extremely desirable. Under a combination of the WEMP optimistic scenario with the resources of the W1-B2 option, the possible annual inflow into the Aral Sea and adjacent areas will amount to 13.3 km^3, which corresponds with the mean annual data over the period 2002 to 2008 (more than 50% of the inflow into the Aral Sea is through the Syrdarya).

The "business as usual" option provides extremely low inflows to the Aral Sea of only 2.5 km^3/year, and the "national" scenario decreases inflow to the areas adjacent to the Aral Sea from 8 km^3/year to 3.3 km^3/year. Table 5.1 is mentioning a deficit of 4.7 km3 / year for the delta's and the sea. However, water availability in the Syrdarya and the Amudarya varies quite markedly according to the various scenarios. By 2035, the situation will be more or less stable in the Syrdarya basin, where according to most scenarios

Table 5.15 Comparing the results of the ASBmm[3] model with other predictions (SIC-ICWC).

Indicator	Option ASBmm			WEMP optimum	IWRUPS
	Optimistic	BAU	National		
Irrigated farming					
Irrigated area, '000 ha	8,500	8,500	9,400	8829.5	
Gross irrigation rate, m^3/ha	9,400	11,500	11,000		
Irrigation water demand, million m^3	79,900	97,750	103,400		86,820
Population, millions	59.0	69.0	77.0	67.0	
Water, m^3/person or l/person/day	0.09/250	0.11/320	0.128/350		
Total consumption	*5,310*	*7,500*	*9,856*		*9,900*
Industry	3,300	3,050	3,500		3,620
Other sectors	1,500	3,500	3,500		3,600
Total	*90,000*	*111,800*	*120,260*	*102,810*	*104,530*
Same plus Afghanistan	95,000	116,800	125,260	113,091*	109,530
Mean annual water resources	126,404	119,274	120,556	126,404	126,404
Water supply to the Aral Sea	39,400	2,474	-4704	13,313	16,674

* Adjusted for the increase in water consumption due to temperature rise.

there will be a water surplus in an average year (of between 2 km^3 and 7.2 km^3) and a water deficit in a dry year (of between 0.8 km^3 and 7.6 km^3). In the Amudarya basin, there will be a water deficit of between 4 km^3 and 9 km^3 in an average year and of between 12 km^3 and 19 km^3 in a dry year!

Of course there is uncertainty in these estimates, but it becomes clear that key attention should go to the Amudarya basin where most of the negative effects are most likely to occur as a result of the

[3]Aral Sea Basin management model. An integrated river basin management model under development by SIC-ICWC and UNESCO-IHE on the basin of an earlier version of this model published in 2003.

Table 5.16 General development indicators for the Aral Sea basin (ASB) states.

Indicator	State	Year							
		1980	**1985**	**1990**	**1995**	**2000**	**2005**	**2007**	**2008**
Population, '000 people	Kazakhstan	2069	2220	2311	2403	2576	2747	2964	3012
	Kyrgyzstan	1813	2062	2364	2505	2726	2908	2976	2998
	Tajikistan	4005	4641	5359	5876	6250	6826	7277	7491
	Turkmenistan	2576	2923	3346	4192	4909	5866	6043	6210
	Uzbekistan	15098	17485	20606	22904	24814	26282	26681	27308
	Total ASB	25561	29331	33986	37880	41275	44629	45941	47019
Population growth, % as compared to previous year	Kazakhstan		101.5	101.3	100.2	100.5	101.3	108.2	101.6
	Kyrgyzstan		103.0	104.1	100.6	101.8	101.4	101.0	100.7
	Tajikistan		103.3	102.1	101.6	100.8	101.2	102.6	102.9
	Turkmenistan		102.5	101.9	103.2	103.2	103.1	100.0	102.8
	Uzbekistan		102.9	102.7	102.0	101.3	100.9	100.7	102.3
	Total ASB		102.6	102.4	101.5	101.5	101.6	102.5	102.1
Irrigated area, % growth with 1980 as a reference year	Kazakhstan		101	108	109	111	103	103	106
	Kyrgyzstan		100	99	101	102	97	96	96
	Tajikistan		106	112	111	112	114	113	119
	Turkmenistan		124	141	182	189	198	201	202
	Uzbekistan		111	117	121	120	119	119	119

	Total ASB		109	115	125	127	126	126	129
Total water withdrawal, '000 m^3	Kazakhstan	12632	11655	11209	9878	7067	7681	7732	6427
	Kyrgyzstan	3370	3092	3339	3270	2851	2706	2838	2739
	Tajikistan	12557	12733	12847	12444	12745	13239	13426	13281
	Turkmenistan	23764	26150	25859	27904	24125	26854	26412	21144
	Uzbekistan	72785	63928	61202	59661	49178	65864	54591	45410
	Total ASB	125108	117558	114456	113156	95965	116345	104999	89000
Including for irrigation, '000 m^3	Kazakhstan	11727	10728	10100	8326	6076	6548	6732	5781
	Kyrgyzstan	3059	2784	2997	2926	2601	2525	2538	2525
	Tajikistan	9839	9844	9905	9646	10152	9715	9161	9210
	Turkmenistan	20823	22891	22113	24036	20008	22715	23179	16250
	Uzbekistan	58788	49280	46103	45212	35687	49875	41006	32540
	Total ASB	104236	95527	91218	90146	74524	91378	82617	66307
Water consumption per hectare, m^3/ha	Kazakhstan	16849.2	15198.1	13430.8	10978.7	7893.4	9166.8	9402.5	7822.7
	Kyrgyzstan	7236.5	6557.3	7162.8	6842.2	6059.0	6144.7	6224.9	6196.1
	Tajikistan	14671.6	13861.0	13198.1	12916.4	13537.8	12730.3	12094.1	11527.7
	Turkmenistan	19280.6	17088.0	14516.2	12220.2	9780.7	10605.6	10671.6	7457.6
	Uzbekistan	15939.9	12062.5	10660.5	10123.6	8039.1	11325.4	9337.3	7409.3
	Total ASB	15895.8	13147.9	11741.4	10775.6	8836.4	10834.5	9783.0	7785.8

(*Continued*)

Table 5.16 (Continued)

Indicator	State	Year							
		1980	1985	1990	1995	2000	2005	2007	2008
Water consumption per capita, m^3/person	Kazakhstan	6105.2	5250.2	4850.1	4110.5	2743.2	2796.1	2608.7	2133.8
	Kyrgyzstan	1859.1	1499.5	1412.2	1305.3	1045.7	930.7	953.7	913.6
	Tajikistan	3135.3	2743.6	2397.3	2117.7	2039.2	1939.4	1845.0	1772.9
	Turkmenistan	9225.3	8946.4	7728.3	6656.5	4914.5	4578.0	4370.7	3404.8
	Uzbekistan	4820.8	3656.1	2970.1	2604.8	1981.9	2506.1	2046.0	1662.9
	Total ASB	4894.5	4008.0	3367.7	2987.2	2325.0	2606.9	2285.5	1892.8
GDP, US$ billions	Kazakhstan	361.0	445.8	1140.0	348.1	365.5	933.7		
	Kyrgyzstan	588.5	603.1	608.8	575.2	500.4	793.1	1035.2	
	Tajikistan	6801.0	8070.0	8793.0	1850.8	1101.4	1595.8	1755.6	1761.6
	Turkmenistan	5137.1	6545.5	433.3	1518.4	3379.2	10882.9		
	Uzbekistan	14184.1	16948.7	20213.0	10157.8	12029.5	9710.2	21849.8	26446.2
	Total ASB	27071.7	32613.1	31188.1	14450.3	17376	23915.68	24640.6	28207.8
Agricultural output,	Kazakhstan	950.0	1110.0	1250.0	370.0	410.0	860.0	910.0	910.0
	Kyrgyzstan	700.0	770.0	1410.0	440.0	410.0	770.0	1190.0	1190.0
	Tajikistan	1880.0	2020.0	1910.0	710.0	490.0	700.0	820.0	790.0
	Turkmenistan	160.0	190.0	210.0	300.0	380.0	510.0	530.0	530.0

	Uzbekistan	9210.0	10150.0	10700.0	4900.0	4590.0	4220.0	6950.0	7530.0
	Total ASB	12900.0	14240.0	15480.0	6720.0	6280.0	7060.0	10400.0	10950.0
Including livestock, US$ billions	Kazakhstan	630.0	750.0	780.0	230.0	260.0	570.0	560.0	560.0
	Kyrgyzstan	370.0	410.0	760.0	220.0	220.0	390.0	600.0	600.0
	Tajikistan	1290.0	1370.0	1250.0	560.0	400.0	490.0	540.0	570.0
	Turkmenistan	140.0	170.0	180.0	260.0	330.0	450.0	460.0	460.0
	Uzbekistan	6530.0	6870.0	5510.0	2520.0	2360.0	2170.0	3830.0	4030.0
	Total ASB	8960	9570	8480	3790	3570	4070	5990	6220
Water productivity in irrigated farming, US$/ha	Kazakhstan	320.0	360.0	470.0	140.0	150.0	290.0	350.0	350.0
	Kyrgyzstan	330.0	360.0	650.0	220.0	190.0	380.0	590.0	590.0
	Tajikistan	590.0	650.0	660.0	150.0	90.0	210.0	280.0	220.0
	Turkmenistan	20.0	20.0	30.0	40.0	50.0	60.0	70.0	70.0
	Uzbekistan	2680.0	3280.0	5190.0	2380.0	2230.0	2050.0	3120.0	3500.0
	Total ASB	3940	4670	7000	2930	2710	2990	4410	4730
Productivity of irrigated area, US$/ha	Kazakhstan	0.05	0.07	0.08	0.03	0.04	0.09	0.08	0.10
	Kyrgyzstan	0.12	0.15	0.25	0.08	0.08	0.15	0.24	0.24
	Tajikistan	0.13	0.14	0.13	0.06	0.04	0.05	0.06	0.06
	Turkmenistan	0.01	0.01	0.01	0.01	0.02	0.02	0.02	0.03
	Uzbekistan	0.11	0.14	0.12	0.06	0.07	0.04	0.09	0.12

melting glaciers, increasing water consumption by Afghanistan and possible manifestations of "hydro egoism". This correlation between future water balances and optimistic options for socioeconomic development requires a revision of the behavior of both the states and society at large. These aspects are very clearly described in a long-term strategic framework written by the Asian Development Bank (ADB 2008). Along with the positive trends and big hopes in the Central Asian region, this document focuses on bottlenecks on the way forward, notably on the low level of mobilization of funds, including community resources, on the uncontrolled priorities of private investment, and on weak fiscal authorities and regulatory instruments. All these factors, along with the insufficiently developed financial system, will widen the gap in income between the urban and rural population and, accordingly, decrease opportunities for rural economic growth. All these shortcomings will be seen in the water sector and the irrigated farming sector as well, where these constraints are particularly felt. The restricted market for investment in long-term projects essentially contributes to the present stagnation in development of the water sector and irrigated farming as a whole. If we take into account the fact that a huge segment of the population has limited access to state credit, the prospects for a way out of this situation are very limited. Unconventional approaches under radical objectives are therefore necessary to find solutions.

In addition to, the present conditions there is a further barrier inhibiting innovations which is the low level and quality of rural education. Although educational institutions are being established in the rural areas at quite high rates, this is not being accompanied by a qualitative improvement in education which would support innovation. This situation is therefore restricting the capacity of the rural public to understand and adapt to state-of-art technologies and to use of new and efficient equipment. Again there are three ways for overcoming these problems under an "optimistic scenario".

1. Economic growth based on a participatory approach. This development strategy should involve all stakeholders in the decision-making process related to the investment and use of public funds. This approach requires improvement in the education system, a basic framework of social security, credit opportunities for all including the underprivileged, and access to fixed assets created during the previous stage of development. Only with financial support for private farmers and rural infrastructure, together with providing farmers with a free choice of which crops to cultivate based on market relations, can the required capacity and conditions for sustainable irrigated farming be created.
2. Wide application of environmentally sound technologies that save natural resources, especially water and energy. Many efficient solutions have been proposed including IWRM, the widespread introduction of extension services, the development of clusters of greenhouses within irrigation schemes, drip irrigation, drainage water reuse, etc. However, to ensure a major impact, the first initiative that should be implemented is the establishment of a transboundary monitoring system with for example SCADA and online meteo in order to show the actual situation regarding water losses in river channels. This is especially critical along the Amudarya. It is of outmost importance to record the actual volumes of water diversion at transboundary gauging stations. This should be based on a 24-hour flow rather than present measurement schemes. The experience learnt from introducing such a system at ten headworks operated by the Syrdarya BWO has clearly shown that improving the accuracy of water measurements by ±2% reveals potential water reserves in the order of millions of cubic meters a year.

3. Regional co-operation as a means to overcome the non-uniform geographic distribution of natural resources and the unequal economic and social conditions in the region.

In conclusion, one should look at the existing trends in a more realistic way (see Table 5.16), on the basis of advanced analysis of data and understand that only scenarios that take into account a balanced shakeholder interest approach will deliver results acceptable to the region over the next 20 years.

So what should be actually done in the sphere of water resource management in the Aral Sea basin? The interstate and national policies of the riparian states in the field of water governance are the main determinant for future success. Water governance at the interstate level should be based on full compliance with the international water laws by all riparian states in the region, especially on the issue of the right of each country to an agreed (and limited) share of water resources and the obligation to prevent damage to downstream countries. Without a doubt there are limits to this right, but it has to be based on agreed norms for water saving and should be aimed at achieving the highest possible productivity. This can be only achieved by means of implementing a co-ordinated regional water policy that should be approved by all national governments in the region. As shown above, the difference between a hydropower and irrigation water release regimes can result in the loss of 7 km^3 in runoff through the Amudarya and 2.2 km^3 in runoff through the Syrdarya. These gaps can be narrowed if riparian countries reach agreements on joint management of river flows along with energy supply provision for the winter season.

The best solution could prove to be multi-year regulation schemes, which can compensate for the abrupt fluctuations in river flows that may arise because of climate change, together with joint management of hydropower plant cascades with appropriate commercial participation in the development of hydropower resources in the upper watersheds of downstream countries. With further joint development of hydropower resources on these rivers, it would be possible to produce hydropower in amounts that will completely meet the needs of all riparian states and generate a surplus both in the summer and in the winter, under the simultaneous implementation of a water release regime that serves irrigation needs and ensures multi-year regulation of the river flows. Another very efficient joint measure for allocating funds and resources would be installation of reliable monitoring systems for surface flows, starting on the Amudarya. National hydro-meteorological services should have free access to the data and the results of any data processing for dissemination among all users. Just improving the accuracy of water measuring on the Amudarya could put into use about 6 km^3 of presently unrecorded water resources.

The use of all return water resources, of which 11 km^3 are in closed desert sinks according to the present scenarios (even under conditions of optimum co-operation between states and sectors), provides a large reserve, especially for environmental needs. For example, 3 km^3 of additional water resources can be provided for environmental purposes by the transfer of the discharge of the Ozerny Collector Drain from the middle stretch of the Amudarya towards the north into the river delta. It is also possible to irrigate some desert areas in Turkmenistan using drainage water (a volume of about 7 km^3 is available) and using the additional water resources that are currently planned to being disposed into the Golden Lake of the Twenty-First Century (a prestige project of a former President of Turkmenistan).

Strengthening water governance at the national level requires observing a few principles, which should be constantly in the minds of decision makers in each riparian country. This should in the first place, be to adopt the attitude of organizing development based on education. Development must be pursued by encouraging communities and by involving young people who will continue to live under conditions of water deficits if we do not change our ways. Central Asia must introduce a special curriculum on "Water and the Future" in the schools and colleges of the region. This must be implemented within an overall framework of awareness building and development of water-saving skills, which should be imparted to the young generation especially.

Another key policy direction is development and approval of national water strategies that specify the agreed priorities for improved water resource use and management. Integrated Water Resources Management (IWRM) within the framework of the IWRM-Fergana project realized a decrease of as much as 25% in water withdrawals for irrigation through a combination of public participation, hydrographic-based water management, integration of new irrigation and drainage practices, and the introduction of financial mechanisms to facilitate rational water resource use. The national water strategies and water codes of each country should now be seeking to replicate these experiences. The data on irrigation water use per unit area provided in Table 5.16 show that in dry years (2000 and 2008) and even in a wet year such as 2007 it is very well possible to achieve an average gross water consumption of 9500 m^3/ha for irrigated farming.

However, it is important to not only reduce irrigation water consumption per unit area, but also simultaneously raise water productivity. The reality of using a target level of 80% potential water productivity was proved in the IWRM-Fergana project, where water productivity for wheat, cotton and other crops was increased by 1.5 times over an area of more than 100,000 hectares. This was achieved even without the introduction of advanced irrigation methods such as drip irrigation or micro-drip (spray) irrigation. Further improvement of water productivity is related to a shift of crop cultivation into greenhouses. In this case, however, it will be of primary importance to establish extension services for farmers and water user associations everywhere. Lastly, the introduction of automatic water monitoring systems is one of the cheapest ways to achieve water saving in key sectors.

Climate change will also force the region to develop skills to adapt to the rising temperatures and the increased water deficits. A program of adaptation to climate change may allow farmers to maintain present irrigation rates under rising air temperatures by taking advantage of a possibly longer growing season in the future.

However, all these approaches will only succeed if Central Asian states are able to ensure sustainable economic progress, because transition to an "optimistic" or, better even, a "rational" scenario for water sector development is possible only on the basis of well-distributed economic growth and greater attention by governments to the water sector and irrigated farming at large. In this respect, the situation in Uzbekistan, the biggest player in the regional water arena, is crucial.

Uzbekistan has no sound long-term socioeconomic development plan reviewed or approved by the government. Nevertheless, the Republican Center of Socio-Economic Studies (RCSES) and the

UNDP have prepared a report that considers long-term development (UNDP and RCSES 2005). This document contains four options for socioeconomic development. The first option follows a forecast by the Hopkins Institute and assumes slow rates of GNP growth (of around 5%) over the whole period up to 2025. This option cannot be called a BAU option because the current rates of GDP growth already considerably exceed this figure. However, the so-called "baseline option" of the RCSES can be adopted as a "BAU option", and assumes the continuation of trends developed over recent years. Average rates of GDP growth are assumed to be 7.2–7.4% up to 2020 (current rates of GDP growth are 10.8%), with industrial production growth at 9.6% up to 2015 and at 11% from 2016 to 2020. Agricultural production growth is assumed to be 4.6% up to 2015 and 4.5% from 2016 to 2020 with services growing by 14% and 15% (current rates of services growth is 11.4%) in the same periods. Option III, the so-called "modernization option", envisages the use of internal opportunities and reserves, increased efficiency of reforms, and considerable technological progress. In this scenario, the GDP growth rates in 2010, 2015, and 2020 reach 8.3%, 9.0% and 9.5% respectively. Industrial production growth is estimated at 11.5%, 12% and 13%. Agricultural production growth is set at 6.1%, 6.2% and 6.2% and services growth at 13.1%, 14%, and 55% respectively.

The increased investment in the water and irrigated farming sectors, financed by economic growth in the "baseline" and "modernization" options, is estimated to allow Uzbekistan to:

- open a credit line for farmers and WUAs in a way that allows the procurement of machinery, fertilizers, seeds and pesticides necessary for farming and facilitates a shift from "direct state governance" to agricultural output based on free market relations and crop diversification;
- increase government financing for improvements in the water sector, including automation and modernization, especially in the field of water application methods and technologies;
- rehabilitate infrastructure in the water and irrigated farming sector.

There is a deep conviction that through government attention, a strengthening of water governance (legal, institutional and financial) along with wide public participation in the process of improved water use efficiency and the co-operation of all riparian states in the region based on mutual respect and taking into account the interests of all, the goal of sustainable use of water resources in the region by 2025 can be achieved.

5.6 Lessons Learned

The heads of the Central Asian states have started to formulate the Aral Sea Basin Program (ASBP) III for 2012–2015. So there are again plans with a start date and an end date, and there are the new directives from the governments of the Central Asian countries. However, what can be expected from this program? Will it be a well-managed exercise leading to better co-operation in the basin or will it fall to the same failures inherent in the ASBP II? The ASBP II is still used by some to demonstrate that the countries themselves were able to increase co-operation. However, co-operation between countries

and donors became almost impossible, because the donors increasingly assessed the program as unclear, vague and lacking priorities, and they were not able to repair this situation.

Now that the ASBP III priorities have been determined by the board of IFAS, there is hope that the development and implementation of the ASBP III will not end in a "tug-of-war" at national and sector levels. One should proceed from the fact that, given the fragmented management and use of water resources in the region, overall water resources should be enough to meet all user requirements for at least the next 20–30 years.

Key to success, without doubt, is improved water governance at all levels (local, regional and national). For a start, it is necessary to harmonize the water policies of all the five and soon to be six (with the inclusion of Afghanistan) countries in terms of elaboration of clear water-allocation rules and procedures, development and approval of river flow regimes, meeting environmental demands and dealing with the combined energy and water needs. There is no doubt about the opportunities for hydropower development, if the hydroscheme operation regimes are taken out of the mono-disciplinary hands of the energy departments. The reservoirs must be operated not only for the benefit of hydropower but for the balanced benefit of all water users, with priority to a base provision for the irrigated agriculture as the most important water user sector in broad socioeconomic terms. The combined capacity of the existing and planned hydropower stations (HEPS) is much larger than the energy demand in the region as a whole. The upper watershed countries have such an excess of energy capacity that it does not make a difference to their own needs when they export energy outside the boundaries of the region. This is important because Pakistan, India and Afghanistan, as energy consumers, are actually more interested in summer electricity (the dry or Kabi season) than in energy during the monsoon period, which coincides with the Central Asia winter. However, downstream countries should get clear guarantees that year-round water release regimes are properly agreed and that they do not serve as a mere instrument of economic and political priorities of the upstream countries. To this end, very detailed and strictly followed agreements and commitments by all parties are needed. The UN Regional Center for Preventive Diplomacy in Central Asia[4] may become the guarantor and centrally agreed instrument for monitoring and follow up in this context. Under such circumstances of agreed joint management of the operational regimes, the downstream countries could even agree to, and take part in, financing the construction of new hydroschemes.

The development and strengthening of user-focused balanced regional water management should be achieved by reducing the numbers of regional institutions and by establishing a single water governance body (which should be more solid and better mandated than, for example, the Mekong River Commission), which would govern the executive water management agencies. These agencies should have representation and participation on the basis of parity and be under the direct monitoring of a regional community body comprising representatives of all national and sector stakeholders

[4]The goal of UNRCCA is to assist and support the governments of Kazakhstan, Kyrgyzstan, Tajikistan, Turkmenistan and Uzbekistan in building their conflict prevention capacities through enhanced dialogue and confidence building measures and in establishing genuine partnership in order to respond to existing threats and emerging challenges in the Central Asian region.

as well as funding agencies (development banks, donors), which would act as a steering committee or independent advisory body for regional management of water and associated resources. Afghanistan should be included (even if only as an observer) from the start. The mission and mandate of this council should be to issue solicited and unsolicited advice on key water and environment management issues and it should be facilitated with all the means necessary to research the background and consequences of this advice. The operating costs of the council should be the joint responsibility of all six countries of the Aral Sea basin.

Water management capacity in Central Asia should continue to be strengthened and supported by a very well-educated and trained workforce of water professionals equipped with all the necessary means of national and regional planning, monitoring and management. A career in water in Central Asia should become attractive again given the long-term development plans for the water sector and the strong and attractive research, education and training infrastructure. This will require strengthening of existing faculties in specialized universities in the region, including upgrading and developing new water and environmental education and research programs. The capacity building efforts must include education and training programs with a regional impact, because a regionally shared framework for water education and research is a key condition for the development of regional policies and programs for integrated water management in the basin.

The work on developing water management planning and implementation concepts and tools must be continued with priority. The development and enhancement of regional, national and local databases, models and planning and decision support tools is crucial in this context. Sharing information through communication within and outside of the sector must be ensured. The regional information system for all water bodies and water associated resources, therefore, should continue to be strengthened. Accountability and transparency of water management, based on well-developed computerized information systems and sound socioeconomic concepts should be the common ethos for modern regional water governance practice. In addition, a regional water strategy is urgently needed for the two basins, which should be based on well-justified and shared water conservation and water use objectives aimed at counteracting growing regional destabilization factors. The strategy should make intensive use of the latest insights and capacities available for the development of integrated river basin management models and include optimization concepts based on the Pareto principle or similar approaches that allow different stakeholder interests to be taken into account within a complex decision framework.

Water governance at the national level should be based on Integrated Water Resource Management (IWRM) policies and projects. The national governance structure should be a single body directly operating under the national government and organized around principles of public participation according to existing administrative and institutional boundaries and hydrographic (dynamic water system based) principles for water management and land management. The national water governance policies should be focused and redirected towards establishing a system of long-term wise water use that embodies water conservation management principles. Apart from being established on IWRM principles, this system should be oriented towards raising the potential productivity of water and land, including the financial tools to cover the costs of water delivery and drainage services, through the joint efforts of water users and the government. It also must be based on the "polluter pays" principle and there

should be wide incorporation of stakeholder initiatives into planning and management procedures. A new "water and education" initiative should be the basis of future "water friendliness", with the aim of ensuring that these attitudes are shared by a large part of the population. Starting from kindergarten onwards, this program would bring a young generation, which will definitely live under conditions of water shortage, to respect water and consider it again a "sacred entity" following in the traditions of their predecessors.

Such a program for water conservation and efficient water use could, for example, include these activities:

- organization of accurate regional, national and local monitoring of water use directly by water management agencies, water user associations, and community representative bodies;
- introduction of water delivery fees using progressive tariffs that also take into account price differentiation principles depending on the source;
- well-managed and monitored joint use of return water and groundwater together with surface water under new and integrated approaches to water allocation;
- application of an updated and automated (telemetric) monitoring, control and management system (such as SCADA) operated at local and regional level.

The second most important issue is agricultural water use reform, which is required to build a new and sustainable physical and institutional infrastructure for agricultural development in the region. Agricultural water use should not be limited to the implementation of IWRM principles. It should be based on principles of ensuring food security under maximum efficiency of water use and, at the same time, provide a decent and sustainable means of living for farmers and agriculture-related industries and services. These are therefore thought to be necessary policy principles:

- establish agricultural extension services, and efficient use of land and water resources based on advanced classification principles;
- sow new drought-resistant and salt-tolerant varieties of wheat, cotton and other crops to increase food security and generate socioeconomic revenues;
- provide maximum incentives to farmers for combined contracted (such as cotton, wheat) and free market (cash) crops, and encourage market-oriented agriculture by providing soft loans for development of greenhouse farming and drip and trickle irrigation of orchards, vineyards, vegetables and cucurbits;
- equip all water management organizations, water user associations and other water users with equipment to measure and monitor water use over time;
- organize remote data collection of land reclamation activities and land and crop growth conditions in combination with an information service system for farmers.

The key to peaceful co-existence in Central Asia, however, will remain with the implementation of joint plans and operations for balanced water allocation between water for hydropower and water for agriculture throughout the seasons.

Epilogue

And then, finally, there will be time for new thought, ideas and solutions. It is likely that the old idea of transferring part of the flow of the Siberian rivers would only be seen useful by future generations when the resources of their own economy have recovered and Central Asia has returned to the a sustainable agriculture based on traditions of the past.

References

1. ADB. 2008. *STRATEGY2020—The Long-Term Strategic Framework of the Asian Development Bank 2008–2020*, ADB, Manila, ISBN 978-971-561-680-5, p. 35.
2. Ahmad, M., Wasiq, M. 2004. *Water Resource Development in Northern Afghanistan and its implementation for Amudarya River*. World Bank. www.cawater-info.net/Afghanistan.htm
3. Analytics, 2007/1 (37), Almaty, p. 16. (in Russian).
4. Bourne, J.K. 2009. "The End of Plenty—special report on the global food crisis", *National Geographic*, June, 26–59.
5. Cholpankulov, E. et al. Factors of changes productivity of agricultural crops in conditions of climate changes, SANIIGMI, bulletin 5, Tashkent, 2002, pp. 36–45 (on Russian). (in Russian).
6. Dowling, M. & Wignaraja, G. 2005. "Turning the corner: the economic revival of Central Asia", *Central Asia and Caucasus*, No. 6 (36).
7. Dukhovny, V. & Sorokin, A. 2007. *Assessment of Rogun Reservoir's impacts on the Amudarya flow regime*. Tashkent, SIC ICWC. p. 119.
8. Dukhovny, V.A. (editor-in-chief). 2003. *Drainage in the Aral Sea Basin for Sustainable Development*. SIC ICWC. p. 316. http//www.cawater-infor.net/library/index.htm
9. Dukhovny, V.A. (editor-in-chief). 2004. *The Strategic Planning and Sustainable Water Resources Development in Central Asia*. SIC ICWC. Tashkent. p. 116. (in Russian).
10. Dukhovny, V.A. (editor-in-chief). 2008b. *Regional Model for Integrated Water Resources Management in Twinned River Basins*. The RiverTwin project. SIC ICWC. Tashkent. p. 214.
11. Dukhovny, V.A. 2007. Water and globalization: case study of Central Asia, *Irrigation and Drainage*, 56, 489–507, www.interscience.willey.com
12. Dukhovny, V.A., Sokolov, V.I. 2002. *Assessment of Water Resources in North Afghanistan, Their use and Impacts on the Aral Sea Basin*. SIC ICWC, Tashkent. p. 42. www.cawater-info.net/Afghanistan.htm
13. Dukhovny, V.A., Sorokin, A.G., de Schutter, J. & Maskey, S. 2009. *Concept Paper for Development of Upgraded ASBMM*, SIC ICWC, p. 14.
14. Fuchinobe H., Tsukatani T, Toderich K.N. Afghanistan's revival: irrigation on the left and right Bank of the Amudarya, Kyoto University, 2002. p. 47.
15. Fuchinoe, N., Tsukatani, T., Toderich, K.N., Kyoto University, p. 47.
16. Global Insight. 2006. February, http //www.global.sight.com
17. http//www.cawater-info.net/library/eng/carewib/summit_ifas.pdf
18. http//www.cawater-info.net/library/rus/carewib/summit_ifas.pdf
19. Ibatullin, S. et al. 2009. *Impacts of Climate Change on Water Resources in Central Asia*. Almaty.

20. KISI. 2008. *Kazakhstan in the modern world: realities and prospects*. Almaty, pp. 102–103, p. 109. (in Russian).
21. Koh, T. et al. 2009. Asia's Next Challenge: Securing the Region's Water Future, report prepared by the Group Water Security in Asia.
22. "Our excited mind has reached the boiling point." *Kommersant-Vlast*, February, 23, 2009. pp. 42–45.
23. Peyrouse, S. 2007. "The hydroelectric sector in Central Asia and the growing role of China", the Central Asia-Caucasus Institute, 195 # 1653–4272, pp. 131–148, China Eurasia Forum, Vol. 5, No. 2.
24. Poslavsky, V.V., 1983. Problems of Irrigation in Central Asia, Tashkent, Publishing House "Fan", p. 229 (in Russian).
25. Ruziev, M. & Prihodko, V. 2007. "Implementation of the Aral Sean bottom socioeconomic model: an assessment of the opportunities to be gained through regional economic integration" in Wouter, P. et al. *Implementing IWRM in Central Asia*, pp. 105–123, Springer (in Russian).
26. Sayfulin, R. *New Independent Countries in Central Asia: Problems of Security, Interstate Co-operation and Partnership*. http//eurasishome.org/xml/t/expertixml?lang = ru&nic = expert&pid = 1517. (in Russian).
27. Shiklomanov, I. (ed.) 2008. *Russian water resources and their use*. St. Petersburg, GGI. ISBN-978-5-98147-006-6. p. 587.
28. Stitienko, B. (ed.). 2007. *International Economic Relations*. Collected papers. Moscow, IFRA-M, p. 35–44. (in Russian).
29. Sultanov, B. (ed.) 2008. *Kazakhstan in the framework of global economic processes*. Collected papers. Publishing House under the administration of President of the Republic of Kazakhstan, p. 19. (in Russian).
30. The IWRM-Fergana Project: Water Governance. (in Russian).
31. The review "Adaptation to Climate Change in Central Asian Countries" was carried out by G. Stulina and presented at the FAO seminar held in Budapest in November 2009.
32. The Second National Report of the Republic of Uzbekistan on the UN Framework Convention on Climate Change. UNEP, 2008. (in Russian).
33. UNDP and RCSES. 2005. *Uzbekistan Human Development Report*, Tashkent.
34. UNESCO 2000, Water-related vision for the Aral Sea Basin for the year 2025, UNESCO, edited by Professor J. Bogardi p. 237.
35. Vasileva, V. 2007. Repeating the destiny of the Caspian Sea was predicted for the Aral Sea. www.tn.ru. June 7, 2007. (in Russian).
36. Verkhogorov, D. *Panj Hydroenergy in Afghanistan*. www.cawater-info.net/Afghanistan. (in Russian).
37. Water and Energy Resources of Central Asia: Problems of Use and Development. April 2008. p. 31. www.eabr.org/media/img/rus/publications/Analytical reports/obzor.water
38. Yakubov, M. & Manthrilake, H. 2009. Water for food as food thought: case study on applying the podium model to Uzbekistan, *Irrigation and Drainage*, 58, pp. 17–37. (in Russian).
39. ____. 2007. *Water is the life-and-death resource for the future of Uzbekistan. UN Office*, Tashkent. p. 127.
40. Marieke de Groen, Makhmud Ruziev, Alexsander Tuchin, Anatoly Sorokin, Valeriy Prikhodko. The Aral Sea Basin Management Model, for awareness raising and strategic decision support. ASBmm Final Report, SIC-ICWC. November 2002.

Subject Index

For Product Safety Concerns and Information please contact our EU representative GPSR@taylorandfrancis.com Taylor & Francis Verlag GmbH, Kaufingerstraße 24, 80331 München, Germany

Batch number: 10733572

Printed by Printforce, the Netherlands